CONSERVATION
OF
METALS

CONSERVATION
OF
OF METALS

Problems in the treatment of metal-organic and metal-inorganic composite objects

INTERNATIONAL RESTORER SEMINAR

Veszprém, Hungary

1–10 July 1989

Editor: **Márta JÁRÓ**

Literary advisers: **Harvey H. SHENKER and Gábor SEIDL**

Cover design and lay out: **István PÉTERFFY**

Publisher: **István ÉRI**

ISBN 963 555 770 1

Contents

Editorial

In keeping with the tradition of publishing the Proceedings of the International Restorer Seminars we present this volume – the seventh in the series. It differs, however, from the former ones in format and make-up.

The changed international situation together with the proliferation of conferences, call for a change. Thus, this seventh volume is perhaps the last in the series, its publication is meant as a full stop (or an exclamation mark).

The format has changed but not the principle of editing. Papers have not been made uniform or revised other than to ensure grammatical correctness or to help to clarify the meaning or to prevent misunderstanding. In addition to the texts of the papers delivered by the speakers of the conference, contributions submitted by those who were not able to attend the conference have also been included. Submitted posters have also been published as well as preliminary reports intended as a means of opening a debate on an intricate problem of restoration.

Customarily, the conference program, and the list of participants are included; a list of the conference publications up till now complements the Proceedings.

In his Introduction, István Éri, chairman of the International Restorer Seminars, surveys the background, and the aims and achievements of the seminars since their inception in 1976. The photos in the brief, illustrated chronicle that follows recall certain events and venues of the seminars – it is hoped that these will serve as a pleasant reminder.

We think that these conferences will be followed by others: refashioned, renewed, may be not in Veszprém but in another town of Hungary but in the traditional "Veszprém mood".

It is with this hope that I should like you to read and enjoy this volume.

The Editor

International Restorer Seminars in Veszprém Review of subjects and publications

1st Seminar 22–29 June, 1976

Subject: Situation and problems of the training of restorers

Publication: edited by T. Kiss, Budapest, 1976

2nd Seminar 15–27 July, 1978

Subject: Problems of the completion of art objects

Publication: edited by Á. Szemere, Budapest, 1979

3rd Seminar 11–20 July, 1981

Subject: Problems of completion, ethics and scientific investigation in the restoration

Publication: edited by M. Járó, Budapest, 1982

4th Seminar 2–10 July, 1983

Subject: Restorer's – museologist's – scientist's common responsibility in the protection of museum objects
The restored object as a source of historical and technological information
Conservation-restoration of church textiles and painted flags
Investigation of museum objects and materials used in conservation-restoration

Publication: edited by Á. Timár-Balázsy, Budapest, 1984, 2 volumes

5th Seminar 29 June–9 July, 1985

Subject: Common responsibilities and tasks of the restorer, scientist and museologist in storing and exhibiting the objects of art, with particular regard to the problems of exhibit installations and show-cases
Questions relating to the treatment of mural paintings and mural painting fragments discovered during excavations; conservation in situ and removal; restoration and reconstruction; approaches to display and storage
Problems of re-restoration, with particular regard to damages caused by acts of aggression (i.e. vandalism, theft), by elemental disasters and/or by improper storage
Round-table conference: Present situation and problems of restorers' and conservators' training; comparisons; practices, methodologies
Miscellaneous Themes: Techniques and results of scientific examinations

Publication: edited by I. Éri and G. Sárközy, Budapest, 1985, 2 volumes

6th Seminar 13–23 July, 1987

Subject: Conservation – restoration of leather and wood
Training of restorers

Publication: edited by I. Éri and G. Sárközy, Budapest, 1988

ÉRI István

International Restorer Seminars in Veszprém

The International Restorer Seminars organized in Hungary have had seven meetings in the fourteen-year period from 1976 to 1989. This series of events seems to have ended – for various reasons. For them to be continued, a different approach would be needed with different objectives.

Number seven in the series is most certainly not the attribute of a round anniversary – being solely of numerological importance. But due to the exciting political changes in Eastern Europe and personal changes in the organisers life, we tend to summarize a concluded period, with biennial international meeting as landmarks, with events constituting a coherent process. No bias is intended but if indeed there is one it is perhaps justifiable that we evaluate these seminars from the viewpoint of the host country and the institution, that undertook the organizing work – though obviously international aspects were taken into consideration.

Several attempts have been made to explain why it was just Hungary that took on the task of launching and of continually organizing this series of events. An explanation seems to be needed for ourselves, rather than for participants from abroad. The point was whether the level of the Hungarian restoration work could be matched with that abroad; is it worth while for an institution with a small staff to devote so much time and effort; will this or that seminar be sufficiently fruitful for foreign and Hungarian participants alike?

It was not until 1974 that the institution organizing the International Restorer Seminars was really established. In the initial stage of its organization was it not premature to appear on the international scene? As its initiator, responsible for this series, after one and a half decades I am convinced that this "opening" was simply inevitable, a vital condition for us.

In Hungary, restoration activities were quite isolated, characterized by a professional lag of decades. Links between institutes and between people were virtually non-existent, special literature barely acquainted with. The need to introduce an up-to-date education system had become pressing; the need to obtain the relevant experience imposed the need to find international connections.

The number of restorers active in Hungarian museums by the early '70s was rather low in relation to the work to be done. With regard to qualifications, except restorers concerned with fine arts objects, there were but a few with high-school or university degrees.

Further problems arose from the lack of even a passive knowledge of foreign languages. Last but not least: within the hierarchy of the staff of public collections, professional acknowledgement of restorers was not very high, this manifested itself in a number of ways but in particular in salaries, resulting in a lack of professional self-esteem, with no recognition of common interests.

The situation strongly changed by now. In public collections (including archives and libraries where, recently, restorers have been given "franchise") the number of restorers has doubled. Nearly 50% are university graduates. Gradually, step-wise training has become the rule: skilled workers – skilled restorers – graduate restorers, in courses of one, two, then four or five years. Professional acknowledgement has improved. This has come about not only as a result of applying the same salary scales as for museologists, but also as a result of the establishment of self-contained restorer sections within public collections, in general, together with the acknowledgement of the importance of preserving objets d'arts. Annual national conferences helped information exchange between specialized professional groups, verbal or written publication of individual achievements (through an independent periodical).

Command of foreign languages has become the general rule permitting – a limited number of – study trips abroad for varying periods. The activities of Hungarian restorers in international organizations has intensified.

Thus, the aimed-at process of drawing closer has started – although not without failures. One should not be ashamed of recognizing that international contacts in the form of seminars primarily served our own interests. While initially it was almost impossible for our specialists to get abroad, for various reasons, this shortcoming had to be offset by experts of abroad, by exchanging experience with guests. The possibility to demonstrate Hungarian achievements – often of a high standard – by means of programmes, lectures, demonstrations, and workshop visits was often of considerable help in promoting both international acknowledgement and the self -esteem of Hungarian specialists. In organizing seminars, however, another aspect has also become overwhelming. When considering the professional isolation of restorers in Hungary, it was impossible not to perceive that of other east-European colleagues, together with – here and there – similar symptoms. Any possible doubts we may have had concerning the proficiency of the International Restorer Seminars tended to be largely dispersed by the statements of those coming from neighbouring countries. By now, it is a commonplace that Veszprém was virtually the only meeting place for restorers from the FRG and the GDR; and this, in turn, was largely responsible for the fact that the largest groups of participants belonged to these countries. But also for restorers from Czechoslovakia, Poland, and, to a lesser degree, Bulgaria and Romania, these events were highly important in establishing first-hand international contacts. (It is a great pity that eminent Soviet specialists were virtually unable to get there.) Of course, these cultural, professional diplomatic achievements were recognized by Western visitors, too. Partly spontaneous, partly official statements, and active cooperation by eminent authorities encouraged "bridge building".

Peculiar political difficulties were apparent at the latest seminar in 1989; this meant very limited participation of specialists from neighbouring countries – just before the bursting of the dam old-timer cultural policy officials of certain countries still tried to hamper international cooperation.

Within about six months, with the historical events in Prague, Berlin, Sofia, Bucharest, things have become quite different, another factor hinting at the need to modify the original goals. Nevertheless, just because borders are less restricted it does not necessarily mean that economic difficulties are eased.

Such a biased evaluation of the goals for our seminars seems to me wrong. Our organizational work served for more than simply to draw closer, to establish links. Peculiar forms of events also played a part. In addition to strict programmes of attending and discussing lectures, various additional events helped in making acquaintances, establishing friendships, gaining experience. With regard to the efficiency of these I do not feel that I am competent to evaluate them. The friendships and professional relationships that have arisen among about seven hundred participants of the seven seminars are best known to the people themselves, but those meeting again at different venues, on different occasions and those visiting or assisting each other do, I am sure, guess how much is due to the Veszprém meetings.

"Professional events with a human touch" – was how one prominent participant put it. Truly democratic coexistence was realized between specialists of different qualifications, age groups, standing. The opposite was experienced at several, excessively formal events. At the Veszprém meetings, no isolation or segregation could be found. It seems me the mood of the participants simply abnegated any hierarchical division. Of course, apart from personal aspects this "melting down" has come about from professional aspects, too. Free exchange of ideas has been hampered solely by language difficulties – although synchronous interpretation, at least for official programs, was of some help.

Last but not least, the seminars had a certain – hopefully other than forced – propagandistic or, better, educational function. Favourable features were the venue of the seminars, the Balaton region, the tours and the concluding meetings in Budapest, all of which permitted the organizers to give a more or less true picture of Hungary. Hungarians are generally said to be hospitable people; the Hungarian organizers and participants endeavoured to merit this epithet. In fact the very site was attractive to newcomers. The majority were in Hungary for the first time but the return of numerous seminar participants also testifies to the attraction of local features: or was it only professional aspects that led to repetitive appearances? This is the most reassuring, empathetic reverberation of the International Restorer Seminars.

Having been responsible for the seven seminars, I do hope that in the world-wide, socially and professionally ever more intensive process of easing up and rapprochement – especially compared with fifteen years ago – we have also been able to do something towards realizing common tasks and goals.

Illustrated Chronicle

After having recapitulated the goals and achievements of the International Restorer Seminars in Veszprém, it was felt that it would be of interest to provide a short, illustrated report.

Veszprém, a picturesque town in a delightful setting, has some 60 thousand inhabitants. One of the oldest Hungarian settlements, it is situated about 120 km sout–west of Budapest, and is a mere 15 km from Lake Balaton the "Hungarian sea" the largest lake in Central Europe.

Initially, the seminars were held in the auditorium of the Chemical University of Veszprém

Since 1981, the Seminars have been held in the seat of the Academic Committee of Veszprém (a regional committee of the Hungarian Academy of Sciences), with lectures being organized not only in the auditorium and minor lecture halls but sometimes also in the courtyard.

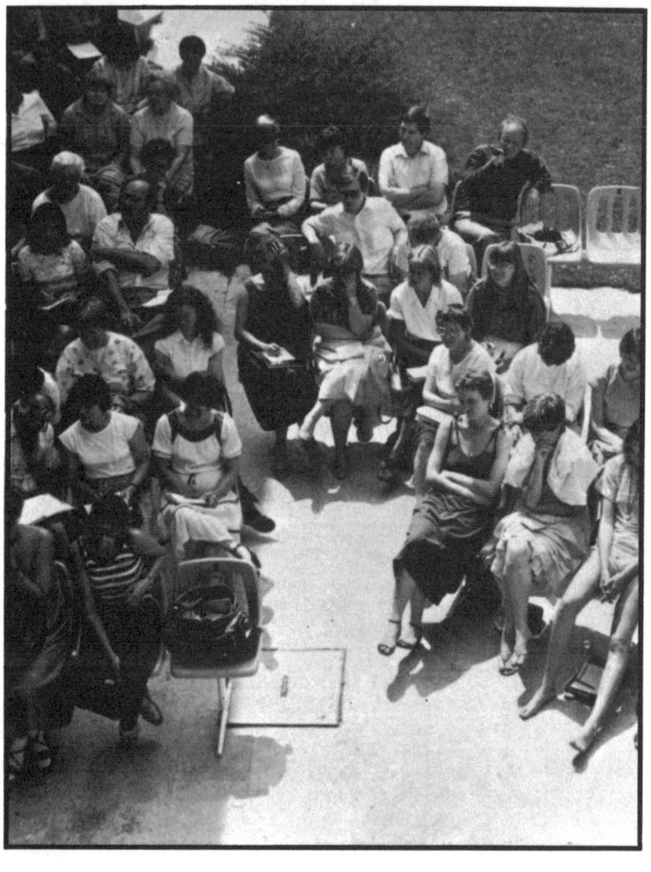

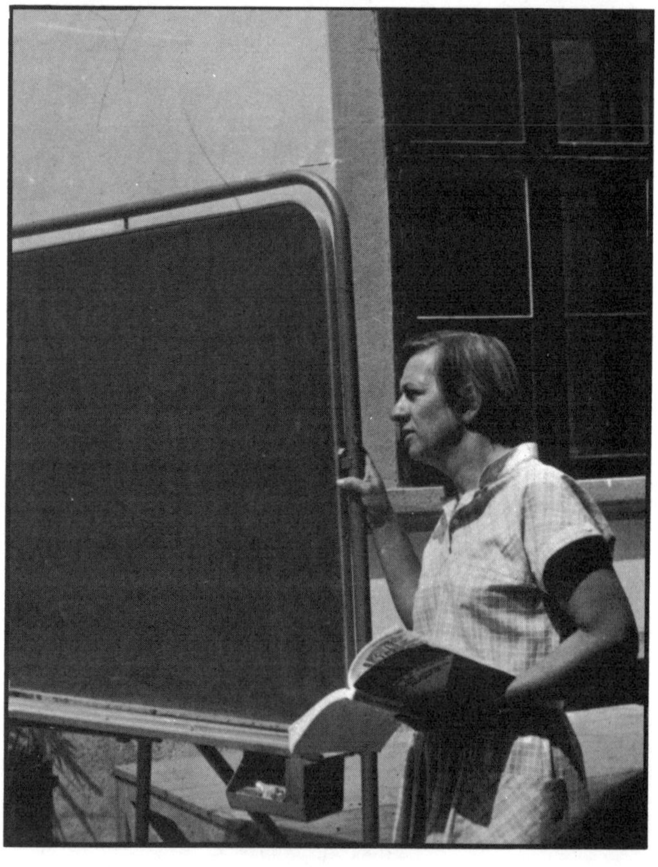

Occasions for entertainments, and for helping people to become better acquainted were offered – in addition to during discussions, and intermissions – at receptions, and at the dinner table.

Excursions helped the participants to get to know something about the Balatons region, with its monuments and its museums: participants could also look for ways to amuse themselves and – last but not least – rest after their labours.

Here, participants are receiving a musical welcome from the balcony of a press house on Szent György mountain. Naturally, in the wine cellar, of the house one takes advantage of the occasion and tastes the local wine.

In front of the press-house, certificates acknowledging attendance are distributed as the closing event of the Seventh International Restorer Seminar.

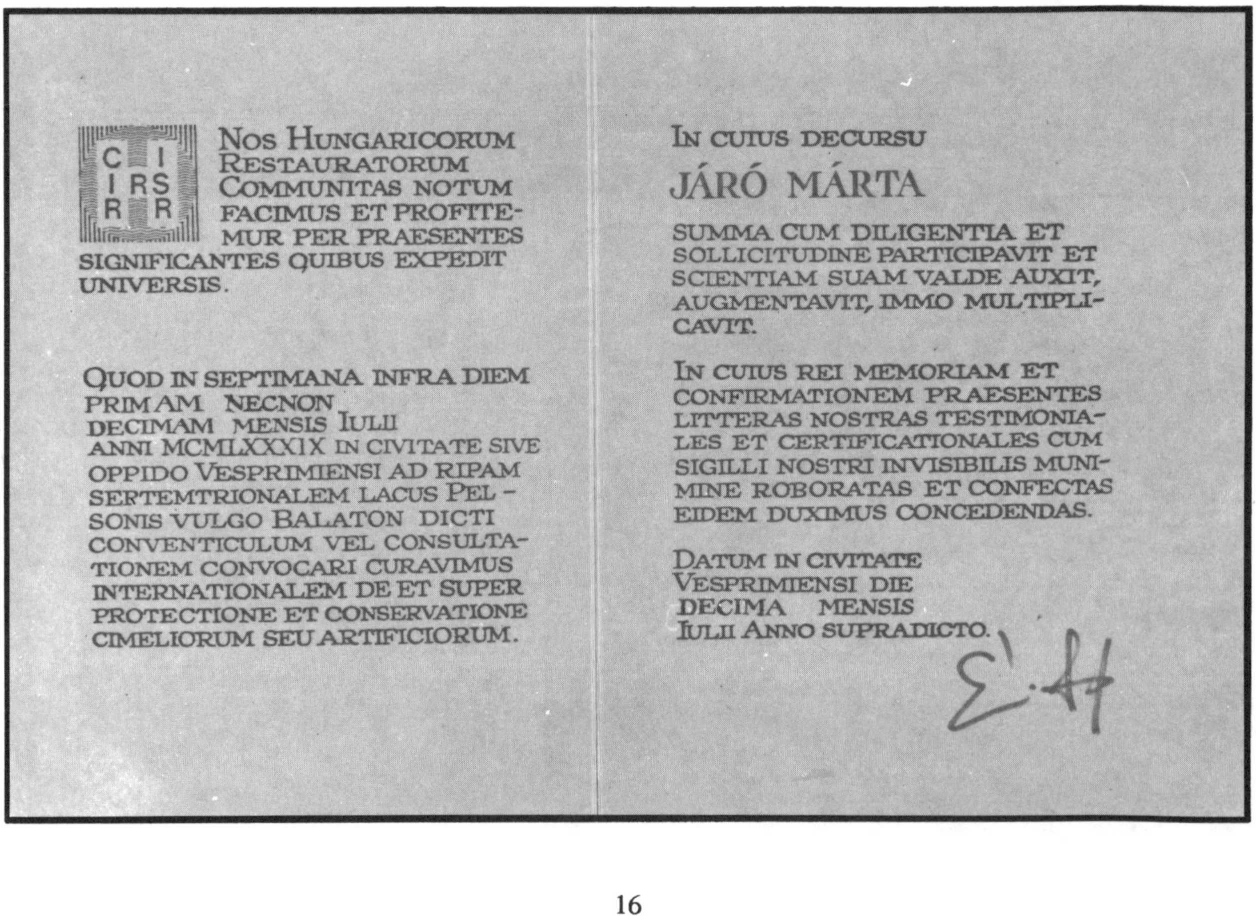

NOS HUNGARICORUM RESTAURATORUM COMMUNITAS NOTUM FACIMUS ET PROFITEMUR PER PRAESENTES SIGNIFICANTES QUIBUS EXPEDIT UNIVERSIS.

QUOD IN SEPTIMANA INFRA DIEM PRIMAM NECNON DECIMAM MENSIS IULII ANNI MCMLXXXIX IN CIVITATE SIVE OPPIDO VESPRIMIENSI AD RIPAM SERTEMTRIONALEM LACUS PELSONIS VULGO BALATON DICTI CONVENTICULUM VEL CONSULTATIONEM CONVOCARI CURAVIMUS INTERNATIONALEM DE ET SUPER PROTECTIONE ET CONSERVATIONE CIMELIORUM SEU ARTIFICIORUM.

IN CUIUS DECURSU

JÁRÓ MÁRTA

SUMMA CUM DILIGENTIA ET SOLLICITUDINE PARTICIPAVIT ET SCIENTIAM SUAM VALDE AUXIT, AUGMENTAVIT, IMMO MULTIPLICAVIT.

IN CUIUS REI MEMORIAM ET CONFIRMATIONEM PRAESENTES LITTERAS NOSTRAS TESTIMONIALES ET CERTIFICATIONALES CUM SIGILLI NOSTRI INVISIBILIS MUNIMINE ROBORATAS ET CONFECTAS EIDEM DUXIMUS CONCEDENDAS.

DATUM IN CIVITATE VESPRIMIENSI DIE DECIMA MENSIS IULII ANNO SUPRADICTO.

Program

July, 3

Éri István
Malliet, Joseph
Opening speech

Jedrzejewska, Hanna
Ethical problems in the conservation of composite metal objects

Roberts, Frances Christine
Organic remains on archaeological iron and bronze in sandy soil

Jensen, Niels Eric
1001 iron nails in wood

Oswald, Niklaus
Konservierung von Bodenfunden aus Eisen und organischem Material – eine Projektarbeit des Schweizerischen Nationalfonds

Schweizer, François – **Houriet**, Claude
Probleme bei der Konservierung und Restaurierung alter wissenschaftlicher Instruments

Weker, Wladislaw
Selected examples of treatments of metal-nonmetal composite archaeological objects (poster)

Weker, Wladislaw
Determination of the concentration of chloride ions extracted from corroded iron artefacts (poster)

Roberts, Frances Christine
Lifting freshly excavated fragile objects (poster)

July, 4

Indictor, Norman – **Ballard**, Mary
The effects of aging on textiles that contain metal: implication for analysis

Darrah, Josephine A.
The microscopical and analytical examination of three types of metal thread

Gondár Erzsébet – **Jankovich** Katalin
Későbronzkori bronzkarperecen lévő textilmaradvány vizsgálata
(Examination of the textile remains on a bronze bracelet from the late bronze age)

Tímár-Balázsy Ágnes
Deterioration process of fibers in different chemical actions

Howell, David
Experiments with chemical cleaning of metal threads

Járó Márta
Corrosion of metal threads after treatment

Flury-Lemberg, Mechtild
Die Kasel des hl. Ladislaus von Ungarn

Chocha, Krystina
The conservation of the XVIIIth century embroideries and tassels from the royal castle collection. Cleaning of metal threads

Sípos Enikő – **Gondár** Erzsébet
Különböző kezelőszerek hatása fémszálas textíliákra (poszter)
(Effect of different treatments on textiles with metal threads)

July, 5

Brunnert, Stephan
Die Glimmentladung in der Restaurierung

Välja, Helmut
Konservierung einer Weissgerberamtslade aus Jahre 1667

Nissel, Dieter
Zur Verantwortung des Restaurators für die Sicherung quellenkundlicher Details – dargestellt am Beispiel des mittelalterlichen Geschützes "Faule Magd"

Heath, Diana
The conservation of painted and polychromed metalwork, two case studies from the Victoria and Albert Museum

Taarnskov, Bodil
Examples of conservation of painted metal objects

Miettinen, Seija
The gilding methods

Slesinski, Wladislaw
Entwicklung der Reinigungsmethoden von Silbererzeugnissen

Green, Lorna Rosemary
A re-evaluation of lead conservation techniques

Milkova, Bogdanka – **Jossiffova**, Zvezdalina
Ein neues Verfahren zur Konservationsbehandlung von musealen Gegenständen aus Eisen

Kovács Petronella
Problems of conserving an 18th century child's coffin of wood coated with silk

July, 6

Hradetzky, Sylvia
Möglichkeiten der Herstellung von Stützkonstruktionen für archäologische Bronzeobjekte

Bakayné Perjés Judit
Technikai megfigyelések meroving-kori bronz leleteken
(Technical study of Merovingian bronze objects)

Leitner – Szapáry, Barbara
Restaurierung und Konservierung von barocken Wandappliken – Unterlegen und Kleben von zerbrochenen sowie ergänzten Teilen

Csere Judit
Középkori körmeneti kereszt restaurálása
(Restoration of a medieval processional cross)

Kozlowsky, Andrzej – **Sekowski**, Stefan
Kombinierte Metall-Lackschichten als Denkmalschutz, 14 Jährige Erfahrungen

July, 7

Báthy Géza
Az ozorai törökkori ötvöslelet restaurálása
(Restoration of the Ozora goldsmith's find from the period of the Turkish occupation of Hungary)

Séd Gábor
A fémrestaurálás oktatása Magyarországon
(Training of metal restorers in Hungary)

Sugishita, Ryuitiro
Metallurgical training for conservation specialists in Japan, traditional and scientific

Olivieri, Lucilla
Les restaurateurs de métal: hier, aujourd'hui, demain

Cubasch, C. Thomas
Ausbildung, Fach-Information

Ashley-Smith, Jonathan
A new conservation course at the Victoria and Albert Museum

Pennec, Stéphane
La conservation des objects du Titanic

Reports of the session leaders

Closing speech

List of participants

The Staff

Chairman:

ÉRI István

Secretary:

JÁRÓ Márta

Organisers:

DALLOS Katalin
ERDŐHALMI-TÖRÖK Klára
KISS-BENDEFY Márta
KOCZKA Mária
TIMAFFY Gyöngyvér
TIMÁR-BALÁZSY Ágnes

Interpreters

HERBOLY Edit
(Français)

DÖMÖTÖR Albert
(Deutsch)

KOSZTIC Szilárdné
(English)

KOZMA Kitty
(Français)

MAGDA Sári
(English)

PALOTÁS Antal
(Français)

PAPP Nándor
(English)

SCHMIDT Péter
(Deutsch)

ZALÁN Péter
(Deutsch)

Lecturers only

JANKOVICH Katalin Dr.
Archaeologist
MTA Régészeti Intézet
Uri u. 49.
H-1014 Budapest

KOZLOWSKI, Andrzej
Scientific Director
Institut für Präzisionsmechanik
Duchnicka 3
PL-00-967 Warszawa

OLIVIERI, Lucilla
Chief Restorer
Soprintendenza Archeologica di Roma
Piazza della Finanze, 1
I-00186 Roma

Participants

ADOMAT-SIMON, Annette
Conservator
Sønderjyllands Amts
Konserveringsvoerksted
Fabriksvej 17-21
DK 6510 Gram

ASHLEY SMITH, Jonathan
Head of Department
Victoria and Albert Museum,
Conservation Department
Cromwell Road, South Kensington
GB-SW7 2RL London

BAKAYNÉ PERJÉS Judit
Restorer
Központi Múzeumi Igazgatóság
Könyves Kálmán krt. 40.
H-1087 Budapest

BALLARD, Mary W.
Conservator
CAL/MSC Smithsonian Institution
USA-20560 Washington D.C.

BARBI ALONSO, Victor
Rua da Pastora 45-2
SP-36210 Vigo (Pontevedra)

BAUMGARTNER, Walter
Conservator
Museum für Völkerkunde
Neue Hofburg
A-1014 Wien

BÁTHY Géza
Head of Conservation Department
Magyar Nemzeti Múzeum
Múzeum krt. 14-16.
H-1088 Budapest

BÍRÓ László
Restorer
Georgikon Majormúzeum
Bercsényi u. 67.
H-8360 Keszthely

BLUMER, Rolf-Dieter
Restorer
Landesdenkmalamt
Baden-Württemberg Silberburgstr. 193.
D-7000 Stuttgart

BONGRAND, Montaine
Textile Restorer
11, rue Duban
F-75016 Paris

T. BRUDER Katalin
Restorer
Magyar Nemzeti Múzeum
Múzeum krt. 14-16.
H-1088 Budapest

BRUNNERT, Stephan
Conservator
Westfälisches Museumsamt
Schwelingsstr. 5
D-44 Münster 1

CARBONNEAUX-LEPRÊTRE, Elisabeth
Restorer
Musée de Normandie
Château
F-1400 Caën

CARRERA RAMIREZ, Fernando
Rua da Pastora 45, 2
SP-36210 Vigo (Pontevedra)

CHOCHA, Krystyna
Textile Restorer
Zamek Krolewski w Warszawe
Pl. Zamkowy 4.
PL-00277 Warszawa

CROOME, Hazel
Conservator
National Museum of Science and Technology,
Conservation Laboratory
P.O.Box 9724, Ottawa Terminal
CD-K1G 5A3 Ottawa, Ontario

CUBASH, C. Thomas
Verlag der Apfel
Gumpendorfer Str. 12.
A-1060 Wien

CSERE Judit
Restorer
Központi Múzeumi Igazgatóság
Könyves Kálmán krt. 40.
H-1087 Budapest

CZOPF Jánosné
Restorer
Janus Pannonius Múzeum
Fazekas Mihály u. 16.
H-7632 Pécs

DARRAH, Josephine A.
Conservation Scientist
Victoria and Albert Museum,
Conservation Department
Cromwell Road, South Kensington
GB-SW7 2RL London

DEY, Anuradha
Restorer
Instituto Centrale per il Restauro
Piazza S.Francesco di Paola 9.
I-00184 Roma

DIEPOLD, Brigitte
Restorer
Mälesskirchestr. 1.
D-8000 München 81

ERDŐHALMINÉ TÖRÖK Klára
Chemist, Conservator
Központi Múzeumi Igazgatóság
Könyves Kálmán krt. 40.
H-1087 Budapest

ÉRI Gyöngyi
Art Historian, Director
NOVOTRADE,
International Art Exhibition
Pozsonyi u. 23.
H-1136 Budapest

ÉRI István
Deputy Director
Központi Múzeumi Igazgatóság
Könyves Kálmán krt. 40.
H-1087 Budapest

FALK, Peter
Restorer
Museum für Volkskunde
Laudongasse 17-19.
A-1080 Wien

FLURY-LEMBERG, Mechtild
Head of Conservation Department
Abegg-Stiftung, Bern
CH-3132 Riggisberg

FÖLBER, Barbara
Restorer
Museum für Völkerkunde Leipzig
Täubchenweg 2
DDR-7010 Leipzig

GONDÁR Erzsébet
Chemist
INNOVATEXT
Gyömrői út 86.
H-1103 Budapest

GONZALEZ AIBERTOS, Teresa Prof.
Santissima Trinidad 4 18
SP-28010 Madrid

GONZALEZ HERNANDEZ, Jorges Antonio
Restorer
CENCREM
Convento de Santa Clara,
10100 Habana Vieja, Cuba

GORICAN, Danilo
Restorer
Pokrajinski Muzej Ptuj
Na gradu 1
Yu-62250 Ptuj

GRANT, Tara
Conservator
Canadian Conservation Institute
1030 Innes Road
CD-K1A OC8 Ottawa, Ontario

GREEN, Lorna Rosemary
Conservation Scientist
The British Museum,
Department of Conservation
Great Russel St.
GB-WC18 3DG London

HAGEMAN, Carmen Prof.
Club 13
SP-48930 Las Arenas (Vizcaya)

HARASZTI Gabriella
Restorer
Intercisa Múzeum
Lenin tér 10.
H-2400 Dunaújváros

HATALAY Géza
Restorer
Szépművészeti Múzeum
Dózsa György u. 40.
H-1146 Budapest

HEATH, Diana Margaret
Conservator
Victoria and Albert Museum,
Metalwork Conservation Department
Cromwell Road, South Kensington
GB-SW7 2RL London

HOWELL, David John
Conservator
Textile Conservation Studios
Apt. 11A, Hampton Court Palace
GB-KT8 9AU East Molesey, Surrey

HRADETZKY, Sylvia
Restorer
Staatliche Museen zu Berlin
Bodestrasse 1-3.
DDR-1020 Berlin

JÁNOSKA Péter
Deputy Director
Központi Múzeumi Igazgatóság
Kinizsi u. 39.
H-1092 Budapest

JÁRÓ Márta Dr.
Chemist, Conservator
Központi Múzeumi Igazgatóság
Könyves Kálmán krt. 40.
H-1087 Budapest

JAVIER, de Miguel Ortego
Restorer
Equipo-7, Restauración de Obras
de Arte
c/Henao, 16—1 izqda.
SP-48009 Bilbao

JEDRZEJEWSKA, Hanna Dr.
Scientist-Restorer
Solec 109A apt. 39.
PL-00382 Warszawa

JENSEN, Niels Erik
Conservator
Randers Kulturhistorisk Museum
Stemannsgade 2
DK-8900 Randers

JOHN-CSERKUTI, Esther
Restorer
Kunsthistorisches Museum
Waffensammlung
Heldenplatz, Neue Burg
A-1010 Wien

KARL, Hannelore Prof.
Hochschule für Angewandte Kunst
Oskar Kokoschkaplatz 2
A-1010 Wien

KISSNÉ BENDEFY Márta
Chemist, Conservator
Központi Múzeumi Igazgatóság
Könyves Kálmán krt. 40.
H-1087 Budapest

KISSNÉ KOROMPAI Bernadette
Head of Conservation Department
Dobó István Vármúzeum
Vár 1.
H-3300 Eger

KNAUS, Petra
Restorer
Staatliche Museen Heilbronn
Abt. Vor-und Frühgeschichte
Deutschhof
Eichgasse 1
D-7100 Heilbronn

KOVÁCS Petronella
Restorer
Központi Múzeumi Igazgatóság
Könyves Kálmán krt. 40.
H-1087 Budapest

KRZEMINSKI, Ryszard
Head of Department
Restorer Workshop
PL-Zagórza

KWARCIAK, Jacek
Archaeologist
P.K.Z.
Ul. Miodowa 4
PL-Kraków

LEITNER-SZAPÁRY, Barbara
Restorer
Bubendorf 85
A-3354 Wolfsbach

LOEDING, Dominique
Textile Restorer
Dorotheen Str. 19.
D-60 Hamburg

MALLIET, Joseph
Architect
ICCROM
Via di S. Michele, 13.
I-00153 Roma

MARQUEZ VIEIRA, Ignacio
c/Ordonez No. 12-14. Tui
(Pontevedra)
Spain

MENDE, Susanne
Restorer
Rüdesheimerstr. 8.
D-6200 Wiesbaden

MÉZES, Árpád
Restorer
SRA Levoca
Bernolakova
CS Levoca

MEYER, Ute
Restorer
Landesdenkmalamt
Baden-Württemberg,
Aussenstelle Tübingen
Schloss Fünfeckturm
D-7400 Tübingen

MIETTINEN, Seija
Conservator for Gilding
Museokatu 28 A 2
SF-00100 Helsinki

MILIC, Zoran
Conservator
Narodni Muzej
Presernova 20
YU-61000 Ljubljana

MILKOVA, Bogdanka
Conservator
National Polytechnical Museum
Samnuil st. 48.
S-1000 Sofia

MILLEI Ilona
Head of Conservation Department
Janus Pannonius Múzeum
Pollack Mihály u. 13.
H-7633 Pécs

MITTELSTEIN SCHEID, Vivica
Restorer
Hosfelds Katernberg 10.
D-5600 Wuppertal 1

MORGÓS András Dr.
Head of Conservation Department
Központi Múzeumi Igazgatóság
Könyves Kálmán krt. 40.
H-1087 Budapest

MÜLLAUER, Edith
Restorer
Urban-Loritzplatz 5/48
A-1070 Wien

NIEMEYER, Barbara
Restorer
Wilhelmstr. 15
D-6502 Mainz-Kostheim

NISSEL, Dieter
Art Historian, Restorer
Armeemuseum der DDR
PSF 11839
DDR-8060 Dresden

OLÁH Éva
Restorer
Munkácsy Mihály Múzeum
Széchenyi u. 9.
H-5600 Békéscsaba

OSWALD, Niklaus Dr.
Chemist
Schweizerisches Landesmuseum,
Chemisch-physikalisches
Laboratorium
Postfach 3263
CH-8031 Zürich

PENNEC, Stéphane
Restorer
97. rue Oberkampf
F-75011 Paris

PETIA, Nikolova
Restorer
National Museum of History
Vitosha 2
B-1000 Sofia

PIEHL, Manfred
Restorer
Museum für Vor-und
Frühgeschichte
Karmelitergasse 1
D-6000 Frankfurt/Main

PONTZ, Klaudia
Conservator
The Textile Conservation Centre
Apt. 22, Hampton Court Palace
GB-KT8 9AU East Molesey, Surrey

POYER, Martina
Restorer
Kunsthistorisches Museum,
Waffensammlung
Heldenplatz, Neue Burg
A-1010 Wien

PYKÄLÄ-AHO, Päivi Helena
Conservator
Museum Administration
P1 913
SF-00101 Helsinki

ROBERTS, Frances Christine
Conservator
Konserveringscentret
Vestergade 5-7
DK-687 Ølgod

RUDZINSKI, Krzysztof
Historian
P.K.Z.
Ul. Solec 104.
PL-00382 Warszawa

SALE, Don
Conservator
Tate Gallery
Millbank
GB-SW1P 4RG London

SANCHEZ EGUILUZ, Jabier
Restorer
Museo Arqueologico Etnografico
e Historico Vasco
Rue Maria Muñoz
SP-48008 Bilbao

SCHMATZER, Egon
Restorer
Kunsthistorisches Museum,
Waffensammlung
Heldenplatz, Neue Burg
A-1010 Wien

SCHMIDT, György
Restorer
Hendorf 111
A-8380 Jennersdorf

SCHWEIZER, François
Chemist, Conservator
Musée d'art et d'histoire,
Laboratoire
9-11 Rue du Clos
CH-1207 Genève

SÉD Gábor
Restorer
Központi Múzeumi Igazgatóság
Könyves Kálmán krt. 40.
H-1087 Budapest

SEKOWSKI, Stefan Dr.
Chemist
Institut für Präzisionsmechanik
Duchnicka 3
PL-00967 Warszawa

SÍPOS Enikő
Textile Restorer
Magyar Nemzeti Múzeum
Károlyi Mihály u. 16.
H-1055 Budapest

SIRKIÄ, Susanna
Conservator
Museum Administration
Pl 913
SF-00101 Helsinki

SLESINSKI, Wladyslaw Prof. Dr.
Restorer
Akademia Sztuk Pieknych
Smolensk 9
PL-91-108 Krakow

SUGISHITA, Ryuitiro
University Professor
Tokyo National University of
Fine Arts and Music
Ueno Koen 12-8, Taito-ku
J-110 Tokyo

SZALAY Sarolta
Restorer
Közlekedési Múzeum
Városliget krt. 11.
H-1146 Budapest

TAARNSKOV, Bodil
Conservator
Nationalmuseet,
Bevaringssektionen
Brede
DK-2800 Lyngby

TEXIER, Annick
Chemist Technician
Laboratoire de Recherche des
Monuments Historiques
29, rue de Paris
F-77420 Champs sur Marne

TOLERTON, Julie-Anne
Conservator
Ulster Folk and Transport Museum
BT18 OEU, Cultra, Holywood
N. Ireland

TÍMÁRNÉ BALÁZSY Ágnes Dr.
Chemist, Conservator
Központi Múzeumi Igazgatóság
Könyves Kálmán krt. 40.
H-1087 Budapest

TRAVIS, Julia
Restorer
Apt. 11A, Hampton Court Palace
GB-KT8 9AU East Molesey, Surrey

VAK, Bettina
Restorer
Mittersteig 2b,15
A-1050 Wien

VÁMOSI Ilona
Restorer
Rippl-Rónai Múzeum
Május 1. u. 10.
H-7400 Kaposvár

VÄLJA, Helmut
Chemist
Restaureerimiskeskus
Pikk 2
SU-200001 Tallinn

VIRTANEN, Mervi
Conservator for Gilding
Hiisiö
SF-47490 Mankala

WEKER, Wladislaw
Conservator
State Archaeological Museum
in Warsaw
Ul. Dluga 52.
PL-00-950 Warszawa

ZAUNER-MEYERHOFER, Silvia
Restorer
Alliiertenstr. 5/9.
A-1020 Wien

ŽIZEK, Ivan
Restorer
Pokrajinski Muzej Ptuj
Na Gradu 1
YU-62250 Ptuj

JEDRZEJEWSKA, Hanna

Ethical problems in the conservation of composite metal objects

Abstract

From the point of view of conservation – composite metal objects can be divided in two groups: viz. where the metal parts can be separated from the artifact and treated individually; where the metallic components must stay in place and be treated in the presence of all other materials. The ethical problems (the preservation of original evidence) are particularly exposed in the first group and need carefully balanced decisions. In the presented paper some basic examples are dealt with.

General considerations

Metals as a material have their own individual chemical properties. They are open to corrosion, changing the original technical and aesthetic functions, and have to be cleaned and protected against further damage. Two basic technical ways are possible here: the metallic element must stay in place and be treated in the presence of all accompanying materials, or it may be removed from the artifact and restored separately. In the first case the metallic component receives only partial care (restricted access) and the other components may react unfavourably to this intervention. In the second case a considerable amount of the original technical evidence may unavoidably be lost.

According to ethical principles, ancient objects are documents from the past and the applied treatments have to comply with this definition as far as possible. There is less danger for the original technical evidence when the metal is cleaned "in situ". Much more destructive is the operation of removal and later replacement.

But there are cases and cases. Every one has its own specific characteristics and requirements. So, ethics here will mean a constant, well balanced compromise between the theoretically desired and the practically possible. The assumption being that *all* was done to achieve the "best possible". Quite often a very difficult problem with composite metal objects.

Cases and comments

As already mentioned, every case is practically unique, but even so it will belong to certain more general groups of artifacts, having similar kinds of technical problems. Among these groups are, for example:

- in textiles, decorative metal threads inwoven into the fabric. The only possibility: cleaning "in situ".
- decorative metal in embroidery (metal threads, sequins, etc.), in most cases removed for cleaning. Often replaced with new elements, or stitched again to a new background.
- metal buttons, galons, fringes, etc. on garments, military uniforms, flags, banners, in the normal course of conservation are usually unstitched, cleaned, repaired, reconstructed, and eventually replaced, or just collected loosely, as such.
- weapons, fire-arms, armour, and other military equipment, have a lot of metal elements, often connected with wood, leather, bone, textiles. There is a general practice to disassemble the objects into their component parts, for treatment, and then join them back again. Eventually, the used or missing parts are reconstructed and replaced in the original. These new parts only rarely receive an informative die-mark.
- old machines, coaches, carriages, other vehicles and installations also are composite metal objects. There is wood, glass, some textiles, different kinds of metal, lots of leather. A good opportunity for disassembling and a large field for ethical considerations.
- there are very many ethical problems in ship-model conservation [1] as demonstrated in a paper from the Maritime Museum in Greenwich, London. Metallic fragments are mixed here with wood, textiles, bone, string, etc.; the preservation of the objects and of the original technical evidence meet here with many controversial points.
- old bookbindings often have quite rich metal ornamentations. The usual principle in the restoration of ancient books and manuscripts is to disassemble the objects into their single components, clean, restore and join again by supposedly "identical methods". It involves also the metal parts.
- ancient furniture is usually decorated with metal (handles, locks, flat surface decoration, figures, inlays, etc.). Also here, only when possible, the particular elements are removed for cleaning and restoration – with the exception of inlays, when these are still well-fitting.
- there are many metallic details in architectural works. The contact of metal (iron, bronze) with stone and mortars in buildings, with water (in fountains), with wood (old doors and windows) usually leads to corrosion and damage. There are statues, balconies, stair-cases, roofs, emb-

lems, memorials, etc. As a principle, they are removed for restoration and eventually — replaced in their original settings (or not!). Metal parts in ecclesiastical architecture, and their restoration, require some specific kinds of consideration.

— another example of composite metal objects is jewellery. Besides metal we have here precious stones, glass, porcelain, leather. There is maybe, less corrosion and more mechanical damage. But there is often the practice of separating the components, and exchanging them for new ones.

— stained-glass windows are another specific case of composite metal objects, with a lot of interventions, maybe not always ethically acceptable.

— metal may serve not only as decoration but also as support for paintings, enamels, and other decorative techniques. Here, the treatment has to be carried out "in situ".

— a specific group are objects in open-air museums. There are decorated doors, windows, roofs, etc., and also original nails, screws etc. During transport to the museum's site and the following rebuilding all these (rusted!) original details are forever lost.

— the biggest, and also the most difficult group of composite metal objects is in the field of archaeology. There are composite metal artifacts found in soil, water (sweet and salty), mud, ice, in caves and in graves. They are big or small, but they always carry valuable technical evidence. Metal is together with textiles, leather, wood, bone, shells, etc. Their treatment needs particular attention, with regard to interpretation as well as treatment.

The presented list is still far from complete, but it just illustrates the enormously large field covered by the definition of "composite metal objects".

Concluding remarks

These may briefly be recapitulated as follows:

— the basis requirement in conservation is to protect artifacts from further deterioration, and make them presentable again, but without harm to historical evidence. Something like the classical problem of "eating one's cake and having it too". One of the solutions is to eat only half of the cake and preserve the other half as evidence,

— with composite objects it is not as easy as that. Before any operations are started there is need for thorough research and a proper classification of all the preserved historical evidence. Some priorities and programme need careful balancing and decisions,

— not all the historical evidence on the object can be saved directly. There is need for samples, analyses, photos, diagrams, etc.,

— very important are decisions as concerns the future of restored composite objects. The different components may need rather different environmental conditions (humidity, temperature, light, sensitivity to pollution), so it is necessary to decide on optimal parameters,

— there are no perfect techniques of treatment. There is need of good information and direct exchange of views. But is also helps to hear (and read) not only about what leads to success, but also what *went wrong* and why. A thing only rarely disclosed...

Reference

Mc Morrow, J., **Lewis,** G.M., **Gregson,** C.: Ethics in Ship-Model Conservation, ICOM Conference, The Netherlands, National Maritime Museum, October 1987, London, UK.

OSWALD, Niklaus

Konservierung von Bodenfunden aus Eisen und organischem Material - eine Projektarbeit des Schweizerischen Nationalfonds[1,2]

Abstract

1. In archeological findings of iron material, the Akaganeitphase is mainly responsible of the deterioration. The entire removal of remaining chlorides is extremely difficult. The reason for that fact is:
a) Chloride, locked in the Hollandit lattice of Akaganeit is coordinated directly to the iron atom: a complete removal will provide a phase-transformation.
b) The chloride concentration has a widely inhomogeneous distribution: high concentrations will be located in little hollows in the near zone of the intact iron kernel. The approach of chemical reagent to the hidden chlorides is hardly handicaped and all reactions will become diffusion controlled.
2. The efficiency of inhibitors to prevent the formation of rust will be directly controlled only in water medium. Under this condition one can measure the provocated variation of the surface potential. To qualify oily systems, like we had to do, one has to use statistical methodes over a long period of time.
3. For iron-organic composites which were not allowed to separated, there is no solution for the moment beeing. All organic material tested will suffer from the examined antirust treatment.

Einleitung

Metallfunde aus dem Boden und Meer sind heikel und vergänglich. Korrosionsschäden sind irreversibel (Redoxreaktionen an der Metalloberfläche) und die Information über den Habitus der Gegenstände geht dabei oft entgültig verloren. Das Konservierungsproblem dieser Objekte ist nur teilweise und punktuell gelöst: ein einheitliches Konzept fehlt. Die bis jetzt bewährten Rezepturen bedingen entweder einen hochtechnisierten apparativen Aufwand (Wasserstoff Plasma[3] oder zeichnen sich durch eine sehr lange Behandlungszeit aus (Sulfitmethode[4]). Ein Schutzüberzug (Anstrich) mit einer längeren Antirostgarantie, d.h. mit einer dauernden Passivierungs- und Inhibitionswirkung fehlt.

Diese Eisenfunde, seit Jahrtausenden ungezwungen schlafend im Boden oder am Grund von Gewässern, erzeugen nach ihrem unverhofften Erwachen, durch ihre Ausgrabung und Freilegung den Konservatoren rostfarbene Alpträume. Die diesen Objekten auferzwungene Behandlung (Konservierung) ist, je nach Methode, Behandlungsintensität und Medikament (Inhibitor), mehr oder weniger erfolgreich – die meisten Patienten liegen in einem Dornröschenschlaf und können nach längerer oder kürzerer Zeit aufwachen und selbstzerstörerisch wieder zu blühen anfangen.

Wir haben uns bei unserer Arbeit über die Konservierung von Eisenfunden folgende Aufgaben gestellt:

1. Was ist der Grund der immer wiederkehrenden Ausblühungen an vermeintlich stabilen Gegenständen?

2. Können neue Methoden gefunden werden, die auch für die Konservierung von Eisenobjekte brauchbar sind, welche in untrennbarer Verbindung mit organischen Materialien (Holz/Leder/Textilien) sind?

Um keine falschen Hoffnungen zu schüren, möchte ich vorwegnehmen, dass Aufgabe 1 mit Erfolg gelöst werden konnte, hingegen die Lösungsansätze für Aufgabe 2, stellt man die Bedingung der Untrennbarkeit der beiden Materialgruppen, nur unbefriedigend waren. Neue Anstrengungen müssen auf diesem Gebiet gemacht werden.

Aufgabe 1

Unser Versuchsobjekte waren eine Hundertschaft von römischen Eisennägeln aus Augusta Raurica. Nägel aus dieser Zeit haben den Vorteil, dass sie in grosser Anzahl verfügbar sind, und deshalb statistisch bewertet werden können. Sie sind ideale Experimentierobjekte, weil sie echt und genau datierbar, aber nich wertvoll sind, und deshalb von den Archäologen à fonds perdu zur Verfügung gestellt werden.

Abb. 1. *Rostoberfläche. (Das Farbphoto befindet sich an der Titelseite.)*

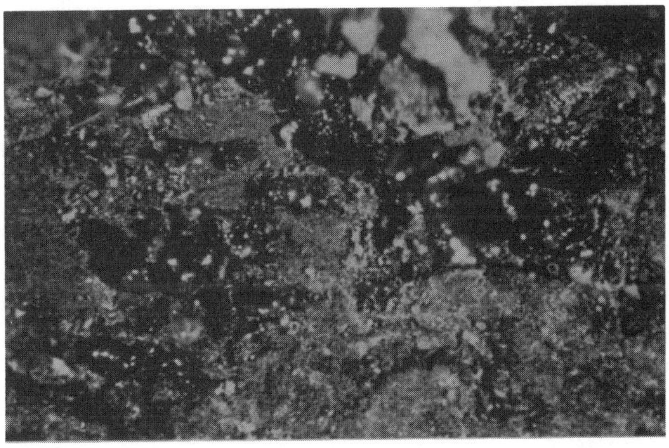

Die abgebildete Oberfläche eines Nagels (Abb. 1.) zeigt sehr schön die verschiedenen möglichen Rostphasen:

rotbraun	Akaganeit	beta-FeOOH
gelb-orange	Lepidocrocit	gamma-FeOOH
gelb	Goethit	alpha-FeOOH
rostrot	Hämatit	alpha-Fe_2O_3
schwarz	Magnetit	Fe_3O_4

Diese verschiedenen Rostphasen unterscheiden sich nicht nur in der Farbe, sondern auch in ihren weiteren physikalischen und chemischen Eigenschaften, wie Dichte, Kristallstruktur, chemische Reaktivität, thermodynamische Stabilität und Adsorbtionsvermögen (spezifische Oberfläche). Eine Auflistung der genauen Werte diesen Eigenschaften und deren Bedeutung führte hier zuweit. Sie können zum Teil im angefügten Literaturnachweis [5, 6, 10] nachgeschaut werden.

Auf Grund unser intensiven Auseinandersetzung mit den Phänomenen des Rostens haben wir folgende drei Thesen aufgestellt:

These 1

Unter den verschiedenen möglichen Rostkomponenten ist beta-FeOOH, Akaganeit, $FeO(OH)_{1-X}Cl_X$ (0<X<0.3) die katalytisch wirkende, agressive Hauptkomponente. Sie ist hauptsächlich verantwortlich für die zerstörenden Ausblühungen.

Tabelle 1 *Theoretische (errechnete) Chloridgehalte von Rostkomponenten in Massenprozenten (%)*

Rostkomponente		min	max
Akaganeit:			
$FeO(OH)_{1-X}\ Cl_X$	$X_{max} = 0.3$	0.0	8.3
$(2-X)H_2OCl_xFe_8O_8(OH)_{8+x}$			
	$X_{max} = 2$	0.0	9.03
Eisenoxichlorid:			
FeOCl			33.0
basische Fe(II)-Chloride			
$XFe(OH)_2\ FeCl_2$	$X_{max} = 3$	18.0	32.7
Green Rust":			
$(4-X)Fe(OH)_2 . XFeO(OH) . FeOCl$			
	$X_{max} = 1.87$	7.5	8.0
Fe(II/III)-Chlorid			
$FeCl_2 . X(H_2O)$	$X_{max} = 4$	35.7	55.9
$FeCl_3 . X(H_2O)$	$X_{max} = 4$	45.5	83.9

Glasgestrahlte Normplättchen aus Eisenblech wurden mit verschiedenen Rostphasen beschichtet (Abb. 2). Die erwähnten Rostphasen wurden entweder gekauft oder synthetisiert und mit der Methode der X-Ray-Diffraktion analysiert. Unter standardisierten Normbedingungen (90% rF, 25 °C, 7 Tage) zeigten die Plättchen deutlich unterschiedliche Verhalten. Im Gegensatz zu den anderen Rostphasen, war das Akaganeitpulver fast nicht mehr zu entfernen, und der Angriff auf das Blech war sehr stark ätzend. Weitere Versuche mit erhöhtem Chloridgehalt und Inhibitorzusätzen, die ich hier nicht eingehend behandeln möchte, zeigten das selbe Bild des agressiven Akaganeit. Die Resultate bestätigen die schon von Keller[5] 1969 beschriebenen Experimente und die von Turgoos[7] 1972 aufgestellte Behauptung, dass Akaganeit schon bei 20% relativer Feuchtigkeit rostkatalytisch wirksam ist.

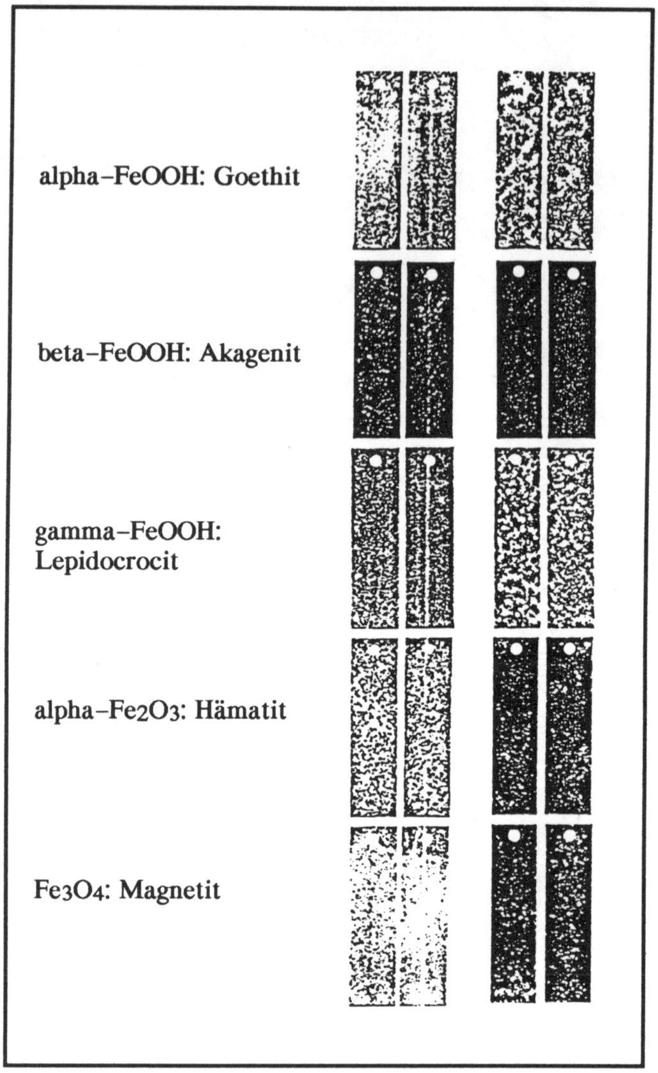

Abb. 2. *Korrosionsversuche mit diversen Rostkomponenten (7 Tage 23 °C 90% rF) rein +0.4% Cl*

These 2

Akaganeit besitzt eine Hollanditstruktur ($BaMn_8O_{16}$): $(2-X)H_2O . Cl_xFe_8O_{8-x}OH_{8+x}$ (0<X<2) und kommt nur zusammen mit Chlorid vor. Eine vollständige Chloridentfernung bedingt eine Phasenumwandlung [6, 10]. (Abb. 3.)

Die ideale tetragonale Elementarzelle (a = 10.5 Å, c = 3.024 Å) besitzt Tunnels von circa 3.5 Å Durchmesser, wie Schneider et al.[12] Mit hydrolytischen Abbauversuchen, Lichtstreumessungen und computersimulierten Kristallwachstumsmodellen zeigen konnte, hat das Chloridion essentiellen Einfluss auf alle wesentlichen Pa-

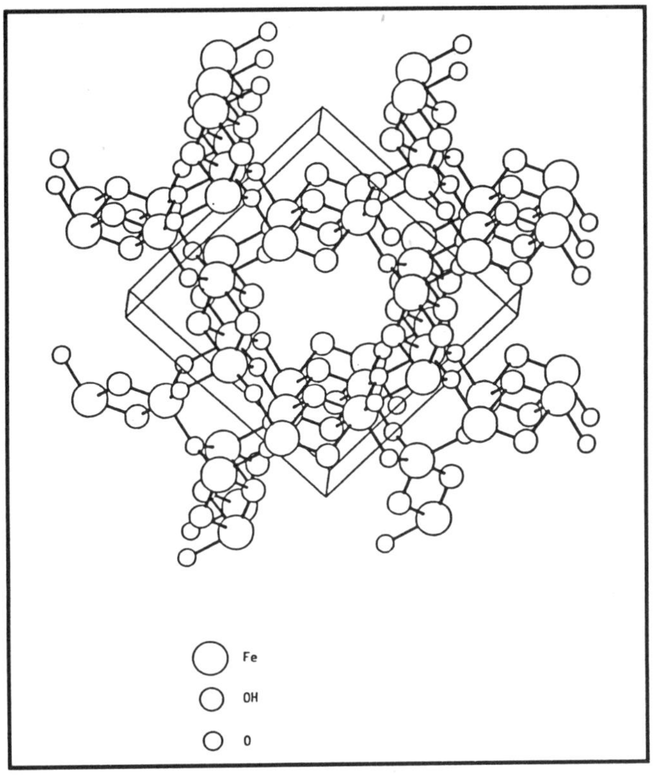

Abb. 3. *Hollandit – Struktur*

rameter. Es existieren pro Elementarzelle 2 Nischen, die eisenkoordiniertes Chlorid (r= 1.81 Å) enthalten können, was zu einem maximalen Chloridgehalt (koordiniertes Chlorid) von 9 Gewichtsprozenten führt.

Die spezifische Oberfläche von 30-60 m^2/g [13] weist auf ein grosses Chloridadsorbtionsvermögen und auf eine katalytisch wirksame Oberflächenstruktur hin. Ein sensibles Gleichgewicht zwischen der Chlorid-Verfügbarkeit (Labilität) und der Chlorid-Adsorbtion an die aktive Oberfläche, so wie der möglichen Koordination ins Hollandit–Gitter, scheint die Reaktivität des Akaganeit zu steuern.

Tabelle 2 *Chloridgehalte von Rostkomponenten in Massenprozenten (%)*

Rostkomponente	Chlorid extrahiert mit HNO_3 warm	Chlorid extrahiert mit HNO_3 kalt
alpha–FeOOH (Goethit)	0.003	0.005
beta–FeOOH (Akaganeit)	5.0	0.03
gamma–FeOOH (Lepidocrocit)	1.2	0.045
alpha–Fe_2O_3 (Hämatit)	<0.001	0.004
Fe_3O_4 (Magnetit)	<0.001	0.001

Auswaschexperimente mit Wasser, warmer und kalter Salpetersäure an verschiedenen Rostphasen, so wie analoge Experimente an rostinduzierten glasgestrahlten Normblechstreiffen, zeigten, die schon 1984 festgestellte Tatsache (Rinuy und Schweizer[4]: wasserlöslicher Anteil = 2/3), dass Chlorid nicht vollständig ausgewaschen werden kann. Bei pH=7 kann das Akaganeit bis etva 1% chloridfrei gewaschen werden; bei höhern pH-Werten werden die Eisenoxid-Hydroxide teilweise und die Ferrioxid-Chloride vollständig hydrolisiert: es entstehen amorphe Fe(III)-Oxid-Hydroxid-Polymere. Mit der vollständigen Entsalzung ist also eine chemische Reaktion verbunden, sei dies eine Phasenumwandlung der beteiligten Eisenoxid-Hydroxiden untereinander, oder zu Hämatit, oder eine Reduktion dieser Komponenten zu Magnetit (Reduktion im Wasserstoff-Plasma). Die vermeindliche Reduktion der Hydroxide mit Sulfit zu Magnetit findet nicht statt. Die rostaktive Grenzkonzentration (critical level) von Chlorid in den eher unkritischen Limiten, geprüft bei Standardbedingungen (90% rF, 25 C, 7 Tage), wurde von uns ebenfall mit Normplättchen ermittelt und betragen $1\text{-}8.10^{-2}$ mg Chlorid pro cm[2] Oberfläche. Die Resultate bestätigen vorgängige Messungen von 8.10^{-2} mg/cm^2 oder 200 ppm (auf Nägel bezogen) [4].

These 3

Chlorid kommt in Rost in zwei unterschiedlichen Formen vor:

a) chemisch nicht ans Eisen gebunden, z.B. als lösliches Salz oder oberflächenadsorbiert am "Rost", speziell am Akaganeit

b) eisenkoordiniert, chemisch gebunden in Akaganeit

Die Chloridverteilung im Gegenstand ist äusserst inhomogen. Es existieren hohe lokale Chloridkonzentrationen in kernnahen Spalten und Rissen. Eine, nach der Entsalzung analytisch gemessenen, sehr geringe Chloridkonzentration (200 ppm) gibt keine Gewähr für eine dauernde Stabilität.

Anhand der Verteilungsanalyse an einem Anschliff eines behandelten Nagels, ermittelt mit einer Elektronenstrahlmikrosonde (X-Ray-Fluoreszenz), zeigt sich deutlich die gehäufte Chloridkonzentration in kernnahen Gebieten (Abb. 4). Diese Resultate stehen im Einklang mit der von North [11] experimentell ermittelten Chloridverteilung eines Ausschnittes einer Kanonenkugel. Die Chloridkonzentration nimmt dort deutlich gegen innen zu.

Aufgabe 2

Heutige angewandte Methoden zur Stabilisierung:

1. Einsalzen

— Lösen der Chloridsalze durch Auswaschen der Objekte mit deionisiertem Wasser
— Lösen der schwerlöslichen Eisenoxide mit Komplexbildnern
— Basische Hydrolyse der Eisenoxide (Sulfitmethode)
— Elektrolytische Freisetzung der Chloridionen

Alle oben erwähnten Methoden werden kombiniert mit einer periodischen oder kontinuierlichen Chloridentfernung (Ionentauscher, Badwechsel)

— Überführen der Chloride in flüchtige Chlorver-

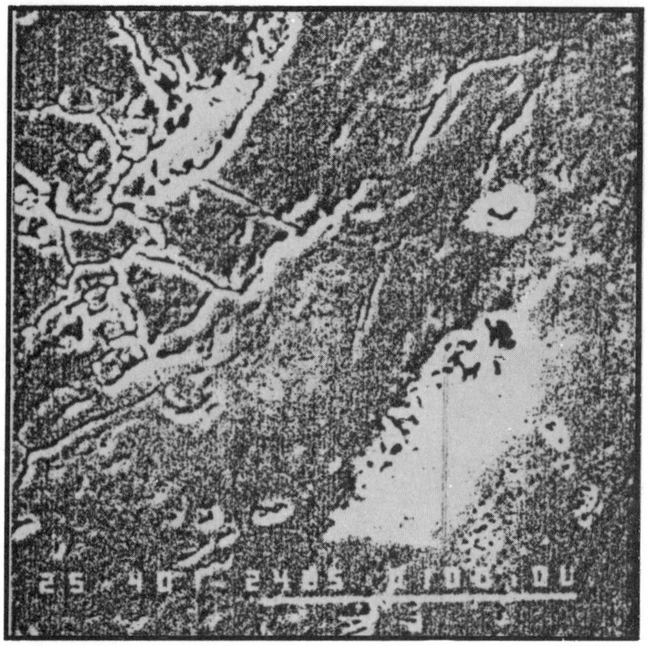

SEI - Bild

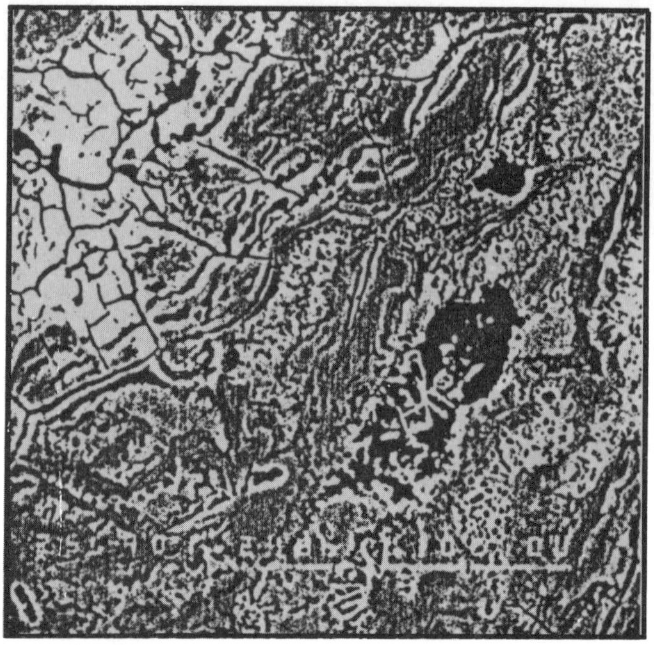

Materialkontrast

Cl - Verteilung

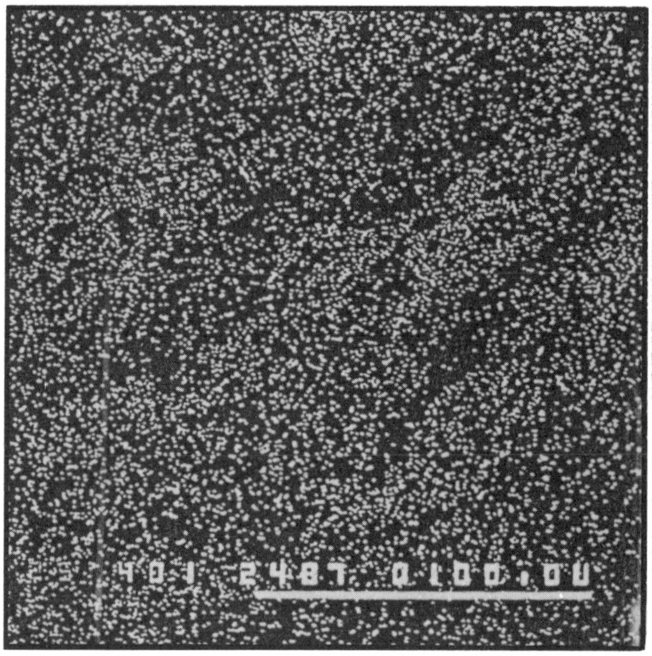

S - Verteilung

Abb. 4. *Römischer Nagel. Vergrösserung : 400-fach*

bindungen: Salzsäuregas (HCl), niedermolekulare organische Chlorverbindungen
— Fällen von schwerlöslichen Chloridsalzen: z.B. Silberchlorid (AgCl)

2. Reduzieren der Eisenoxide
— Elektrolytische Behandlung
— Wasserstoff-Plasma

3. Passivierung der Oberfläche
— Phosphatisierung
— Gerbsäurebehandlung
— Nitridierung
— Behandlung mit Korrosionsinhibitoren

Da uns die Behandlung mit Korrosionsinhibitoren die mildeste und einfachste Methode schien, experimentierten wir mit Nägeln und organischen Stoffen, zusammen und getrennt (Flussdiagramm, Abb. 5).

Holz verschiedener Abbaustufe, Leder mit unterschiedlichem Alter (unterschiedlichem Abbaugrad), Seide, Wolle und Leinen wurden zusammen mit Nägel behandelt und photographiert. Sowohl die erwartete Inhibitorwirkung auf die Nägel, wie auch die Schadwirkung auf das organische Material wurde genaustens dokumentiert. Die beiden unkäuflichen Wirkstoffe auf Aminbasis, W1[6],und W2[6], wurden in verschiedenen Lösungsmitteln (RAG Oel[6] , Sangaiol, Isopropanol, Granoöl, Wasser) angewandt. Zusätzlich erprobten wir auch die Inhibitor- und Schadwirkung von Silberaminkomplex-Lösung [$Ag(NH_3)_2OH$]. Trotz grosser Anstrengungen waren die Ergebnisse unbefriedigend und statistisch schwer auswertbar. Die Resultate bedürfen einer weiteren Laborüberprüfung. Im Ganzen gesehen hat das unkäufliche RAG-Oel, welches zur Nachbehandlung von modernen Eisengegenständen bei PVC-Bränden (PVC = Polyvinylchlorid) eingesetzt wird, bezüglich der Langzeit-Inhibitorwirkung am besten abgeschnitten.

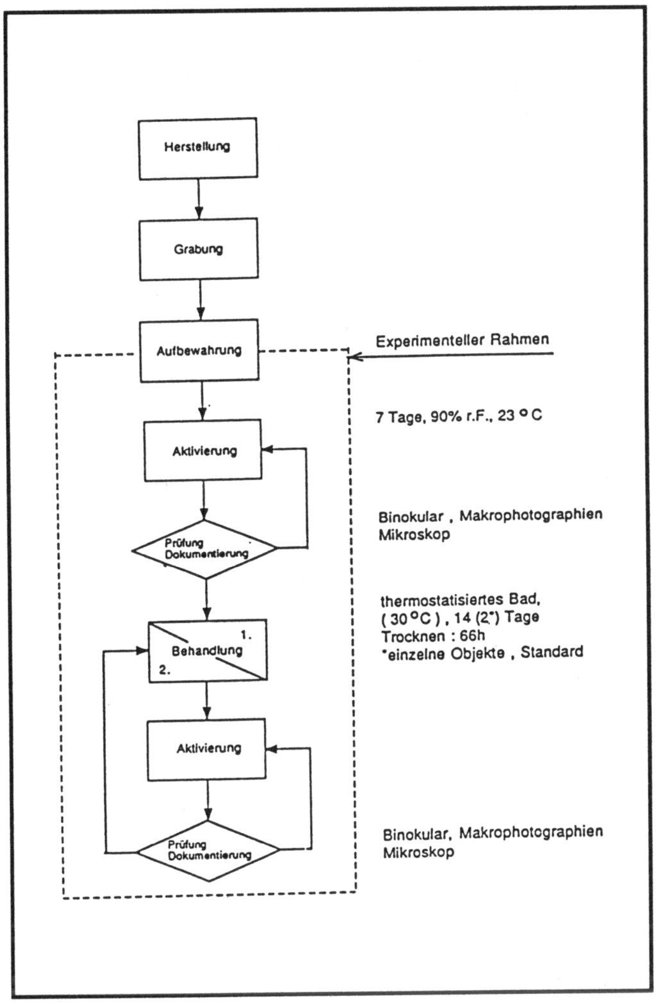

Abb. 5. *Flussdiagramm*

Die mitgeführten organischen Stoffe leiden immer, sowohl strukturell (irreversible Veränderungen im chemischen Aufbau), wie auch äusserlich (Verfärbung, Versprödung), in unterschiedlichem Masse unter allen Behandlungen.

Literatur:

1 **Mühlethaler** B., **Theiler** F., **Oswald** N., **Geiser** R., **Pulver** W., Methoden zur Konservierung von Bodenfunden aus Eisen in Verbindung mit organischen Materialien, Methoden zur Erhaltung von Kulturgütern, 166 (1989), Bern, Hauptverlag, ISBN 3-258/04000-1

2 Originalarbeit: NFP 16, Konservierung von Eisenfunden, EMPA Dübendorf 1989, EMPA Nr. 11 227

3 **Veprek** S. and **Elmer** J.Th., **Eckmann** Ch. and **Jurka-Rajman** M. Restauration and Conservation of Archeological Artefacts by Means of a New Plasma-Chemical Method, J.Electrochem. Soc. 134, (10) 2398 (1987)

4 **Rinuy A.** and **Schweizer** F. Application of Alkaline Sulphite Treatment to Archaeological Iron: a Comparative Study of Different Desalination Methods,
Nat. Maritime Museum Monographs and Reports, **53**, 44 (1982)

5 **Keller** P. Eigenschaften von $(Cl,F.OH)_2$ Fe_6 (O,OH) und Akaganeit, N. Jahrb. Miner., **113**, 30 (1970)

6 **Keller P**. Vorkommen, Entstehung und Phasenumwandlung von Beta-FeOOH in Rost, Werkstoffe und Korrosion, 20, 102 (1969)

7 **Turgoos** S. Post-Excavation Changes in Iron Antiquities, Studies in Conservation, **27**, 97 (1982)

8 RAG-Oel: Reichenberger AG, CH-6038 Gisikon

9 **Watkinson** D. Degree of Mineralisation Significance for Stability and Treatment of Excavated Ironwork,
Studies in Conservation, **28**, 85 (1983)

10 **Merono** D.M., **Morales** J. and **Tirado** J.L. Thermal Behaviour of Synthetic Akaganeit under Different Experimental Conditions, Thermochimia Acta **92**, 525 (1985)

11 **North** N.A. Corrosions Products on Marine Iron,
Studies in Conservation, 26, 75 (1982)

12 **Schneider** W. Comments Inorg. Chem. 3, (4), 205 (1984)

13 **Holm** N.G. New Evidence for Tabular Structure of Beta-Iron(III)-Oxide-Hydroxide, Akaganeit, Origins of Life, 15, 131 (1985)

Adresse des Autors

Schweizerisches Landesmuseum
Chemisch-physikalisches Laboratorium
Postfach 3263
CH-8031 Zürich

SCHWEIZER, François – HOURIET, Claude

Probleme bei der Konservierung und Restaurierung alter wissenschaftlicher Instrumente

1. Einleitung

Das Musée d'art et d'histoire in Genf, besitzt in seiner Zweigstelle, dem Musée d'histoire des sciences, eine reiche und wertvolle Sammlung alter wissenschaftlicher Instrumente. Diese zeugt von der Erfindungskraft der Genfer Gelehrten sowie dem hohen technologischen Stand der heimischen Industrie seit dem späten 17. Jahrhundert. Die Sammlung umfasst die verschiedensten Instrumente wie Mikroskope, astronomische Messgeräte, Fernrohre, Barometer, Chronometer und Feldmessgeräte. Sie wird konservatorisch von unseren Restaurierungswerkstätten sowie dem chemisch-physikalischen Laboratorium betreut.

Der vorliegende Beitrag soll einen Ueberblick über die bei uns entwickelten Konservierungs- und Restaurierungsverfahren geben. Diese wurden in den letzten 15 Jahren immer wieder den neuen Erkenntnissen angepasst. Obwohl jedes Museumsobjekt einen Sonderfall darstellt, scheint es uns nützlich, einige allgemeine Arbeitsmethoden vorzustellen. Dies umso mehr als die Konservierung wissenschaftlicher Instrumente bis anhin nur zu sehr wenigen Publikationen Anlass gegeben hat [1,2]. Eine sehr reiche Literatur liegt hingegen über die historisch-technologischen und wissenschaftlichen Aspekte vor [3,4,5,6].

Für die Konservierung stellen die wissenschaftlichen Instrumente einen Sonderfall dar. Die bestehen oft aus einer Vielzahl von Materialien wie Messing, Eisen, Stahl, Bronze, Glas, Holz, Horn, Elfenbein, Leder, und sogar Papier. Die Metalle mögen zudem an ihren Oberflächen vergoldet, versilbert oder gebläut worden sein.

Für optische Systeme finden wir uns in Anwesenheit von exotischen Produkten wie hauchdünnen Spinnenfäden, die als Fadenkreuze Anwendung fanden.

Eine Besonderheit stellt der berühmte organische Instrumentenlack dar, der seit Jahrhunderten den wissenschaftlichen Instrumenten ihr vornehmes Aussehen verleiht. Auf die glänzende Messingoberfläche aufgebracht, verleiht er ihnen nicht nur den notwendigen mechanischen Schutz, sondern wird zum wesentlichen esthetischen Bestandteil.

Obwohl die wissenschaftlichen Instrumente nie Umweltbedingungen ausgesetzt waren, die archäologischen Objekten vergleichbar sind, bedürfen sie doch konservatorisch-restauratorischer Pflege. Die unsachgemässe Handhabung, die Lagerung unter ungünstigen klimatischen Bedingungen oder die Benutzung im Felde haben zum Verlust des Instrumentenlackes, zur Oxydation der Metalle oder gar zu deren Korrosion, zur Versprödung der organischen Teile oder zur Ablösung der glänzenden Metallschichten auf den Spiegeln geführt. Ganz wesentlich haben aber viele Instrumente unter unsachgemässer Reinigung gelitten. Die weit verbreitete Anwendung von "Metallputzmitteln", wie z.B. Sigolin oder alkalischen Scheuerpasten, hat zu intensiver Korrosion der Teile geführt, die von diesen nach der Reinigung nicht genügend befreit wurden. Diesen "Reinigungsmethoden" ist auch oft der originale Instrumentenlack zum Opfer gefallen, der dann durch einen gewöhnlichen Transparentlack ersetzt wurde.

2. Voruntersuchung und Wahl des Konservierung-Restaurierung Konzeptes

Vor jedem konservatorischen Eingriff wird das Objekt einer Voruntersuchung unterzogen. Diese umfasst eine genaue Beschreibung des Ist-Zustandes mit photographischer und zeichnerischer Dokumentation. Die Funktionsweise des Instrumentes wird untersucht und die Anwesenheit objektfremder oder ersetzter Bestandteile ermittelt. Frühere restauratorische Eingriffe werden festgehalten.

Mit dem zuständigen wissenschaftlichen Leiter der Sammlung sowie dem chem.-physikalischen Laboratorium werden anschliessend folgende Punkte abgeklärt wie z.B.:

— Wo soll das Instument ausgestellt werden?
— Handelt es sich um eine Reinigung für eine photographische Aufnahme?
— Liegen technologische Fragen vor?
— Soll das Instrument wieder funktionstüchtig sein?
— Sollen fehlende Bestandteile ersetzt werden?
— Müssen Analysen der Korrosionsprodukte, der Legierungen oder Metalloberflächen (Vergoldung usw.) durchgeführt werden?
— Wird an der Echtheit gezweifelt?

Aus alle diesen Fragestellungen ergibt sich dann ein Konservierungs- und Restaurierungskonzept. Oft sind zuerst naturwissenschaftliche Analysen durchzuführen. Hat man sich auf ein Konzept geeinigt, so kann die eigentliche Konservierung beginnen.

3. Konservierung und Restaurierung

3.1. Die Zerlegung der Instrumente

In den meisten Fällen ist es wünschenswert und notwendig, das Instrument in seine einzelnen Bestandteile zu zerlegen. Diese Arbeit erfordert viel Fingerspitzengefühl und ein Verständnis für mechanische Konstruktionsprinzipien. Die oxydierten oder korrodierten Teile wie Gewinde lassen sich oft nur mit Hilfe von Schmiermitteln auf graphitbasis (z.B. Molykote; 7) und speziell gefertigten Werkzeugen lösen. Skizzen oder photographische

Aufnahmen sowie die Numerierung der entfernten Bestandteile erlauben dann wieder nach der Konservierung das sachgemässe Zusammensetzen. Das Zerlegen der Objekte hat verschiedene Vorteile:

- die Funktionsweise ist oft erst in diesem Zustand ersichtlich (optische Systeme bei Fernrohren);
- falsch zusammengesetzte optische Systeme werden ersichtlich;
- originale und ersetzte Teile können erkannt werden (z.B. alte oder metrische Gewinde);
- Spuren von originalem Instrumentenlack sind meist im Innern erhalten;
- Reste von alten korrosiven Reinigungsmitteln können entfernt werden.

3.2. Bestandteile aus Messing und ihr Oberflächen-Schutzlack

Wie schon erwähnt, ist ursprünglich die Oberfläche der Bestandteile aus Messing mit einem transparenten Instrumentenlack versehen worden. Dieser besteht im wesentlichen aus Schellack, sowie färbenden natürlichen Substanzen (siehe Anhang I).

Bei der Konservierung muss also die Erhaltung der metallischen Bestandteile immer im Zusammenhang mit der Erhaltung des Schutzlackes gesehen werden. Grundsätzlich unterscheiden wir die folgende Fälle:

- Schutzlack vollkommen erhalten oder nur kleine Fehlstellen; Messingteile meist nur oxydiert.
- Schutzlack nur stellenweise erhalten; Metallteile oxydiert und ev. korrodiert.
- Schutzlack nur noch in Spuren vorhanden; Metallteile meist oxydiert und korrodiert.

Trotz den vielen Zwischenstadien sollen im Folgenden zwei Interventionsarten beschrieben werden.

3.2.1. Minimal Intervention

Hier handelt es sich um Objekte, bei denen der originale Instrumentenlack fast vollständig erhalten ist. Unser Ziel ist es, ihn zu bewahren. Der Arbeitsablauf gliedert sich in folgende Schritte:

1) Die Reinigung der Oberfläche erfolgt lokal mit Hilfe von mit Lösemitteln getränkten Wattestäbchen. Als Lösemittel kommen Wasser, Reinbenzin, White spirit oder Perchloraethylen in Frage (unter dem Abzug arbeiten!).

2) Oxydationen der Metalloberfläche (z.B. Fingerabdrücke) werden durch sanftes Reiben von Hand unter Wasser und mit einem Gemisch von Bimsteinpulver/Kaolinpulver entfernt.

3) Zur Anwendung gelangt auch eine "Reinigungswatte" die unter dem Namen "Metarex" erhältlich ist (8). Unsere Untersuchungen haben gezeigt, dass es sich dabei im Wesentlichen um eine Watte handelt, die mit einem sehr feinen Quarzmehl sowie einem schwerflüchtigen org. Lösemittel belegt ist. Korrosionsversuche an Messingplättchen über mehrere Monate ergaben ausgezeichnete Resultate (siehe Anhang II).

Diese Versuchsreihe zeigte ebenfalls, dass die weit verbreiteten Reinigungsmethoden mit ammoniakalischer Schmierseifenlösung sowie mit Komplexon- oder Rochelle-Salzlösungen (Na-K-Tartrat) oder Reinigungspasten wie Sigolin oder VIM mittlere bis sehr starke Nachkorrosionen erzeugen (siehe Anhang II).

4) Nach dieser Reinigung muss die Oberfläche gut mit einem Wattebausch sowie den unter 1) erwähnten Lösemitteln gereinigt werden.

5) Es besteht immer die Tendenz, die Oberfläche der Instrumente zu stark zu reinigen. Sie wirkt dann zu "neu" und störend. So verzichten wir darauf, dünne braune und schwarze Oxydationsschichten zu entfernen, die nicht korrodierend wirken.

6) Vor der Retouche der Lackfehlstellen stellt sich die Frage, ob die gereinigte Metalloberfläche farblich den oxydierten Teilen angepasst werden soll. Ursprünglich verwendeten wir gelegentlich verdünnte Lösungen von Schwefelleber (K_2S) und stabilisierten die Oberfläche durch Behandlung mit verdünnter Natriumthiocyanat-Lösung. Trotz dieser Vorsichtsmassnahme wurde der umliegende originale Instrumentenlack unterwandert, und die Metalloberfläche verfärbte sich. Eine gewisse Nachdunkelung der Oberfläche mit in Alkohol gelöstem Benzotriazol (2% g/v) hat sich nicht bewährt, da der originale Lack angegriffen wird und quilt.

7) Zur Retouche der Lackfehlstellen verwenden wir drei Produkte:

a) Meist begnügen wir uns damit die Metalloberfläche mit einer feinen Schicht eines mikrokristallinen Wachse zu schützen (9);

b) in anderen Fällen bringen wir eine sehr dünne Schicht eines Acrylharzlackes auf (Paraloid B72 oder Incralack; 10).

Die Beschichtung erfolgt mit einer Mini-Spritzpistole, wie sie im Modellflugzeugbau Anwendung findet. Die Lackschicht ist in diesen Fall extrem fragil.

c) Manchmal verwenden wir auch einen Lack, der dem originalen Instrumentenlack sehr ähnlich ist. Dies besonders, wenn eine mechanische Resistenz gefordert ist. Dieser Lack enthält vor allem Schellack sowie Naturfarbstoffe, und wurde von uns nach langeren Versuchsreihen nach historischen Quellen zusammengestellt. Hinweise zu seiner Zusammensetzung und Herstellung befinden sich im Anhang I.

Er wird wie der Acryllack unter b) appliziert und gleicht sich farblich gut dem originalen Lack an. Alterungsversuche ergaben gute Resultate.

3.2.2. Weitergehende Intervention

Ist der originale Lack abwesend oder nur noch an einigen Stellen vorhanden, die Metalloberfläche jedoch stark oxydiert oder korrodiert, so drängt sich eine weitergehende Behandlung auf. Hier gilt es, der Korrosion Einhalt zu gebieten, und das Instrument vor weiterer Korrosion zu bewahren. Die originalen Lackresten sind aber trotzdem, wenn immer möglich, zu belassen. Der Ablauf unterteilt sich dann in folgende Behandlungsschritte:

1) Das Objekt wird zuerst wie im Abschnitt 3.2.1. unter 1) mit dem Bimsteinpulver / Kaolinpulver Gemenge gereinigt.

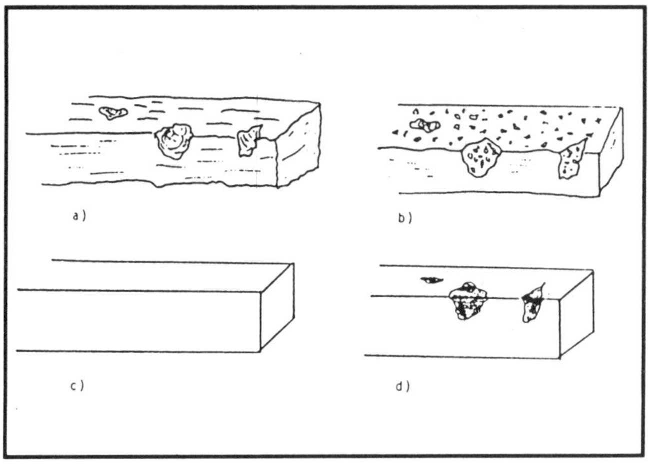

Abb. 1. *Entstehung der schwarzen Flecken auf der Messingoberfläche.*

a) *Beim Guss der Messingstange bilden sich an dessen Oberfläche durch anhaftende Luftblasen feine Hohlräume*

b) *Beim Schleifen der Oberfläche werden diese mit dem Schleifmittel (Kreide, Sandkörner, Tonerde) sowie feinem Messingpulver aufgefüllt.*

c) *Für das Auge erscheint die Oberfläche glatt und glänzend.*

d) *Nach längerer Zeit wandelt sich das Gemenge durch oberflächliche Oxydation des Mesingspulvers in eine schwärzliche Masse um.*

2) Korrosiv wirkende Korrosionsprodukte, wie basische Kupferchloride, müssen entfernt werden. Deren Zusammensetzung hatten wir analytisch abgeklärt.

Die Entfernung wird entweder von Hand mit einem feinen Skalpel, mit dem Mikro-Sandstrahlgerät oder der Reinigungswatte "Metarex" [8] durchgeführt. Kleine dunkle Flecken oder linienförmige dunkle Adern können belassen werden. Wie wir mittels analytischen Untersuchungen zeigen konnten, handelt es sich um ein Gemenge von oxydiertem Messingpulver und Schleifmitteln (Kreide, Ton und Quarz). Dieses Gemenge entstand beim Polieren der gegossenen und gehämmerten Metalloberfläche, die aber noch voll kleinen Rissen und Blasen war. Das Gemenge füllte diese Unebenheiten auf und täuschte eine glänzende Oberfläche vor. Durch die Oxydation erscheint es nun optisch dunkler (Abb. 1)

3) Nach erneuter Reinigung mit Wasser wird eine Entfettung in Perchloraethylen im Ultraschallbad vorgenommen. Dies natürlich nur in den Fällen, bei denen keine originalen Lackreste mehr vorhanden sind.

4) In einem Luft-Trockenschrank bereitet man eine geeignete Aufhängevorrichtung aus Drähten vor und erwärmt den Messingteil auf 120 °C.

5) In einer Spritzkabine wird ebenfalls eine Aufhängevorrichtung eingerichtet. Der MAH-Instrumentenlack wird mittels Vorversuchen farblich richtig eingestellt. Mit der kleinen Spritzpistole wird er bei 50 - 90 °C auf das warme Instrumententeil in feiner Schicht aufgetragen. Zur Wärmehärtung wird dieses im Luft-Trockenschrank bei ca. 120 °C ausgelagert. Der Vorgang wird bis zur Erreichung der gewünschten Dicke und des gewünschten Farbtones wiederholt. Die Schichtdicke ist jedoch geringer als diejenige des originalen Lacks. Der Farbton wird individuell für jedes Instrument eingestellt.

3.2.3. Mechanische Bearbeitung originaler und Bezeichnung neuer Instrumententeile

Verbogene Messingbestandteile werden in der Regel im kalten Zustand zurückgeformt, um den originalen Instrumentenlack nicht zu zerstören. In Ausnahmefällen rechtfertigt es sich, den Bestandteil zu erhitzen. Der Lack ist dann natürlich verloren.

Gebrochene Bestandteile sollen wenn möglich mit einem Zweikomponenten-Kunstharz wieder verleimt werden (z.B. Araldit LY 554 mit Härter HY 554) [12]. In Ausnahmefällen wird mit einem Messinglot gelötet (Castollin 1802).

Fehlende Bestandteile ersetzen wir nur, wenn das Instrument sonst nicht präsentiert werden kann. Das fehlende Bestandteil wird dann nach einem Modell in einer fremden Legierung hergestellt und mit dem Zeichen "MAH" versehen. Besitzen wir keine Vorlage so begnügen wir uns mit einem Ersatzteil aus Plexiglas, der den Verlust des originalen Bestandteiles deutlich ersichtlich macht. Schrauben verwenden wir wenn möglich wieder. Fehlen solche, so werden die neuen ebenfalls mit einem Stempel "MAH" versehen.

3.3. Bestandteile aus Eisen und Stahl

Zu hohe Luftfeuchtigkeiten in Lagerräumen sowie Fingerabdrücke führen oft zu starken Korrosionserscheinungen an Oberflächen aus Eisen und Stahl. Da die Oberflächenbehandlung (bläuen, brünieren) ein wesentliches Element der wissenschaftlichen Instrumente darstellt, müssen Konservierungsmethoden gefunden werden, welche diese bewahren.

Die gängigsten Konservierungsmassnahmen seien kurz zusammengefasst:

1) Nach der Zerlegung des Instrumentes werden die korrodierten Eisen- und Stahlbestandteile in ein nicht korrosiv wirkendes leicht alkalisches Bad eines Spezialreinigers "R 2000" gelegt [13].

2) Nach gründlicher Reinigung mit Wasser wird über Alkohol- und Aceton- Bädern und im Luft-Trockenschrank getrocknet. Anschliessend wird mit Perchloraethylen im Ultraschallbad entfettet.

3) Immer noch anhaftender Rost wird dann von Hand mit einem Skalpell oder mit feinen Zahnarztbürsten, dem Mikrosandstrahlgerät oder feinstem Schmirgel entfernt. Es folgt das Einlegen in ein Korrosionsschutzoel (14) und das Abtrocknen der Oberfläche.

4) Die farbliche Anpassung der gereinigten Stellen ist heikel. Eine Nachahmung originaler Anlaufverfahren vermeiden wir. Versuche zur Färbung mit verdünnter Arsensäure gaben kein überzeugendes Resultat. In Ausnahmefällen verwenden wir ein kommerzielles Produkt, Nigrit, das jedoch nicht immer sehr gute Resultate ergibt.

5) Als Oberflächenschutz kommt ein Acrylharzlack wie (Paraloid B 72 [10] oder ein mikrokristalliner Wachs in Frage [9].

Für Objekte die in Vitrinen ausgestellt werden und somit dem Staub nicht ausgesetzt sind, kann auch ein Korrosionsschutz-Oel verwendet werden [14].

3.4.Versilberte und vergoldete Oberflächen

Während bei vergoldeten Oberflächen die exterm dünne Goldschicht oft durch unsachgemässe Reinigung entfernt worden ist, kann die Anwesenheit einer Feuervergoldung durch den spektralanalytischen Nachweis von Quecksilberspuren mittels Röntgenfluoreszenz bestimmt werden. Dafür verwenden wir ein Spektrometer, das für die Analyse grösserer Objekte speziell konzipiert wurde [15].

Versilberte Oberflächen sind oft stark geschwärzt und, da sie sehr dünn sind, in einem schlechten Zustand. Wir beschränken uns auf eine Reinigung mit organischen Lösemitteln (Reinbenzin, Aceton) und dem Aufbringen eines dünnen Schutzlackes (Paraloid B 72). Auf eine neue Vergoldung oder Versilberung verzichten wir bewusst aus ethischen Gründen.

3.5. Spiegel

Die Konservierung von Licht-Spiegeln bei Mikroskopen und Hohlspiegeln bei Fernrohren gehört zu den schwierigsten der Erhaltung wissenschaftlicher Instrumente.

Literaturangaben zur Materialzusammensetzung dieser Bestandteile haben wir nicht gefunden. Dies bewog uns seit einiger Zeit, ein Analysen-Programm in die Wege zu leiten. Obwohl erst wenige Analysen vorliegen, scheinen sich bereits Tendenzen abzuzeichnen. Hohlspiegel bei Mikroskopen bestehen aus Quecksilber-Zinn Amalgamen. Durch den Verlust von Quecksilber wandeln sich diese Amalgame um, was zu deren Ablösung von der Glasoberfläche führt. Konservatorische Massnahmen ergreifen wir bis anhin keine, solange der Chemismus des Zerfalls nicht besser abgeklärt ist.

3.6. Glas

Die aus Glas hergestellten Linsen der Objektive und Okulare sind meist in relativ gutem Zustand, jedoch verschmutzt. Die Reinigung darf nicht mit Xylol erfolgen, da die Linsen früher mit Kanadabalsam verklebt wurden. Meist beschränken wir uns auf eine Reinigung mit Wasser und einem feinen Hirschleder. Bei sehr schmutzigen Linsen leisten die in Optikergeschäften erhältlichen Spezialpapiere gute Dienste. Schliesslich kann bei extremer Verschmutzung eine sorgfältige Reinigung mit Alkohol, Chloroform oder Aether vorgenommen werden.

3.7. Organische Bestandteile

3.7.1 Holz

Die Behandlung der Bestandteile aus Holz stellt im allgemeinen kein grosses Problem. Die konservatorischen Massnahmen beschränken sich auf eine leichte Reinigung der Oberfläche und dem Aufbringen eines Wachsüberzuges [9].

3.7.2. Horn, Knochen, Elfenbein, Schildkröten- und Walfischhaut

Meist sind auch diese Bestandteile oder Einlagen in einem relativ guten Zustand. Gute Erfahrungen haben wir mit einem Reinigungsmittel gemacht, das unter dem Markennahmen "R 2000" verkauft wird [16]. Es handelt sich grundsätzlich um ein Produkt auf Ochsengalle-Basis. Es wird mit einem feinen Lappen auf die Oberfläche gebracht und anschliessend gut mit lauem Wasser oder Alkohol gewaschen.

3.7.3. Leder

Leder pflegen wir mit handelsüblichen Produkten, die für die Konservierung entwickelt wurden. Es handelt sich um das vor allem im Frankreich verbreitete Produkt "Cire 212" [17] und das aus England stammende "Pliantine" [18].

3.7.4. Papier

Wissenschaftliche Instrumente in Verbindung mit Papier sind eher selten. Treten jedoch Schäden an solchen Objekten auf, so ist der Papierteil vom Rest völlig zu trennen und einem erfahrenen Papierrestauratoren zur Bearbeitung zu überlassen.

4. Ausstellung und Lagerung

In der Ausstellung sind die Instrumente vor direkter Sonneneinstrahlung zu schützen. Liegen dekorierte Leder- oder Papierteile vor, so sollte die Beleuchtung stark reduziert werden (max. 100 Lux).

Die ideale Luftfeuchtigkeit zu finden ist nicht einfach, verlangen doch Metalle und organische Bestandteile verschiedene Werte. Wir haben die Tendenz, die relative Feuchte eher etwas tief zu halten (45-50 %), um Korrosionserscheinungen an den Metallteilen zu verhüten.

Glocken aus Plexiglas oder Glasvitrinen tragen sehr zur Stabilisierung der Objekte bei und vermeiden eine Beschädigung durch die Besucher.

In den Lagerräumen kann durch eine relativ tiefe Temperatur die Umwandlung der Materialien verlangsamt werden. Feuchtigkeitswerte sind wie für die Ausstellung einzustellen.

Wichtig ist, sowohl für die Ausstellung als auch für die Lagerung darauf zu achten, dass keine Fremdmaterialien (Unterlagen aus Kunstschaumstoff, Holz, Leime oder Dekorationsstoffe) korrosive gasförmige Substanzen abgeben [19].

Verdankung

Die Arbeiten zur Konservierung und Restaurierung wissenschaftlicher Instrumente wurden in enger Zusammenarbeit mit verschiedenen Personen ausgeführt, denen wir an dieser Stelle unseren besten Dank aussprechen. Es handelt sich um: Frau M. Archinard, Konservatorin des Musée de l'histoire des sciences; Frau M. Mas, welche die Analysen der Korrosionsprodukte und Metall-Legierungen durchführte und Herrn Stephan Ramseyer, welcher bei der Konservierung half. Dem Direktor des Musée d'art et d'histoire, Herrn Dr. C. Lapaire, danken wir für sein Interesse an dieser Arbeit.

ANHANG I

Herstellung des Instrumenten-Lackes MAH

Die Zusammensetzung dieses Instrumenten-Lackes lehnt sich eng an historische Quellen an [20, 21, 22]. Versuche mit modernen Kunstharzlacken haben nicht zum erhofften Resultat geführt. Diese Ueberzüge sind nicht kratzfest und ergeben farblich unbefriedigende Resultate. Der neu aufgebrachte Instrumentenlack nach historischen Rezepten löst sich im Gegensatz zum originalen Lack leicht in Alkohol. Er kann also vom Instrument wieder entfernt werden.

Der Lack besteht aus drei Teilen

a) dem Grundlack (Schellack)

b) den Farbstoffen (Drachenblut, Curcuma, Safran)

c) Zusatzstoffen wie Spikoel (zur Verzögerung der Trocknung) und Borsäure (Härtung und Verbesserung der Haftfestigkeit am Metall).

Korrosionserscheinungen durch die Borsäure an Messing haben wir auch nach langjähriger Beobachtung nicht feststellen können.

Die Herstellung erfolgt in folgenden Schritten:

Herstellung der Grundlösungen

In einem grossen Becherglas erwärmt man ca. 1 l Aethylalkohol auf 50 °C. Darin löst man Schellack (reine weisse Sorte) bis zur Sättigung (ca. 300–350 gr). Die Lösung wird zuerst durch einen groben Filter (Metallsieb, Nylonstrumpf), anschliessend durch einen Papierfilter in der Porzellannutsche unter Vakuum filtriert (Papier öfters wechseln).

In einer Flasche lässt man das Filtrat absetzen. Nach einiger Zeit bildet sich unten wieder eine Trübung. Für den Gebrauch hebt man mit einer Pipette die obenstehende klare Lösung ab.

Gleich vor dem Gebrauch wird diese Lösung, da sie nun zu dünn ist, auf einem heizbaren Magnetrührer eingedampft, bis sie die für das Spritzen richtige Viskosität besitzt.

Die färbenden Stoffe (Drachenblut, Curcuma, Safran) werden einzeln im Alkohol gelöst und abfiltriert. Die Konzentration richtet sich nach der gewünschten Intensität der Färbung.

Nun erfolgt die Mischung der Komponenten in etwa folgenden Gewichtsverhältnissen:

Alkohol-Schellacklösung	450 gr.
Alkohol-Drachenblutlösung	40 gr.
Alkohol-Safran/Curcuma-Lösung	20 gr.
Spikoel	60 gr.
Terpentinoel	40 gr.
Borsäure hist. (zuerst in Wasser) lösen, dann Alkohol zufügen und das Wasser verdampfen lassen).	3 gr.

Die Viskosität der Mischung kann nun, durch Zusatz von Terpentinoel, nach Belieben eingestellt werden. Vor dem Spritzen erwärmt man die Lösung auf 50–80 °C. Durch Versuche ermittelt man den gewünschten Farbton. Das Objekt wurde inzwischen im Trockenschrank bei ca. 120 °C vorgewärmt. Die Aushärtung des Lacks nach dem Spritzen erfolgt ebenfalls bei ca. 120 °C.

Es empfiehlt sich, verschiedene dünne Schichten aufzubringen und dazwischen immer wieder auszuhärten.

ANHANG II

Versuche zu Reinigungsmitteln für Messing

In der Restaurierungspraxis und im Fachhandel werden eine ganze Reihe Reinigungsmittel für Messing angepriesen. Wir haben einige dieser Mittel näher untersucht. Dabei gingen wir von der Voraussetzung aus, dass immer Reste der Reinigungsmittel auf dem Messing bleiben. Um das Korrosionsverhalten dieser unerwünschten Reste zu ermitteln, wurden saubere Messingplättchen mit diesen Reinigungsmitteln belegt und während 5 Monaten in einer Feuchtkammer bei ca. 80 % rel. Luftfeuchtigkeit ausgelagert. Die Resultate sind in der nachfolgenden Tabelle aufgeführt:

Verhalten von Messing gegenüber Reinigungsmitteln nach fünfmonatiger Lagerung bei 80 % rel. Luftfeuchtigkeit

1) Bimstein + Kreide-Pulver	keine Korrosion
2) Metallreinigungswatte Metarex [8]	" "
3) Ammoniak. Schmierseifelösung	leichte Korrosion
4) Komplexon III (EDTA) 5%	" "
5) Alk. Rochelle-Salzlösung 5 % (Natrium-Kalium-Tartrat)	" "
6) Sigolin Metallreinigungspaste [24]	starke Korrosion
7) Vim Scheuerpulver [25]	" "

Die im ersten Block aufgeführten Reinigungsmittel sind tatsächlich für das Messing völlig ungefährlich. Bei der Metallreinigungswatte METAREX handelt es sich übrigens um sehr feines Quarzmehl, das mit einem org. Lösemittel in Watte aufgenommen ist.

Interessant ist die zweite Gruppe, da die ammoniak. Schmierseifelösung die klassische Reinigungsmethode darstellt. Bleibt ein Tropfen dieser Lösung auf dem Messing liegen, so bildet sich eine lokale Korrosion.

Die dritte Gruppe umfasst Reinigungsmittel, die der Vergangenheit angehören sollten. Während ein Haushaltreinigungsmittel wie VIM kaum mehr benützt wird, erfreut sich Sigolin doch noch in weiten Kreisen grosser Beliebtheit. Die auf der Flasche aufgedruckte Reklame muss als irreführend bezeichnet werden.

Referenzen

1 E. **Gay**: Brass instruments: practical guidelines for repair, Technology and Conservation 6, (3), 38-46, Fall 1981

2 R. **Pearsall**: Collecting and Restoring Scientific Instruments, David and Charles, Newton Abbot, London 1974 (Das Kapitel über Restaurierung ab S. 172 ist nicht zu empfehlen).

3 A. **Turner**: Early Scientific Instruments Europe 1400 - 1800, Sotheby's Publications, London 1987, ISBN 0 85667 319 6

4 G.L.E. **Turner**: Essays on The History of The Microscope, Senecio Publ. Comp. Ltd., Oxford 1980, ISBN 0 906831 01 6

5 R.S. **Clay** and T.H. **Court**: The History of The Microscope, The Holland Press, London 1975

6 G.L.E. **Turner**: Scientific Instruments Nineteenth-Century, Sotheby's Publications, London 1983. ISBN 0 85667 1703

7 **Molykote**: Dow Corning GmbH, München BRD

8 **Metarex**: Estalin AG, 4000 Basel

9 Renaissance Micro-crystalline wax Picreator Renaissance Products, 44 Park View Gardens, GB-London NW4 Vertreter: Archival Aids Ltd., P.O. Box 5, GB-Spondon, Derby DE2 7BP

10 Paraloid B 72 (Copolymerisat Aethyl-Methyl Metacrylat) 5 % in Toluol (Hoechst)
Diethelm Farbenfabriken CH-8603 Brüttisellen
Incralac LA/1: Gleiche Zusammensetzung wie Paraloid, enthält zusätzlich Benzotriaol als Inhibitor
Archival Aids Ltd. P.O. Box 5, Spondon, Derby DE2 7BP, England

11 Röntgenographische Untersuchungen (Candolfi-Kamera, Fe-Röhre) ergaben, dass es sich bei den beobachteten grünlich-blauen Korrosionsprodukten um das basische Kupferchlorid Atacamit, $Cu_2(OH)_3$ CL, handelt (ASTM 25-269).

12 Araldit LY 554 mit Härter HY 554
Ciba-Geigy Basel über Astorit SA, CH-8840 Einsiedeln
Bei Klebungen, die rasch ausgeführt werden müssen, eignet sich besser das Epoxydharz Dercon "5 Minutes Epoxy", Neidharz 1233 Bernex oder Archival Aids (Ref. 9).
Dieses Harz löst sich im Epoxy-Desintegrater (Archival Aids).

13 R 2000
Schober, Fabrik chemischer Artikel, Enzenberg 13, 3428 Duderstadt, BRD

14 Multifix, Universaloel
Ernst Hürlimann, Mineraloelprodukte, CH-8820 Wädenswil

15 Die Apparatur besteht aus einer Kevex Mini-Röntgenröhre (50 kV, lmA), einem Seforad Si(Li)-Detector und einem Tracor Northern Vielkanal- Analysator. Grössere Objekte können direkt analysiert werden.

16 "R 2000" siehe bei 13

17 Cire 212, Microcrystallines Wachs mit Fungiziden
Bibliothèque nationale, Château de Sable, F-72300 Sable-sur-Sarthe

18 Pliantine (British Museum Leather Dressing) enthält Bienenwachs, Lanolin und Zedernoel in Hexan
Archival Aids Ltd., P.O. Box 5, Spondon, Derby DE2 7BP England

19 Ca. 50 gr. des zu testenden Materials werden in einem geschlossenen Glasgefäss mit verschiedenen Metallplättchen (Gold, Silber, Kupfer, Messing, Eisen, Blei) bei 50 °C in einem Trockenschrank ausgelagert. Die Test-Metalle dürfen das Material nicht berühren. Man fügt 1-2 Tropfen Wasser dazu, um eine relative Luftfeuchtigkeit um 50 % zu erhalten. Die Oberfläche der Plättchen wird mit denjenigen einer Null-Probe verglichen.

20 E. **Rouveyre**: Cinq cent soixante recettes et procédés. Flammarion, Paris 1911

21 Ch. **Coffigner**: Couleur et vernis, manuel du peintre, J.B. Ballière, Paris 1921

22 R.J. **Gettens** and G.L. **Stout**: Paintings Materials - A Short Encyclopaedia, Dover Pub., New York 1965

23 **Sigolin**: Metallreinigungsmittel

24 **Vim**: kommerzielles Scheuermittel für Glas, Porzellan, Metalle

Adresse der Autoren:

Laboratoire, Musée d'art et d'histoire
9-11, rue du Clos
CH - 1207 Genève

ROBERTS, Frances C.

Organic remains on archæological iron & bronze

Abstract

Conditions for the preservation of buried artifacts in western Denmark. Work with these archæological materials (metals/organics) and the basic information extracted as a matter of course. Description of collaboration with specialist colleagues to achieve the best analysis of the material, thereby extending our sparse data from the poorly preserved objects.

Introduction

During the last ice age Denmark was only partially covered by the maximum extention of the ice, the line cutting north/south through the Jutland peninsular (Fig. 1).

The preceeding ice age had covered western Jutland with moraine sand. During the last ice age it underwent extreme wind erosion and was crossed by innumerable streams from the ice, thus producing a flat landscape of glacial and alluvial sand.

The area covered by the two counties, Ribe and Ringkøbing, coincides remarkably well to the same line (Fig. 2). This means that the county conservation laboratory which services the more than 20 museums in these two counties recieves most of its archæological finds from a well defined soil/geological area.

Fig. 1 *Denmark during last ice age*

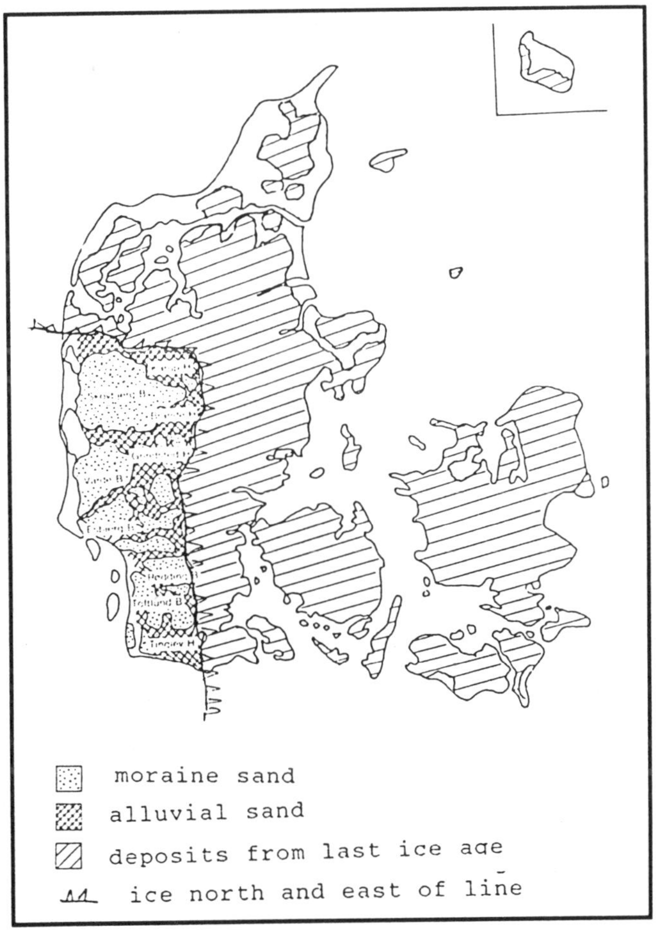

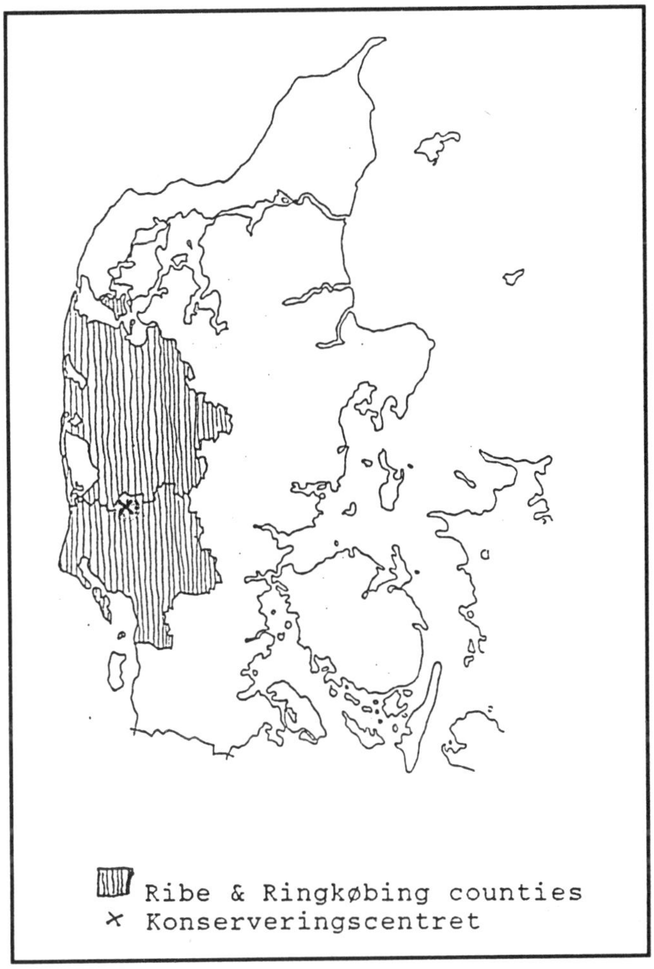

Fig. 2 *Denmark*

In prehistory light soils were popular, being easier to clear and cultivate, so this same area has always been well populated. Archæologically rewarding but problematical for conservation.

State of Preservation

Sandy, porous, acid soils, salt winds from the North sea coast and an excess of rainfall proportional to evaporation have all worked their worst on metal artifacts from prehistory. Generally speaking, the result is total mineralisation within a short space of time.

Bronze

Bronzes remain in three different states of decay:

a) Mineralised with oxides at the core and the usual variety of green corrosion products on and covering the surface, well known from most areas.

b) Uniform, usually very powdery, green corrosion products with little or no cohesion which "disintegrate" on excavation.

c) Objects consisting only of tin salts (oxides and hydroxides) cream to grey in colour and without a hint of copper. The original surface is finely preserved and visible, being without any corrosion crust. Archæologists seeing them for the first time tend to classify them as bone, and they are mostly found among objects made from wrought wire or sheetmetal.

Whatever the state of the "bronze" object, the speed in which corrosion began and developed has preserved organic material in their immediate vicinity. The most common being textile fragments on fibula and buckle, but other materials such as leather, wood around fittings and bone from combs and human remains are widely found. The poisonous effect of the copper salts has hindered microbiological degradation so that the cells themselves remain. Under the microscope wool fibres are green or blackish, they are identifiable but appear to have had bites taken out of them. Though these holes are on extreme small scale, weakness is apparent and results in brittleness. Fibres tend to snap in lengths down to a few millimetres at low levels of stress. Other organic materials retain some degree of original integrity. In one case a bronze seal stamp was found to have preserved a small amount of human flesh from the grave, first thought to be a leather purse!

Iron

Iron objects are found in all stages of decay, such being the complex nature of iron, though most usually as hollow shells of corrosion or magnetite cores with mountainous corrosion products. Where organic materials have been in intimate contact with iron, one tends to find that the metal object is so leached as to be hollow and that the resulting salts have invaded the cell structure, forming a pseudofossilised remnent of organic material. It would seem that the salts, being led away through the cells, are unable to concentrate to form worts or irregular disfiguring corrosion. The cellular material is not usually preserved, but rather replaced, by the salts. Most common finds include textile patterns, leather and especially wood.

Organic remains

Without contact with metals next to nothing can be found in the light soils, a mere change of colour in the sand, even heavy oak beams disappear. Urban digs and bog sites do offer up a considerable amount of material, but this is beyond the scope of the present paper.

Treatment

It should be clear that a large percentage of the objects from digs in western Jutland are in a sorry state, but also contain a potential of information concerning organics. A useful method, and one much resorted to in our area, is to lift objects or groups or objects in earth blocks using paraffin wax (the sort used to make cheap candles) as a support. Melted (m.p. ca. 55-65°C) wax is brushed on in several layers according to the size of the block. An X-ray in the lab facilitates work; and with a warm spatula it is a simple matter to remove the lid and excavate the fragile object.

Treatment is relatively "simple" though time consuming. Careful removal of the earth and extraneous matter is in most cases the most important element, followed by examination and a description of the find (see next paragraph). Where necessary, we have begun to consolidate with an acrylic colloidal dispersion, Acrysol WS-24 from Rohm and Hass Company, sold in Denmark under the name Primal WS-24. It has a pleasant, low viscosity and, being on a water basis, is less of a safety hazard than consolidents based on solvents. We have become more restrictive in our use of consolidents, prefering to offer strength for the possibility to re-examine organic remains at a later date. Consolidation can make this difficult or even impossible, but is in some cases unavoidable when very fragile and degraded objects are needed for exhibition.

Standard observations

Textiles are examined on four points, as a matter of course (Figs. 3 and 4).

Fig. 3 *Three bronze fibuli with textile remains*

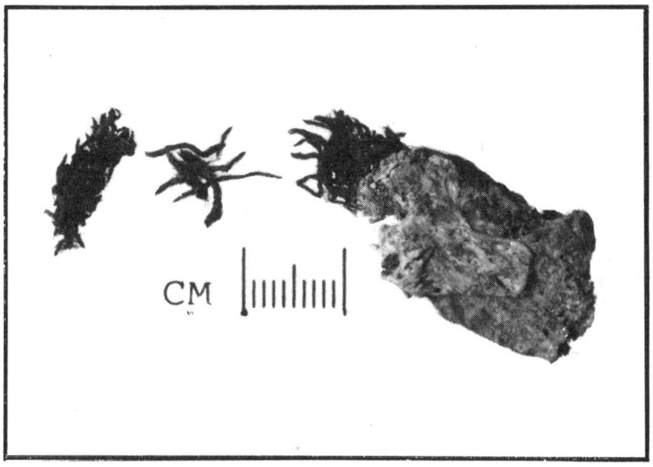

Fig. 4 *Wool fibres from fibula*

a) Whether they are S or Z spun and/or twined threads. Each thread in both warp and weft is defined, as patterns were made, as they are today, by weaving combinations of S & Z threads. It might sound like a great task, but the average size of the pieces found around fibuli is 1- 4 cm^2 .

b) Which weaving technique(s) have been used. Tabby, various twills, tablet weave, sprang or other type. Selvages and borders are noted where they appear. Small fragments seldom give up adequate information, though surprising amounts are sometimes found in very compact pieces.

c) Number of threads per centimeter, in both directions. This indicates the finesses or courseness of the cloth. It is important to record how long the piece measured was, if there is only a little fragment (e. g. 14/18 threads per cm Z/Z, 2/2 kipper, measured over 5/10 mm).

d) Textile material if possible (wool, linen...)

Most of the textile fragments are only of scientific value, and would not be exhibited, thoung the material is always of interest to the public, if it is accompanied by sufficient additional information.

Wood is examined for type of wood, direction of grain and the presence of more than one piece of wood (e. g. nails passing through two planks).

Leather is interesting in connection with iron, a surface may be preserved which facilitates identification of the type of skin used.

Other materials are described.

Fig. 5 *Reconstruction of waggon casing from viking grave*

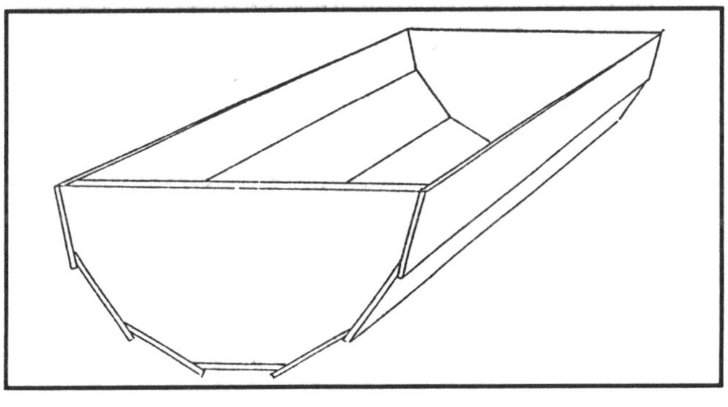

Fig. 6 *Iron mount from waggon*

Collaboration

No objects conservator has the time to work in depth with the results hereby obtained, there is always plenty more work to be done. The conclusions are also within the domain of the archæologist, but she will also have many different areas of work and may not be able to spend the time on analysis of wood or textiles.

We are lucky, in Denmark, to have specialists in these two fields, archæologists who have spent many years investigating and analysing wood and textiles from European prehistory.

Lise Bender Jørgensen's book "Prehistoric Scandinavian Textiles" is an extremely detailed and thorough analysis, being the first volume of a larger work covering the whole of Europe. The book covers the investigation of 1,322 grave finds, collected from museums in Denmark, Schleswig-Holstein, Sweden and Norway. This puts the small fragments we find into a broarder context, whereas each piece alone would be of limited interest.

Claus Malmros, from the National Museum in Copenhagen has worked on our material on numerous occasions, using his rather special combination of archæology and natural history.

From the rows of nails and remains of the gables in a viking age grave, he has described the waggon casing which had been used as a coffin. The two gable ends were

Fig. 7 *Nail from waggon*

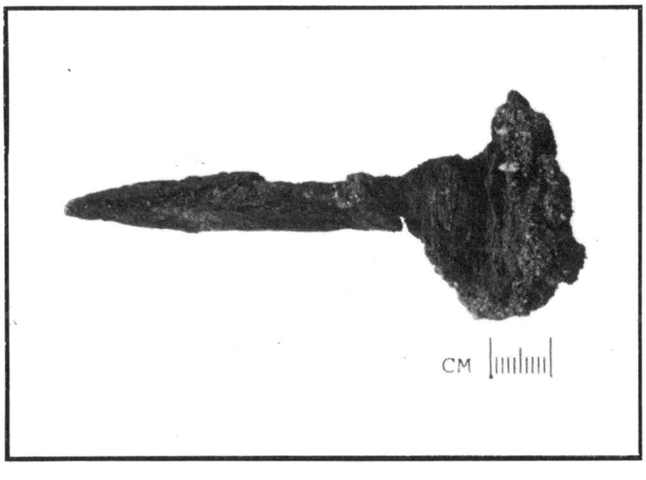

taken up as blocks which, after being X-rayed, were excavated by Claus. The relative positions of the gable mountings could be reconstructed, as well as the position of the timbers and the wood employed (Fig. 5). Had the mountings been excavated on the site and brought into conservation in the usual way, important information would have been lost. The mounts and nails in themselves are not too inspiring (Figs. 6 and 7).

Other projects have given us a detailed knowledge of the viking period's use of different types of wood for tools and weapon parts; and a total analysis of all the knife shafts from an iron age burial ground.

Conclusion

Stress has been laid on the importance of metal objects found on excavations ever since archæology became a science. As a result, organic remains have often been overlooked and even stripped from the object without documentation. Though we must remove some organic remains, it should be a matter of routine to describe and photograph them first, and in many cases learn to admit that the information in the organic layer is of greater importance than the sight of the degraded metal object. Routine use of radiography in conservation makes matters so much simpler, a thorough X-ray documentation often gives the clearest picture of a very degraded metal object.

Acknowledgements

Thanks to Rita Fredsgård Nielsen for photographs.

Litterature

Forhistoriske textiler i Skandinavien.
Prehistoric Scandinavian textiles. (danish and english text),
390 pp. By Lise Benger Jørgensen.
Publisher Det Kongelige Nordiske Oldskriftselskab, Copenhagen, 1986.
Vikingernes brug af træ.(danish text english summery) article,
8 pp.
Nationalmuseets Arbejdsmark 1987.
By **Malmros** Claus
Publisher Nationalmuseet Denmark

Author's adress:

Konserveringscentret for Museerne
i Ribe & Ringkjøbing
Amter Vestergrade 5-7,
DK-6870 Ølgod,

JENSEN, Niels E.

1001 iron nails in wood

An irritating problem which crops up time after time is that of the iron nail knocked into a wooden object. This phenomenon is most usually found to be a secondary addition to the object, an early repair. Its purpose has been to strengthen an original joint which had become instable in the course of time (loose) through use.

The great disadvantage to this method is that, apart from initially keeping the fragments together, it does nothing to improve the stability of the joint (tenon and mortise). On the contrary, it introduces a number of new problems.

A nail will corrode (rust) and the corrosion occupies a good deal more space than the iron from which it came. This expansion causes ruptures and splitting in the wood. Rust will also wander into the very pores of the wood, staining an area around the nail. What is more, difficulties in removing the nail can cause even more harm to the object.

If the nail is to be removed for any of the above reasons, or if the further treatment of the object makes disassembling necessary, then the question is: "how to do it?"

In many cases it may seem that a nail is only accessible from its head, which lies flush with the object's surface, and as a rule one is tempted to drag it out by the head, or even use drills, etc., thereby causing further damage to the wood surface. As long as the nail is part of a construction, not just embedded in a plank, it is possible to avoid this damage, as will be shown in the following:

Figures **1** & **2** illustrate a classical situation – that of a loose chair leg which has been repaired with a couple of nails.

The first nail "**A**" is knocked through the side of the chair leg, and through the length of the tenon in the seat rail. The other nail, "**B**", has been made to penetrate the tenon cross wise, so as to lock it in position in the mortice.

If we tried to pull the nails out by their heads, the result could easily be severe damage to the top of the leg, and the nails might well break in the wood.

By taking the entire construction of the chair into consideration, it is possible to find an alternative method.

Fig. 1 *Chair leg, front section*

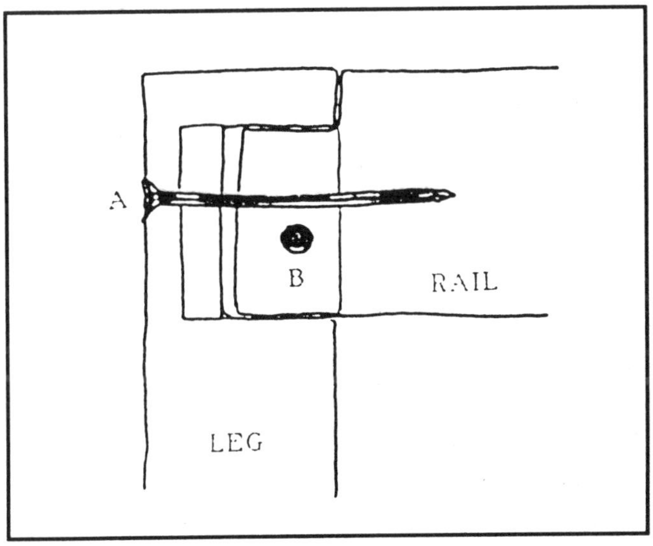

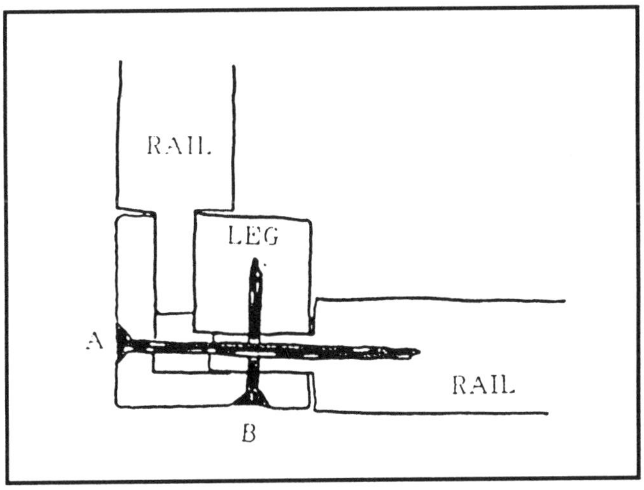

Fig. 2 *Horizontal section*

Nail "**A**": Begin by removing the head of the nail. Make a depression in the centre of the head with a blow from a pointed punch. Using a metal drill of slightly larger diameter than the nail, but less than that of the head, drill down until the body of the nail is reached. (The remainder of the head falls away as a little ring.) (See Figure 3) Now only nail "**B**" prevents the joint from being taken apart, nail "**A**" will follow the rail when the chair is disassembled.

Nail "**B**" is more inaccessible and cannot be removed without some damage to the tenon. It is removed from the leg after the chair is taken apart.

By means of slim wooden wedges driven from above into the joint, between leg and rail on both sides of the tenon, the joint is slowly opened and nail "**B**" cuts its

Fig. 3 *Nail head remover with metal drill*

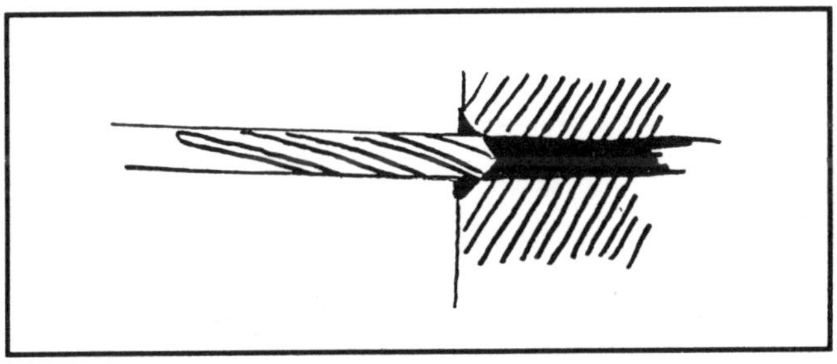

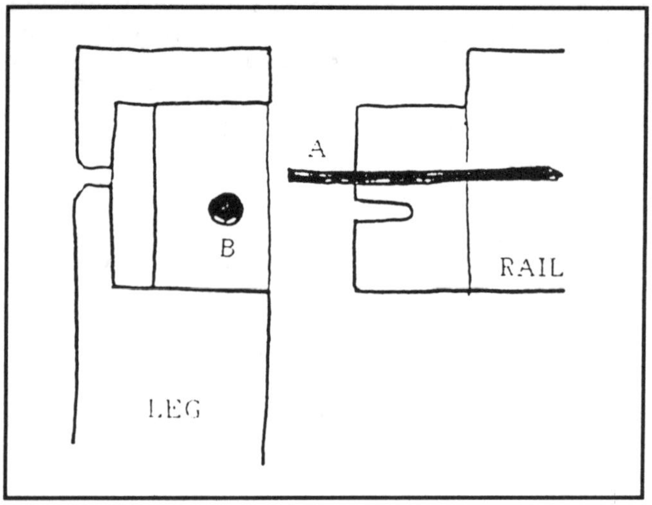

Fig. 4 *Chair leg with seat rail detached front section*

way through the length of the the tenon. (See Figures 4 & 5.)

The joint is free and the damage is restricted to a minimum of loss in the tenon. Nail **"B"** is still in situ, but is now accessible to treatment from inside the mortise. A number of sharp taps with a punch may loosen rust adhesion to the wood and a pair of pliers in the mortise can be used to lever it up.

With the chair taken apart, nail "A" can now be extracted from the rail. Fix the rail in a bench vice and loosen the rust adhesion by sharp blows with a hammer, whilst holding the nail with a flat jawed pliers to prevent it from bending. The nail is then extracted with a pair of pincers (preferably blunt ones – they should not cut through the metal, only hold it). Removal by shock treatment, hitting the pincers upwards with a hammer, is more gentle than leverage.

Fig. 5 *Chair leg with seat rail detached, horizontal section*

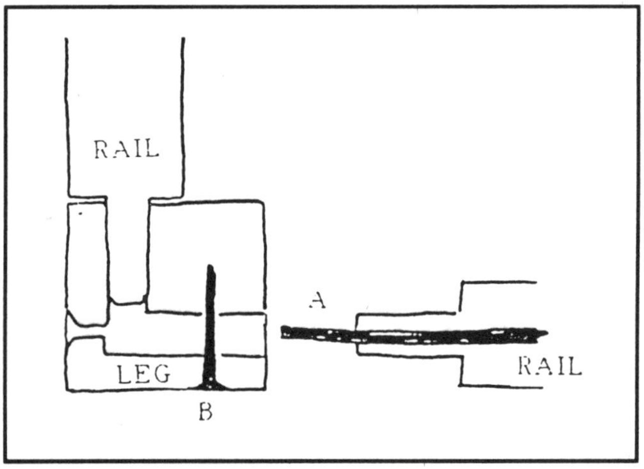

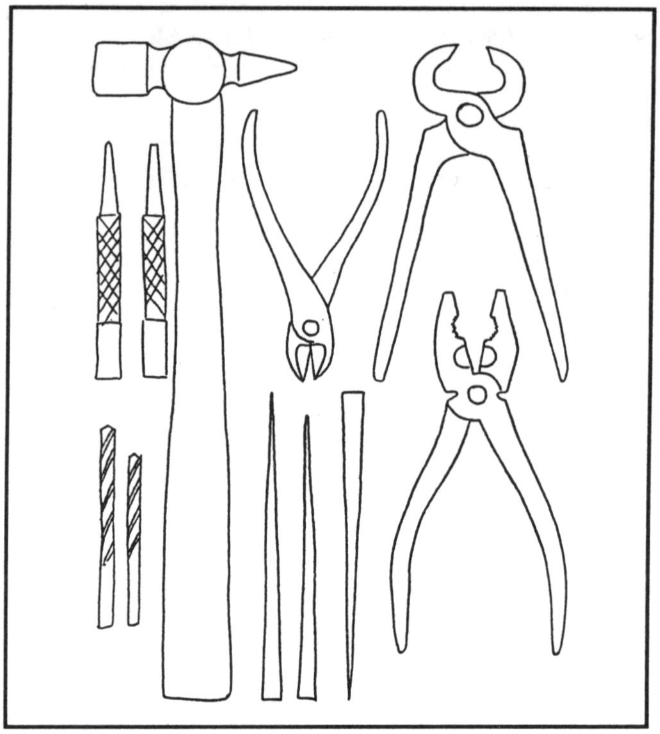

Fig. 6 *Tools*
Hammer, Punch with butt end
Punch with pointed end
Metal drills & hand-drill
A carpenters pincers
A pair of pliers with side cutters
A pair of pliers with flat jaws
A number of wooden wedges (beechwood)

General remarks

1. Before starting, plan out the overall strategy of the operation.

2. Follow the principle of least resistance disassemble loose joints, but delay difficult operations until the object is so "simplified" that the component parts can be held in the bench vice for treatment.

3. The amount of corrosion decides whether it is possible to loosen the nail by shock treatment. If not, (where the nail is not on the surface of the object e.g. "A" being removed from the tenon) holes drilled on both sides, hugging the nail, remove some of the rust, paraffin is dripped into the holes and the process of extraction continued.

4. Some damage is unavoidable, the point is to be able to decide where and how much there is to be.

Author's address:

Randers Kulturhistorisk Museum
Stemannsgade 2
DK - 8600 Randers

WEKER, Wladyslaw Henryk

Selected example of treatment of metal-nonmetal archaeological objects

Abstract

Examination and treatment of an archaeological find composed of bronze, pottery and amber are considered.

The object was found on excavation as a hard clod of earth (size about 5x8x3 cm), with some bronze wire rings covered with green deposits, sticking out if it. On one of the rings a round bead could be recognized (Fig. 1). The find was brought to the laboratory for further examination and treatment.

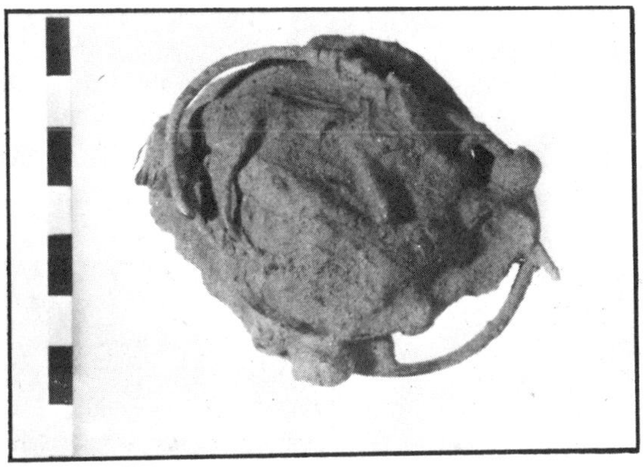

Fig. 1 *Object before treatment*

The first step was to have it radiographed. The X-ray photographs (Fig. 2) show clearly the presence of three complete rings (of different size) and fragments of a fourth. In addition, three spiral ornaments became visible on the two smaller rings, as well as two (metal) beads and a loop of wire on two parts of the rings. The round bead that could be observed on the surface

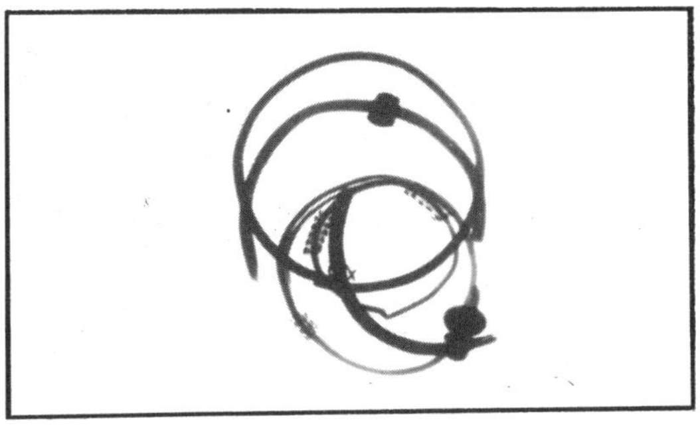

Fig. 2 *X-ray photograph*

(Fig. 1) did not show on the radiogram and was obviously a non-metallic element. This in turn suggested the possible presence of other such beads inside the clod.

The clod was put in a Petri dish and immersed in distilled water for a few hours. The earth became soft and could be carefully removed with the help of a scalpel and a needle (Figs. 3, 4, 5), revealing a broken fragment of pottery, complete and fragmentary bronze rings, and some broken amber beads (Fig. 6).

The bronze rings were passing through holes in the pottery. All these fragments were carefully washed with water and dried.

Then, the metal pieces were put into a wet chamber for two days to check the eventual tendencies to corrosion. There were none and the next conservation operations could be carried out.

The pottery fragment was probably already cracked and it separated into two pieces during the final

Fig. 3 *Cleaning of earth with H_2O*

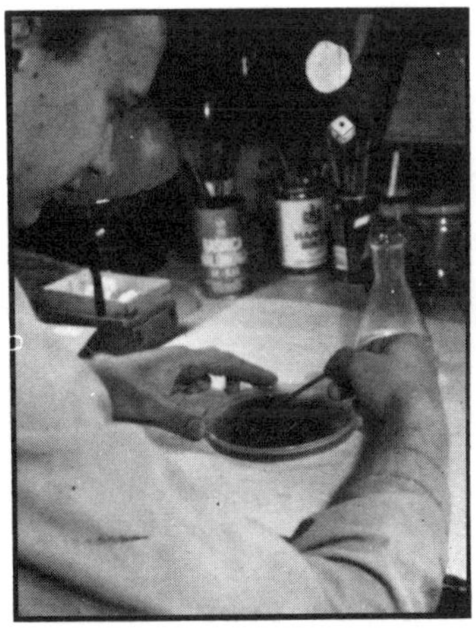

Fig 4 *Successive steps of cleaning*

Fig. 5 *Successive steps of cleaning*

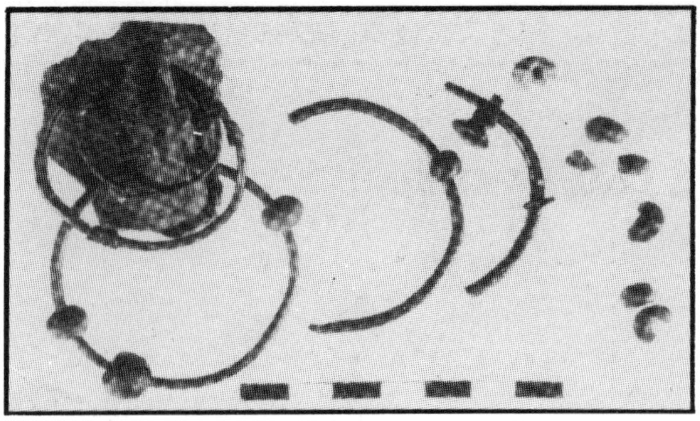

Fig. 6 *Object after removal of earth*

cleaning procedure. The fragments of broken amber beads were strong and well preserved as concerns their substance.

All the discovered elements were carefully classified and cleaned (Fig. 7), and treated according to need:

- The two ceramic fragments, after washing, were joined with poly-vinyl-acetate (PVA) in 40 % solution in alcohol.

- the bronze rings (fragments) were mechanically cleaned from corrosion deposits and immersed in benzotriazole (BTA) solution (1 % in alcohol) for 24 hours to

Fig. 7 *Object in the course of final cleaning*

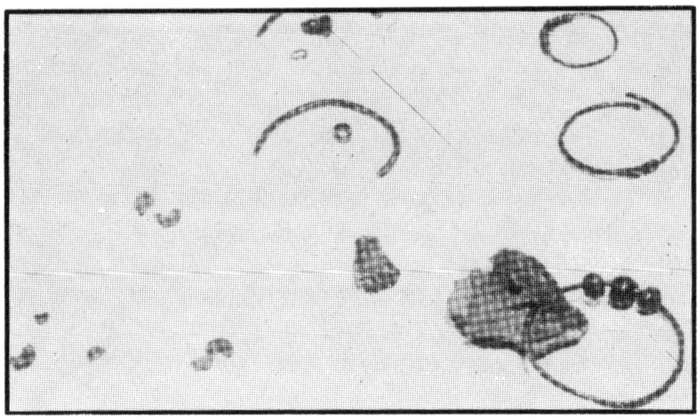

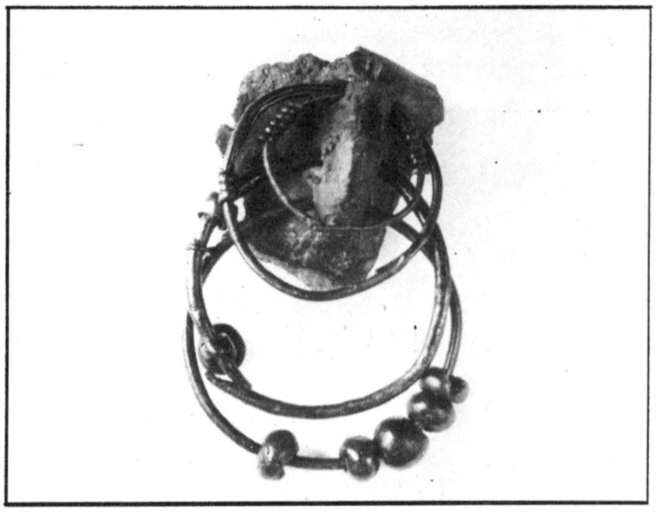

Fig. 8 *Object after cleaning and restoration*

inhibit future corrosion processes. Finally they were coated with a nitrocellulose lacquer for protection. The two bronze fragments belonging to the same broken ring were strong enough to be joined together with a tin solder (glues were too weak to be effective).

- as concerns amber, it became possible to set together two beads from seven broken fragments. They were joined with PVA (as above). All the amber pieces were coated with a nitrocellulose lacquer for protection.

- the bronze rings were inserted again in their respective holes in the ceramic fragment (Fig. 8).

What was this find originally? Most probably it was an ornament of a big pottery vessel.

The conservation treatment was carried out by the restorer Tomasz Mayer.

Author's address

State Archaeological Museum in Warsaw
Ul. Dluga 52
PL-00950 Warszawa

KOVÁCS Petronella

Problems of conserving an 18th century child's coffin of wood coated with silk (Preliminary report)

Previous to monument reconstruction of the Roman Catholic church in Sárospatak, Vera Gervers-Molnár – commissioned by the Hungarian National Museum and the National Monuments' Inspectorate – carried out archaeological excavations inside, and in the direct neighbourhood of the building in order to disclose bulding periods and earlier levels of the building[1].

The child's coffin from 1772 – the object of concern – was found in crypt XV, and had been left untreated for about 25 years after its disclosure, under poor storage conditions[2].

The pinewood coffin was originally coated with grege silk fixed to the wood with iron nails with copper buttonheads, also used for studding the name and year on the coffin [3]. (Fig. 1)

Primary destruction processes in the coffin may have been due to the 94% R.H. atmosphere in the crypts – as entered in the record of excavations[4]. As a consequence, the wood has undergone brown rot, hydrolytic decomposition of the silk started, and components – iron, copper and tin solder – of metallic nails formed local galvanic cells inducing marked deterioration. Dissolved metallic salts stained the fabric, at the same time hampering somewhat its decomposition.

Variation of climatic conditions is known to be rather harmful to any art object. Actually, the excavated objects had moved from an atmosphere of high relative humidity to rather dry air much affecting the condition of both silk and wood. Both having dried out, the silk has become frail, and the wood crumbly. The bottom of the coffin has broken down. (Figs. 2-4)

Such was the condition of the child's coffin when it was handed over to me for restoration. What is more, about two thirds of the silk coating was missing, some of the nail heads had fallen off due to heavy corrosion, and some had become lost; the remaining ones were very weak.

Fig. 2 a *Board from the bottom of the coffin.*
b *and* **c** *Side-boards from the lower part of the coffin*

The condition of the object imposed the following needs:

— to reinforce the crumbly, unsound wood;
— to conserve the silk, to make it more material-like, and to clean it;
— to stop active corrosion processes in the metal nails, then to replace missing parts – of course not more than structurally required for the object.

After having prepared written, photographic and drawing documentation recording the condition[5], the material of the coffin – exhibiting, of course, all the

Fig. 1 *Upper part of the coffin with silk fragments and nails with copper buttonhead*

Fig. 3 *Side-boards*

features of a composite object – was subjected to tests for cleaning and conserving the nails and the silk. Trials were made with ion exchangers, complexing agents and with conventional textile and metal detergents, both in paste and foam form [6]. Unfortunately, however, no satisfactory result was achieved: no detergent could be found, that was suitable for treating one component material without doing harm to the other; and, of course, the poor condition of the wood had to be kept in mind. This meant that liquid detergents were out of the question. Hitherto trials could not decide on the mode of treatment to be followed: either to leave the object un-

Fig. 4 *Buttonheads fallen down from the coffin*

decomposed, seeking for efficient but harmless agents and methods, or to decompose the coffin to the various materials each to be conserved by the appropriate conserving agent. According to the actual stade of our tests, this latter seems to be adoptable. Ligneous material would be strengthened by imbibing it with methyl-- methacrylate or methyl- and butylmethacrylate, and polymerization, then the cleaned silk and metal parts applied on a proper base would be fastened again on the wooden support.

Adopting this solution means that we have to face technical and – of course – ethical problems of decomposing and re-assembly.

Notes

1 Vera **Gervers-Molnár:** Burial Monuments in Sárospatak, Művészettörténeti Füzetek 14, Akadémiai Kiadó, Budapest, 1983 (In Hungarian)

2 For description of the crypt, see the quoted work by **Gervers-Molnár**, p. 84, note 26, and p. 85, note 29. Also three other coffins removed from this crypt were left untreated, to perish with time, leaving no traces.

3 This had been an age-old custom as testified by the will of Anna **Monaki-Andrássy** dated November lst, 1651: "No silver nails should be studden on my coffin's cover but all the nails I have had made should be used to stud my name, forming a blazon at my head, the others should be given away". **Béla Radvánszky**, Hungarian Family Life and Household in the 16th and 17th Centuries, III. Budapest, 1879, pp. 328 (In Hungarian)
In the case of the particular child's coffin, birth and data relating to death could be found in the register of the parish church of Sárospatak.

4 Climatic conditions in crypts have been referred to by Mária V. **Ember** in: "16th – 17th Century Garments for Crypts in Sárospatak", Folia Archeologica XIX, 1968, p. 151 (In Hungarian)

5 Drawing of the coffin was also needed for recording the inscription by marking spots of the heavily corroded pinheads. To our greatest delight, it was the same as found by **Gervers-Molnár**, See Ref. 1. p. 85. note 29.

6 Detailed description of the investigations will be published after the restoration.

Author's address:

National Centre of Museums
Könyves Kálmán krt. 40.
H-1087 Budapest

CSERE Judit

Process of restoring a medieval cross of procession

(Preliminary report)

Originally, it was intended to embrace all aspects related to the restoration on this Medieval cross, with all its problems. In cleaning, however, it has become clear that the subject embraces a far wider set of problems than had originally been presumed.

The object – a cross of Italian origin from about the 14th century – belongs to the Christian Museum in Esztergom. It was purchased in 1884 from the Schüttgen collection in Cologne by Dr. Béla Czobor on commission by Cardinal Archbishop János Simor for his collection bequeathed to the Christian Museum founded by himself.

The upright and horizontal members (the bars) of the cross are 44.5 and 36 cm respectively. The wooden core supports images of Jesus and saints, and their symbols embossed in copper plates fixed on with copper nails. (Figs. 1-3)

The face is composed of five figures; the back, of nine images. The sides are coated with strips ornamented with foliations folded over at 0.5 cm by face plates.

The crucifix, as the primary symbol of the Christian faith, represents Christ's crucifixion, victory over sin and death. The whole surface of the object was coated with a dusty, dark brown, rather thick, greasy, waxy corrosion layer. The plates and their ornaments were quite deformed, damaged. About one third of the nails were missing. I supposed – though there was no evidence to support this – that the plates might have been silvered or gilded.

Here and there side cover strips were deficient or, rather, damaged. At these spots, also the wooden core appeared to be damaged, to an indefinite degree.

Before starting restoration, our team deliberated on the restoration procedure suited to the object. The first problem to be decided was whether cleaning and treatment should be carried out in the original condition or whether the object should be disassembled. Some were against, but the majority were for disassembly. This was my view too: namely what ever agent is used for cleaning, the possibility for it to have access to the wooden core across defects and/or injuries of the plates cannot be excluded, while no perfect cleaning is possible at overlappings. Moreover the backs of plates being in contact with wood or organic matter were supposedly in a poorer condition than the outside. Without disassembly, defor-

Fig. 1 *Face of the procession cross*

Fig. 2 *Back of the procession cross*

Fig. 3 *Copper plates fixed on with copper nails*

mations cannot adequately be rectified, deficiencies replaced, timber disinfected let alone strengthened.

As a conclusion, the object had to disassembled. Nails were rather difficult to remove since some lacked heads to be held. Those with regular lentiform heads contacted wide areas of the plate, corrosion had thickened their shafts almost preventing them from being pulled out. In the end, they could be pulled out with a side-nipper-type nail extractor without damaging the surface.

After stripping, the plates from the two sides of the Crucifix showed their true condition. Their reverses were coated with a thick layer of patina, and had a marked chloride efflorescence. Beyond that, it became obvious that the object had already been repaired (Fig. 4a and 4b). Most of the plates were copper lined and tin soldered. All this resulted from the high embossment of the thin copper plate which had made it rather thin and rigid to become the most delicate and fragile part of the surface.

The core is of limewood, fastened at the cross-rabbeting by a wooden pin. The horizontal bar was broken in the middle, just as was earlier the upright bar that had been glued vertically and cross-wise. Its substance was weakened by wormbiting by longhorn beetle and Anobia. The ends of the bars exhibited rusty remnants of former iron nails.

Cleaning started by removing dusty, greasy impurities by means of a distilled water solution of fatty alcohol sulphate, pH 7. The resinous layer could not be dissolved before cleaning the greasy, dusty surface, it being inaccessible to watery detergent solutions. The resinous matter coating the whole surface of the object, polymerized due to oxidation, has fixed the top grease and dust layer and, by its acid groups, copper salts of the basic metal in complex form. Because of ageing, some resins could hardly or not at all be dissolved. Bases caused the resin to swell making it mechanically or chemically removable.

At that, my supposition was verified: namely that – it being an ecclesiastic object – the face and the back plates were fire-gilded, the side strips silvered. Precious metal coats subsisted only spotwise. Microscopy showed them not to adhere organically to the supporting metal–

Fig. 4 a *Repaired parts of the copper plate*

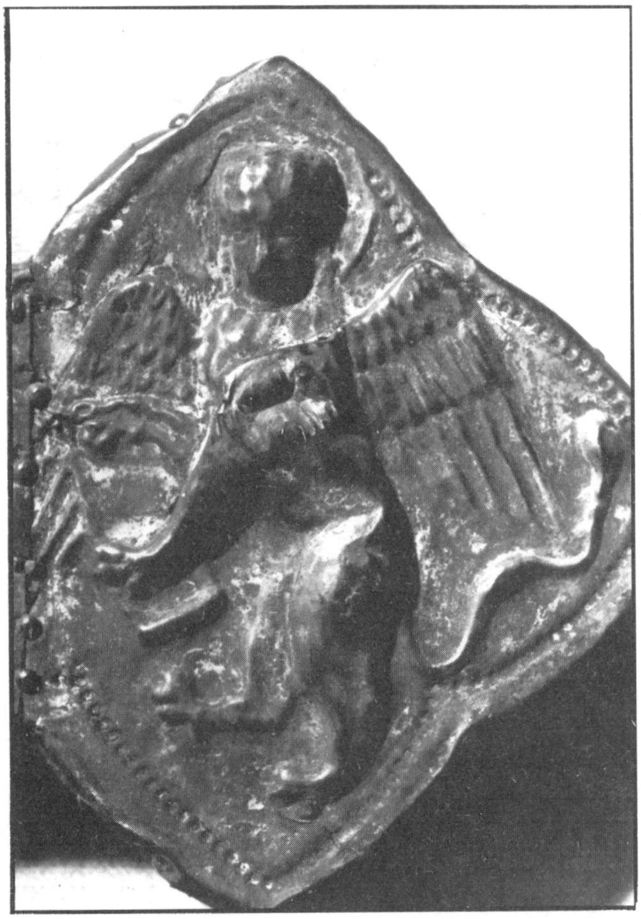

Fig. 4 b *Repaired parts of the copper plate*

itself a composite matter – an alloy of copper and zinc. In acidic conditions, copper loses its zinc due to chemical and electrochemical processes, zinc becomes dissolved from its surface layers and a reddish surface is left. Due to oxidation and other chemical reactions, a surface corrosion layer typical of copper develops. In the actual case, almost every modification occurs: viz. monovalent reddish brown copper oxide, bivalent black copper oxide and sulphide, basic copper carbonate, as well as copper chloride.

Coloured copper compounds can be dissolved in basic or in acid media.

The next problem was the tin soldering present in repaired parts. Tin is amphoteric, equally soluble in acids and in bases.

All these counteract the selection of a proven treatment agent known to us, such that, when removing the corrosion product of the supporting metal, it does not dissolve the layer between cooper and gilding or silvering, but saves the tin solder in the repaired parts. Therefore such a treatment is only possible by section–wise separating every corrosion layer.

For the continuation of the work we need some more research work.

Author's address:

National Centre of Museums
Könyves Kálmán krt. 40.
H-1087 Budapest

VÄLJA, Helmut

Wie die Lade der Weißgerberzunft restauriert wurde

In der dreißiger Jahren des 17. Jahrhunderts fingen die großen Zünften in Tallinn an, sich aufzulösen. Die Meister verschiedener Gewerbe bildeten ihre eigenen Berufsorganisationen.

Die Idee, eine Organisation für Weißgerber zu gründen, entstand in den 50-er Jahren des 17. Jahrhunderts. Wegen des Pestes und aus verschiedener anderen Gründen wurde die Satzung des Gewerbes erst im Jahre 1668 endgültig zusammengestellt und vom Magistrat offiziell bestätigt.

Am Innendeckel der Lade steht aber die Jahreszahl 1667. Zu dieser Zeit hat sich die Zunft wahrscheinlich schon ausgebildet und von der Lade wurde Gebrauch gemacht. Das die Lade wirklich aus dieser Zeit stammt, das beweist sowohl die Gestaltung der Schmiedearbeiten als auch die Uniform aus der Schwedenzeit, die die Männer auf den Bemalungen tragen.

Die mit Eisen beschlagene Lade war ursprünglich mit einem Schloß ausgestattet, das mit Hilfe von drei Schlüsseln geöffnet werden konnte. Das Schloß selbst ist im Laufe der Zeit verlorengegangen, aber sein Vorhandensein beweisen 3 Schlüssellöcher in der Vorderwand.

Die hölzernen Teile der Lade haben sich ziemlich gut erhalten, ebenso das Messingschild auf dem Deckel. Als der jahrhundertealte Schmutz beseitigt wurde, kamen an den Wänden der Lade Bemalungen hervor, die am Innendeckel auch die Metallteile schmücken. Gerade mit der Reinigung der Metallteile gab es die meisten Problemen.

Die eisernen Leisten waren ursprünglich verlötet und mit Kreidegrundfarbe bedeckt. Darauf war die Temperabemalung aufgetragen worden. Im Laufe der Zeit entstanden wegen der schlechten Aufbewahrungsbedingungen Rostschäden auf den Metallteilen. Ein Teil des Rostschäden ist als Ergebniß eines sehr langwierigen Schädigungsprozesses entstanden und hat sich mit den Schichten der Grundfarbe und der Bemalung vermischt. Ein Teil der Rostschäden ist aber als Ergebniß eines schnellen Prozesses entstanden und der Rost bildet auf der Oberfläche der Bemalung mit bloßem Augen erkennbare Kügelchen oder pilzförmige Gebilde.

Auf Grund der durchgeführten Analysen kann man sagen, daß in den Korrosionprodukten die Chloriden fehlen - also bestand auch keine Notwendigkeit, sie zu beseitigen. Um möglichst wenig die Grundfarbe und die Bemalungen zu schädigen, beschlossen wir die Rostkügelchen und "-Pilze" mit Hilfe des Ultraschallskalpells zu beseitigen. Dazu haben wir ein sowjetisches Gerät Y3X-201 benutzt, dessen Schwingungsfrequenz 44 kHz ist. Bei der Benutzung eines gewöhnlichen Skalpells besteht immer die Gefahr, daß sich große Teile der Grundfarbe und der Bemalungen loslösen. Das Ultraschallskalpell ermöglicht uns aber, die "Rostpilze" sozusagen am Fuße abzuschneiden.

Die ganze Fläche, die mit Rostschäden bedeckt war, wurde mit 0,5% (prozentiger) Lösung von Phosphorsäure bearbeitet. Dieser Vorgang wurde nach und nach in kleinen Teilen durchgeführt. Nach dem Austrocknen wurde die bearbeitete Fläche mit 3% Spirituslack bedeckt.

Das Messingschild der Lade wurde in 5% alkalischen Komplexonlösung bearbeitet. Nach der Reinigung wurde das Schild in eine Lösung von 3% (prozentigen) Bensotriazol spiritus getaucht, dann getrocknet und mit einer dünnen polymeren Schutzschicht bedeckt.

Größtenteils waren auch die Nägel, die man bei der Herstellung der Lade gebraucht hat, gerostet. Der lose Rost wurde mit dem Ultraschallskalpell beseitigt. Die Nägel wurden mit Tanninlösung, der man ein wenig Phosphorsäure hinzugefügt hat, konserviert.

Um die Teile der Nägel, die tief im Holz sitzen, haltbar zu machen, wurde unter die Nagelköpfe Tanninlösung geimpft. Dabei gebrauchten wir eine medizinische Injektionsspritze mit einer feinen Nadel.

Die gestrichen Holzteile wurden von dem Oberflächenschmutz mit einem leicht alkalischen Waschmittel gereinigt. Der Farbenauftrag wurde mit einer aufgewärmten 10% Lösung von Polyvinylalkohol befestigt. Unter die losen Farbteilchen wurde vorausgehend eine kleine Alkoholmenge geleitet, damit das Polyvinylalkohol besser resorbiert werden könne. Weil im Laufe der Zeit sich die Struktur der Farbe geändert hat und die Farbe den früheren Glanz verloren hat, entstand die Notwendigkeit sie zu erneuern. Zu diesem Zweck betupften wir den Farbenauftrag mit Tampons, die mit Äthylzellosolv durchgetränkt waren.

Als Ergebniß einer sorgfältigen Arbeit entstand aus einer unansehnlichen Kiste ein ordentliches Museumsgegenstand.

Die Adresse des Autors:

Restaureerimiskeskus
Pikk 2
SU-200001 Tallinn

DARRAH, Josephine A.

The microscopical and analytical examination of three types of metal thread

Abstract

The materials and methods of manufacture of metal threads and filaments are studied microscopically and by energy-dispersive x-ray fluorescence spectroscopy. Three discrete groups are discussed: 17th - 20th century silvered and gilt copper; late 18th early 19th century silver and silver-gilt and Far Eastern paper membrane. Some suggestions are made for the cleaning and conservation of metal threads.

Introduction

In recent years an increasing amount of information has become available about the materials and technology of metal filaments and threads. Barker[1] and Braun Ronsdorf[2] have published general surveys. Darrah[3], Járó[7] and Stodulski et al.[10, 11] have presented general surveys and analyses. Medieval European thread has been studied and analysed in greater detail by Hoke [4], Hoke and Petraschek-Heim[5], Indictor et al.[6] and by Járó[8]. Járó and Gondár[9] have written on European medieval membrane threads.

The materials and methods of manufacture of three discrete groups of metal threads about which little information is available are described here:

1. Silvered and gilt copper threads, c. 1670-1988.
2. English silver and silver-gilt threads from a single source dated 1760-1821.
3. Far-Eastern paper membrane threads, 17th-20th century.

It is hoped that this information may help to date and assign a provenance to textiles of unknown origin.

Methods of Examination

Samples of thread or filament 5-20 mm long were taken from textiles undergoing conservation treatment. When no sample could be taken the metal was analysed in situ and no further examination was undertaken. Very small samples were divided into four parts.

A. A short section of thread was unwrapped, the filament flattened, and mounted with the separated core fibres and a length of whole thread in Canada Balsam on a microscope slide. Filaments were similarly mounted.

B. A thread or filament was mounted in embedding resin. A cross-section was prepared by grinding on abrasive papers. A final polish was obtained using Durmax Alumina on a polishing cloth.

The samples were examined under reflected light using a Leitz Orthoplan microscope at magnifications of x100, x175 and x350. The separated fibres were examined in transmitted light and crossed polarizers at magnifications of x125 and x320. Some cross-sections were stained with saffranin to aid examination of the core fibres and membrane.

Microscopic examination allowed the identification of the core fibres and their colour; the type of membrane and ground colour of Far-Eastern threads; the main components of the cast, drawn and rolled metal filament and the extent and sometimes the cause of deterioration and corrosion.

C. The thread or filament was attached to metal-free adhesive tape and analysed by energy dispersive x-ray fluorescence spectroscopy (EDXRF). This nondestructive technique gives a semi-quantitative analysis of the metallic elements. The results are given in the form of a scan showing the elements present as a series of identified peaks and a printout identifying and listing the area covered by each peak. The x-rays penetrate the surface of the metal for a short distance. The surface gilding or silvering is overrepresented. Alloys are represented in their true proportions.

To compare the content of the metal filaments a simple ratio was calculated for each analysis using the area of the main peak only for each element ($K\alpha$ for silver and copper, $L\alpha$ for gold). These are shown in Tables 1-3. The ratios constitute a simple basis for comparison. In section 2 a method of estimating the true proportions of silver and copper in silver threads is given.

1. Copper based thread and filament (Nuremberg type)

In 16th and 17th century Europe when textiles with silver and silver-gilt decoration were available only to the wealthiest and most fashionable members of society their production was still controlled to a great extent by the Guilds. Large numbers of small workshops produced cast, drawn and rolled filament and thread spun onto a silk core. In the 18th century as industrial manufacture grew and both population and the numbers aspiring to belong to fashionable society increased a new, cheaper metal filament was made in Germany and came to be called Nuremberg thread [1, 3]. The metal was copper, thinly silvered or gilt. It was cast, drawn and rolled in the same manner as silver and silver-gilt filament. Much less precious metal was required and it must have been far cheaper to produce.

In the course of the survey of metal threads carried out at the VAM thirty-four copped based examples have

No 136 *A silvered copper thread spun (S twist) onto silk, c. 1670 Magn. x875*

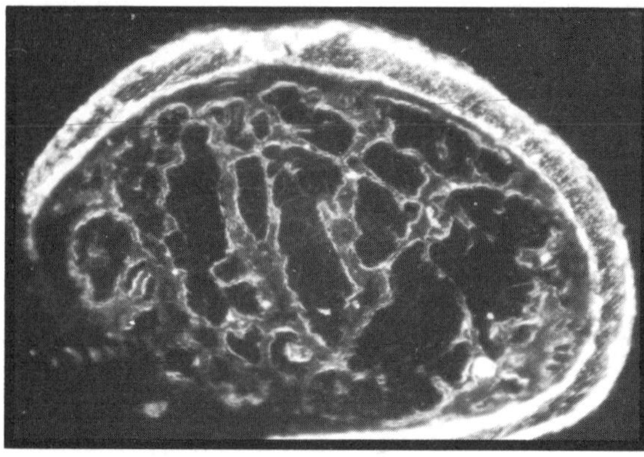

No 139 A silvered copper filament spun (S twist) onto undyed linen Magn. x1750. Cross section

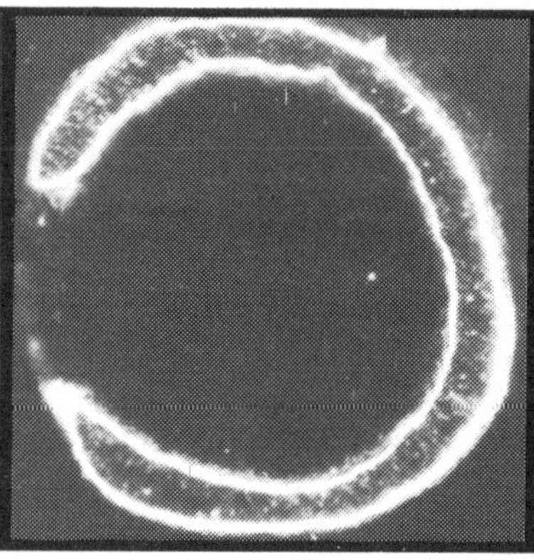

No 136 A fine early silvered copper filament spun (S twist) onto silk, c. 1670 Magn. x1750. Cross section

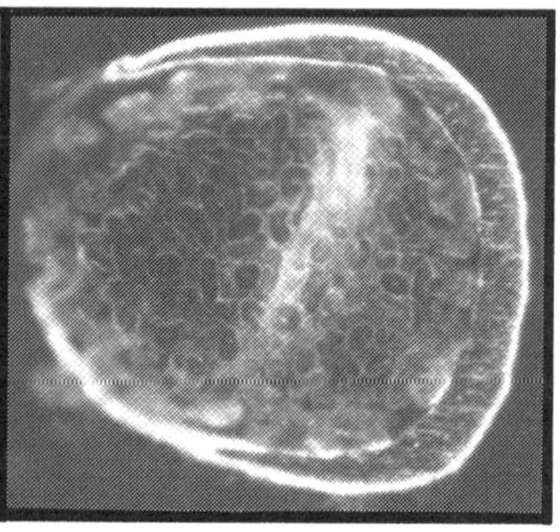

No 336 A silvered copper filament spun (Z twist) onto undyed cotton, 18 C Magn. x1750. Cross section

been identified, one tenth of the total. The earliest dates from 1670 and the latest from 1988, most are western European but a few came from the Indian sub-continent and one from China.

The copper-based threads and filaments are usually in good condition because copper is relatively resistant to corrosion under normal conditions. Silver and silver-gilt metal may be completely blackened and badly corroded by air pollutants attacking the silver.

Copper filament with a relatively heavy silvering or gilding is not recognizable with the naked eye. At low magnifications (x10 - x30) the brown-pink metal may be seen showing through thin or rubbed areas of the coating. Poorly gilt or silvered copper may be recognized by its overall brownish-pink tone.

Most of the 18th and 19th century samples examined were spun onto undyed cotton cores. Silver and silver-gilt threads were almost always spun onto silk cores and cotton has never been seen.

Some of the samples listed in Table 1 have been mentioned in a previous paper[3] (Nos 4-104). I shall discuss below some of the new examples and those of particular interest from the conservation aspect.

The majority of European samples date from 1730-1820. One interesting example does not fit into this group. The silk wall-hangings, c.1670 in the Queen's Closet at Ham House, Surrey, are rather rare survivals of this type of wall covering. The central crimson panels are silk woven with silver-gilt thread. The borders are blue and white silk woven with silvered-copper thread. The

Table 1 *Copper based filaments (Nuremberg type)*

#	Origin–Date	Type	Core, colour	Ratio based on EDXRF analysis			
				Au	Ag	Cu	Trace
136	Eng. c.1670	C.D.R.	Silk, white		t	+	
92	Eng. c.1730	C.D.R.	None	t		+	Ni zn
93	Eng. c.1730	C.D.R.	Cotton, white	t		+	Zn
94	Eng. c.1730	C.D.R.	None	t		+	Zn
95	Eng. c.1730	C.D.R.	Cotton, white	t		+	Ni Cr Zn
74	Ital. c.1734	C.D.R.	None	t		+	Ni
75	Ital. c.1734	C.D.R.	None	t		+	Ni
104	Eng. mid. 18c	C.D.R.	None	t		+	Ni Zn
65	Eng. 1760	C.D.R.	None	1		256	Ni Zn
4	Ital. 18c	C.D.R.	None		1	896	Ni
25	German 18c	C.D.R.	None		1	507	
26	German 18c	C.D.R.	Cotton, white		t	+	
71	German 18c	C.D.R.	None		1	248	Ni
72	German 18c	C.D.R.	Cotton, white		1	245	Ni
328	? 18-19c	Drawn	Cotton, yellow	1	0.4	159	Ni Cr Zn-0.7
306	Eng. 18-19c	C.D.R.	None	1	0.15	172	Ni Cr
336	? 18-19c	C.D.R.	Cotton, white		1	337	Ni Cr
337	? 18-19c	C.D.R.	None		1	842	Ni Cr
310	Eng. 18-19c	C.D.R	Cotton, white	t	1	112	Ni Cr Sn-2.9
311	Eng. 18-19c	C.D.R.	Cotton, white	1	1.4	325	Ni Cr Sn-4.0
139	French? late 18c	C.D.R.	Linen white		t	+	Ni Cr Zn (Ca Fe)
287	Eng. 18-19c	C.D.R.	None	1		366	Ni Cr
288	Eng. 1819	C.D.R.	None	1	+	206	Ni Cr
176	French 1810-20	C.D.R.	None	1	0.2	94	Ni Cr
205	Indian 18/19c	Foil		t	1	90	Ni As Pb
206	Indian 18/19c	Foil (glazed)			1	237	As
28	Indian 19c	C.D.R.	Spangle		1	1065	
88	Indian 19c	C.D.R.				+	Pb
89	Indian 19c	C.D.R.	None		+		
191	Chin. mid 19c	C.D.R.	Spangle	t	+		
200	Eng. c.1920	?	?	t	+		Fe
35	Ital.17 or 20c	? Electroplate	Cotton, yellow	1	0.4	227	
342	Bangladesh 1988	C.D.R.	Cotton, yellow	(glaze)t	1	561	Ni Cr
343	Bangladesh 1988	C.D.R.	Cotton, yellow	(glaze)t	1	484	Ni

C.D.R. = cast, drawn and rolled

t = trace

No 26 *Silvered copper thread spun (Z twist) onto cotton. Blackened lacquer breaking away from the surface 18C Magn. x500*

thread is minutely fine and is spun onto an undyed silk core, the silver is tarnished black. This silvered-copper thread is sixty years earlier than any other seen, it is extraordinarily fine and is spun onto silk. It is not of the Nuremberg group and so far its the only example of its type seen.

Several samples of gilded copper metal are dated to the 1730's, four from lace and fringe on an English dress (Nos 92-95) and two from the curtains of a Venetian Marionette theatre (Nos 74-75). These are in good condition, even after many years exposure in the case of the curtains. Those on the dress could have been new judging by their appearance.

Three examples from the early 19th century proved to contain small amounts of silver as well as gold. It is not known whether this was applied to the copper before the gold or whether a silver gold alloy was used; the latter is more likely. No 176 was from a French empire dress, c. 1810-1820, made of net with spots of gilt filament; No 288 from a gauze of 1819 from the Warner Pattern Book (see section 2) and No 306 from an undated silk brocade.

A heavy passementarie cord, No 328, of yellow spun with two strands of gilt wire also contained some silver. The copper contained sufficient zinc for this be a brass wire.

Two unusual examples, No 310-311, unfortunately of unknown provenance and date, taken from a fragment of brocade (18th or 19th century) proved to contain a considerable amount of tin and were silvered. This suggests that the base metal was a copper-tin alloy.

A single gilded-copper spangle was found on a chinese robe of the mid-19th century, No 191. Chinese metal filaments are of very different construction (see section 3) so this spangle may have been imported.

The metal thread in an Italian cut velvet (No 35) said to be 17th century, proved to be gold on one side and silver on the other, suggesting beaten and cut metal, but the base was copper. It was spun onto bright yellow cotton. The date of this textile had already been questioned on art historical grounds and the metal thread adds further doubt. It is perhaps a 20th century fake or copy. The filament cannot have been cast, drawn and rolled and may be electroplated although it could be beaten and cut metal made to deliberately deceive.

Silvered copper threads may be tarnished but the copper is usually in good condition. However, two examples of metal lace, probably 18th century German, Nos 25-26, 71-72, and a 19th century Indian Kalaga, Nos 28 and 88-89, were badly damaged. Nos 25-26 were blackened and flaking with areas of the metal falling away. The white cotton core of the thread was stained green with copper corrosion products. Under the microscope it could be seen that the lace had been lacquered. The lacquer had attacked the copper, blackening in the process and breaking away from the metal lifting the silver from the surface as it did so. The second lace, Nos 71-72 has also been lacquered but the heavier silvering seemed to have protected the metal to some extent. The metal of an Indian Kalaga was completely black with deteriorated lacquer. A trace of silver was found on the spangle but none could be identified on the filaments suggesting that the original silver must have been exceedingly thin.

One sample, No 139, came from excavations at Porchester Castle, in the South of England, which was used as a prison during the Napoleonic wars. The woven braid, much corroded from two hundred years burial, may have originated from an army uniform. The thread had a linen core, the only example seen. Uniforms are a rich source of metal threads but they form no part of the VAM collections.

It is interesting that the Warner Pattern Books (see section 2) contain very few copper-based filaments. The lightweight silk and gauze fashion fabrics of the periode 1760-1820 must have had a very short periode of use and might be expected to contain the cheaper filaments. Of ninety samples examined only three were copper-based. No 65 was dated 1760, Nos 287-288 were dated 1819 and, one silvered and one gilt, were woven into the same fabric. Perhaps the ban on the use of copper-based filament in England in the 18th century, designed to protect the English producers, was effective most of the time.

Finally a modern example, Nos 342-343 purchased in Bangladesh this year and apparently brilliantly gilt was not what it seemed. The copper filament was silvered, cast drawn and rolled and on a yellow cotton core and therefore similar to the earlier examples. It's gold appearance was produced by an applied yellow glaze or lacquer on the surface.

The copper-based filaments in Europe generally form a discrete group dating from c. 1730 to c. 1820. Other 19th century examples are Indian and Chinese. The earliest example of c. 1670 was on silk core, the Nuremberg threads c. 1730-1820 are on white cotton and in one case on linen. Modern examples are on bright yellow cotton.

2. The Warner Pattern Books

In the 18th and 19th centuries the textile manufactures Warners kept sample books of all their products. These books, now at the VAM, contain dated silk fashion fabrics, many woven with metal filament. One book, running from 1809 to 1821 contains only fine silk and gauzes woven with metal filament. Samples were taken from each year to determine whether the filament was of consistent or variable metal. The results are shown in Tables 2 and 3.

Standard samples of silver-copper alloys had been analysed by EDXRF and Atomic Absorption spectroscopy in the VAM laboratory. The results of the quanti-

Table 2 *Silver Filaments* *Warner Pattern Book T.3730-1972*

#	Year	Au	Ag	Cu	Other	%Au
66	1760	t	18	1	Zn	t
11	1799	-	42	1	Zn	-
12	1800	-	45	1	-	-
83	1805	t	46	1	Zn	t
126	1808	t	101	1	-	t
220	1809	-	81	1	Zn Fe	-
223	1810	t	113	1	Zn Cr	t
224	1810	t	104	1	Zn Cr	t
226A	1810	0.72	106	1	Zn Cr	0.7
227	1810	t	92	1	Zn Cr	t
127	1811	t	130	1	-	t
231	1811	t	167	1	Zn Cr	t
232	1811	t	141	1	Zn	t
234	1811	t	91	1	Zn	t
235	1811	t	>100	t	Zn	t
236	1811	t	100	t	Zn Cr Fe	t
237	1811	t	364	1	Zn Cr	t
240	1811	0.62	1.62	1	Zn Cr	0.4
241	1811	-	184	1	-	-
244	1811-12	0.65	186	1	Zn Cr	0.3
245	1811-12	t	164	1	Zn Cr	t
246	1812	t	78	1	Zn Cr	t
248	1812	-	30	1	Fe Ca	-
250	1812	0.49	77	1	Zn Cr Pb As	0.6
251	1812	0.44	166	1	-	0.3
253	1813	t	205	1	Zn Cr Pb	t
255	1813	t	150	1	Zn Cr Pb	t
256	1813	0.51	137	1	Zn Cr Pb	0.4
257	1813	0.98	180	1	Zn Cr Pb As	0.5
258	1813	1.05	204	1	Zn Cr Pb As	0.5
260	1814	0.08	18. 9	1	Zn Cr	0.4
261	1814	0.10	21.0	1	Zn Cr	0.4
262	1814	0.53	100	1	Zn Cr	0.5
266	1815	t	76	1	Zn Cr	t
267	1815	t	107	1	Zn	t
268	1815	t	48	1	Zn Cr	t
270	1815	0.18	55	1	Zn Cr	0.3
128	1816	t	88	1	-	t
272	1816	0.44	89	1	Zn Cr	0.5
273	1816	0.71	187	1	Zn Cr	0.4
275	816	0.48	103	1	Zn Cr	0.5
277	1816	t	62	1	Zn Cr	t
280	1817	t	128	1	Zn	t
281	1818	0.75	120	1	Zn Cr Fe	0.6
286	1818	t	143	1	Zn Cr Fe	t
290	1819	t	99	1	Zn Cr	t
292	1821	0.40	94	1	Zn Cr	0.4
293	1821	t	109	1	Zn Cr	t
294	1821	0.53	135	1	Zn Cr Fe	0.4
297	1821	t	116	1	Zn Cr	t
299	1821	t	47	1	Zn Cr Fe	t

Table 3 *Silver-gilt Filaments* *Warner Pattern Book T.373 1972*

#	Year	Au	Ag	Cu	Other	%Au
5	1790	0.6	5.7	1	Zn	8.2
13	1802	0.4	4.0	1	Zn	7.4
229	1807	0.83	6.9	1	Zn Cr	9.5
221	1810	0.91	7.2	1	Zn Cr	10.0
222	1810	1.01	8.3	1	Zn Cr	9.8
225	1810	1.06	7.5	1	Zn Cr	11.1
226	1810	1.09	7.2	1	Zn Cr	11.6
228	1810	1.14	8.5	1	Zn Cr	10.7
230	1811	0.99	7.3	1	Zn Cr	10.7
233	1811	0.93	9.4	1		8.2
238	1811	0.39	9.1	1	Cr	3.7
298	1811	0.63	11.2	1	Zn Cr	4.9
239	1811	1.10	6.3	1	13.1	
242	1811	1.06	10.3	1	8.6	
243	1811-12	1.08	9.7	1	Fe	9.1
247	1812	1.28	10.7	1	Fe	9.9
249	1812	1.37	10.4	1	10.7	
252	1813	0.86	7.6	1	Cr	9.1
254	1813	0.63	6.6	1	Zn Cr Fe	9.5
259	1814	0.77	8.0	1	7.9	
263	1814	0.55	6.6	1	6.7	
264	1814	0.59	5.5	1	Zn Cr	8.4
265	1814	0.71	6.8	1	Zn Cr	8.3
269	1815	0.75	7.7	1	Zn Cr	7.9
271	1815	0.63	6.4	1	Zn Cr	7.9
274	1816	0.20	4.1	1	Zn Cr	3.7
276	1816	0.55	8.0	1	Zn Cr	5.8
278	1816	0.74	8.4	1	Zn Cr	7.4
279	1817	0.50	5.8	1	Zn Cr	6.8
282	1818	0.69	8.1	1	Zn Cr	7.0
283	1818	0.77	8.1	1	Zn Cr	7.7
284	1818	0.61	8.6	1	Zn Cr	6.0
285	1818	0.58	5.8	1	Zn Cr	7.9
289	1819	0.62	5.5	1	Zn Cr	8.7
291	1821	0.63	7.4	1	Zn Cr	7.0
295	1821	0.68	7.5	1	Zn Cr	7.4
296	1821	0.73	9.0	1	Zn Cr	6.8

No 304 *Gilded paper membrane filament. Tibet 18C. Magn. x875.*

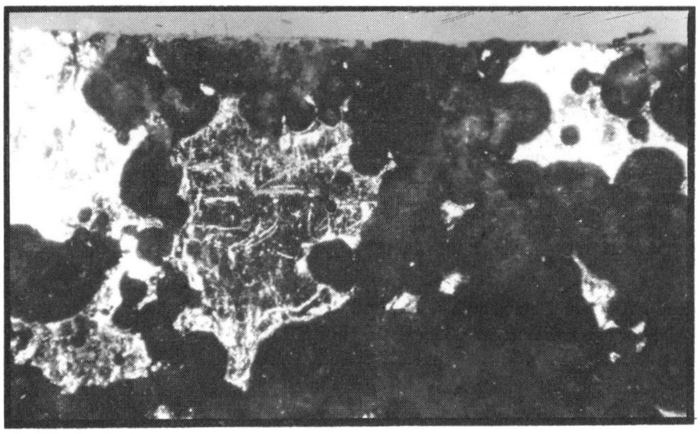

No 108 *Tinned paper membrane filament. Progressive circles of corrosion. Japan 19C Magn. x500.*

tative A.A.S. were used to produce a true copper-silver ratio. The EDXRF results were prepared as in the analytical Tables 1, 2 and 3. The ratios were plotted against the known silver content of the standards (see Fig. 1).

The EDXRF line closely follows that of the A A line but is consistently lower by a factor of about 2.5 (this is due to basing the ratio only on the copper Kα lines of silver and copper). When very little copper is present it becomes impossible to determine the concentration with any accuracy. Figure 1 may be used to estimate the true silver: copper ratio of the filaments listed in Table 2.

It has been said that silver metal contained 12% copper[1]. This does not seem to be so in western Europe, at least after the 16th century. Most 17th and 18th samples contain only 2-5% of copper. The early 19th samples of the Warner Pattern Book contain less than 1%. Only four early samples Nos 66, 11, 12 and 83 seem to contain a little more with occasional later occurrences such as Nos 260 and 261. Whether this very small addition of copper is general in the 19th century or is characteristic of Warners supplier is not known as few other 19th century samples have been examined. The Warner filament almost always contains a little gold, traces of zinc and chromium and in a few samples of 1812-1813 lead and arsenic.

The graph cannot be used to determine the silver content of silver-gilt thread. The gold being on the surface is over represented. However it does seem that there is more copper in the silver-gilt filament. The silver and silver-gilt filaments in the Warner samples are very consistent over a period of twelve years. In only three cases (see section 1) out of a total of ninety was copper based filament found.

No 177 *Silvered paper membrane filament. Overall tartnish. China, early 19C. Magn. x875.*

3. Far Eastern Membrane

The metal threads and filaments of China and Japan, see table 4, are constructed quite differently from those of Europe. A sheet of paper, probably of mulberry fibre, was coated with a ground on which metal leaf was laid.

When gold leaf was used the ground was red, similar in appearance to European bole. Silver and tin leaf were laid onto a colourless or brown ground. The gold leaf was almost always alloyed with a little copper and traces of zinc and silver are also sometimes present. Gold filament retains its brillance as there is no corrosion. Silver leaf seems to have been rarely used and that seen was completely corroded and black. It may contain a little copper and gold. One Tibetan sample, No 304, contained similar quantities of silver and gold; it was a paler colour than the purer gold leaf and was slightly tarnished. Tin leaf seems to have first been used in Japan [7] in the 16th or 17th century. All the examples seen in this survey have been 19th century. The tin has usually corroded to a dull grey–brown colour. Examples of 19th century tin membrane in a collection of decorative cords and papers at the VAM [3] which have always been protected have retained their bright white metallic finish.

No 302 *Gilded paper membrane filament, red ochre and clay ground. Tibet, 17C. Magn x1750 Cross section*

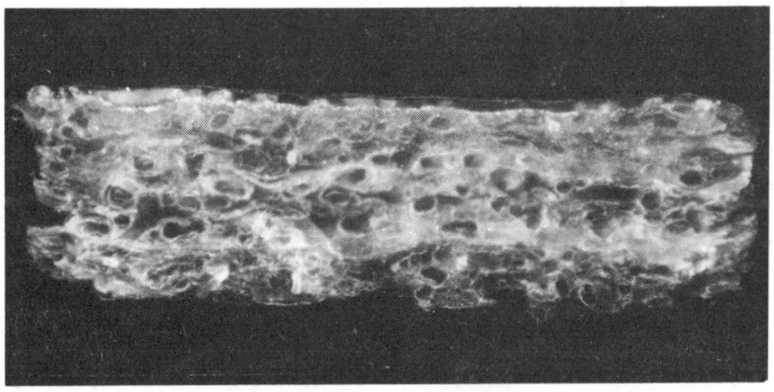

Table 4 *Far Eastern Membranes*

#	Orgin, Date	Form, Core	Base	EDXRF Analysis			
				Metal	Trace	Ground	Trace
121	C.Asia 14-15c	Filament	Leather	Au	Cu	Ca Fe	K
178	Chin. 16-17c	Filament	Paper	Au	Cu	Fe	K Ca
180	Chin. 17c	Thread, silk	Paper	Au	Cu Ag	Fe K Ca	
345	Chin. 17c	Thread, silk	Paper	Au	Cu Ag	Fe	K Ca
302	Tib. 17c	Filament	Paper	Au	Cu Ag	Fe	K Ca Ti
153	Chin.	Thread, silk	Paper	Au	Cu	Fe Pb	K Ca Ti
154	Chin.	Thread, silk	Paper	Au	Cu	Fe	K Ca
155	Chin.	Thread, silk	Paper	Au	Cu	Fe	K Ca
195	Chin. early 18c			Au	Cu	Fe	K Ca Ti Pb
196	Chin. early 18c			Au	Cu	Fe Pb	K Ca Ti
346	Chin. 18c	Thread, silk	Paper	Au	Cu Ag	Fe Pb	K Ca
303	Tib. 18c	Filament	Paper	Ag	Cu	Fe	K Ca Ti Hg
304	Tib.	Filament	Paper	Au Ag	Cu	Fe	K Ca Ti
100	Chin. 18c	Thread, silk	Paper	Au	Cu Ag	Fe	K CA Ti Pb
101	Chin. 18c	Thread, ramie	Paper	Au	Cu Zn	Fe	K Ca Ti
102	Chin. 18c	Thread, ramie	Paper	Au Zn	Cu Ag	Fe	K Ca Ti
103	Chin. 18c	Thread, ramie	Paper	Au	Cu Zn	Fe	K Ca Ti
179	Chin. late 18c	Filament	Paper	Au	Cu Zn	Fe	K Ca Ti
192	Chin. late 18c			Ag	Cu Au		
193	Chin. late 18c			Au	Ag	Fe	Mn Pb
197	Chin. late 18c			Au	Cu Ag	Fe	Mn Ti
177	Chin. early 19c	Filament	Paper	Ag	Cu	Fe	K Ca Hg Mn
86	Chin. 19c	Filament	Paper	Au	Cu (Zn)	Fe	K Ca
190	Chin. mid 19c			Au	Cu Ag	Fe Pb	
194	Chin. late 19c			Au	Cu Ag	Fe Pb	Ca
185	Chin. 1988	Filament	Paper	Au	Cu		Ca
186	Chin. 1988	Filament	Paper	Au	Cu Ag		Ca
187	Chin. 1988	Filament	Paper	Au	Cu Ag		Ca
188	Chin. 1988	Filament	Paper	Ag	Ca		
85	Japan 19c	Filament	Paper	Au	Cu (Zn)	Fe	K Ca
6	Japan 19c	Filament	Paper	Au	Cu (Zn)	K Ca Fe	
34	Japan 19c	Thread, ? core	Paper	Sn	Cu (Zn)	K Ca Fe	
156	Japan 19c	Filament	Paper	Sn	Cu Cr Zn	Fe	Ca Ba
157	Japan 19c	Filament	Paper	Sn	Cu	Fe	Ca Ba
158	Japan 19c	Filament	Paper	Sn	Cu Cr Zn	Fe	Ca Ba
159	Japan 19c	Filament	Paper	Sn	Cr Zn Cu	Fe Ca	Va Pb
160	Japan 19c	Filament	Paper	Sn	Zn Cu	Ca Fe	Ba Pb
161	Japan 19c	Filament	Paper	Sn	Cu Cr Zn	Fe	Ca Ba
162	Japan 19c	Filament	Paper	Sn	Zn Cu	Fe Ca	
163	Japan 19c	Filament	Paper	Sn	Cu Zn		Fe Ca

The grounds were prepared from clays or kaolin coloured with iron oxides. EDXRF analysis indicated that iron was usually the major element present with traces of potassium, calcium and titanium or barium. A few samples contained some lead, perhaps red lead, and manganese and mercury (vermilion) were occasionally identified.

The majority of the samples examined were filaments woven into fine silk fabrics used in scroll, screen or than-ka borders. Embroidery threads were prepared by spinning the filaments onto a silk core usually dyed yellow. The cores of three heavy 18th century embroidery threads, Nos 101-103, were identified as ramie, a vegetable fibre. Without exception the filaments were spun using a Z twist on a core which also had a slight Z twist, (European threads are spun with as S twist).

Modern membrane filaments are made using a different method; in 1988 a colleague saw it being made in Hong Kong. The mulberry paper was coated with shellac and the gold or silver leaf was laid onto it while it was still tacky. When it was dry the sheet was cut into fine filaments leaving the borders of the sheet whole. It was sold as a sheet (No 185). The gold metal contains a trace of copper and/or silver. The trace of calcium probably results from a filler in the paper. The lack of a red ground may make the filaments less brilliant in colour. However the clay grounds tend to be brittle and to flake so the shellac adhesive may produce a more flexible and resistant filament.

Three Tibetan samples from than-ka borders have been examined and were of the same construction as the Chinese and Japanese filaments. One was corroded silver, No 303, one gold, No 302, and one a silver-gold alloy No 304.

Only one sample from Central Asia has been examined, No 121. The textile, a silk lampas from a chasuble from Regensberg Cathedral, is of the 14th or 15th century. It is of the type that is believed to have been made in China before the introduction of paper membrane. Gold leaf was laid onto a thin sheet of leather rather than the gut used in medieval middle-eastern membrane. Although no ground could be seen calcium and iron were identified during analysis.

Fig. 1 *Copper: Silver ratio (EDXRF based on area of Kα peaks only)/known % silver content of standards*

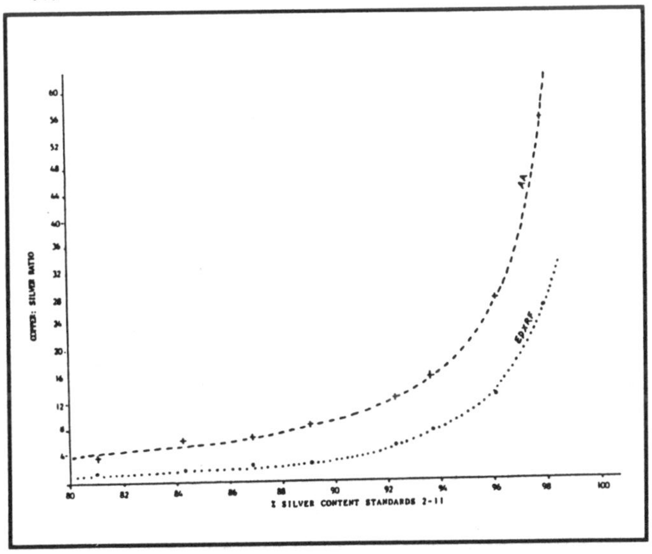

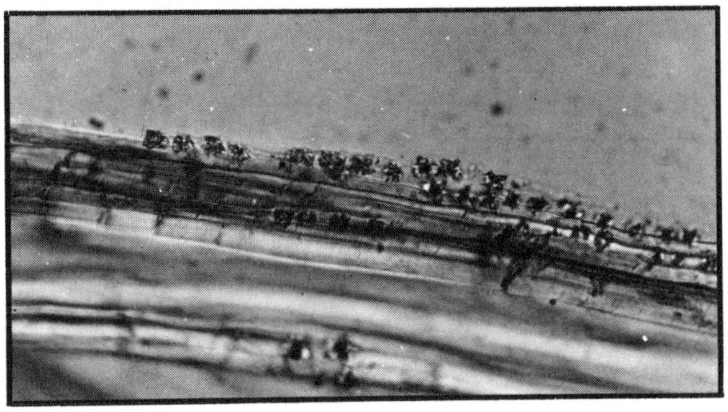

No 101 *Ramie fibres with typical cluster crystals from core of gilded paper membrane threat. China 18c Magn x1600.*

4. Conservation and Cleaning

A paper given at this conference by Howell reports on the cleaning of metal threads. I shall discuss here only the copper-based and Far-Eastern threads.

Copper-based filament is usually in relatively good condition except for tarnish on the silvered samples. Cleaning copper-based filament should be restricted to lifting surface dirt and dust with methylated spirits. The acids and alkalis, including ammonia, contained in some commercial and laboratory metal cleaners which have been used in the past will attack copper and may stain the fibre cores and adjacent fabric with dissolved copper salts. Copper filaments have been seriously corroded by the application of lacquers, possibly nitrocelluloses, a technique used until quite recently in conservation studios. No attempt has been made to remove this lacquer.

Far-Eastern membrane creates very different problems. Silver and tin leaf is usually severely corroded and little can be done to improve its appearance. Gold membrane may be dirty and need simple cleaning. The greatest care is required as any pressure on the surface may crack the brittle ground and cause the break up of the whole. Dry-cleaning is probably the only safe way to clean the silks and embroideries in which these membranes occur. The VAM cleans fragile silk fabrics of this type using white spirit on a vacuum table which causes the least damage, though a little metal may be lost using this technique.

Preventative conservation must be taken into consideration for the display and storage of all fragile textiles but those with silver and silver-gilt thread are perhaps even more susceptible to damage by air pollutants and volatile substances given off by many materials. Wood, chipboard and hardboard, paper, adhesives, paints and modern textile finishers and some dyes can cause tarnish on silver. Many museums are now attending to this problem by testing the materials to be used in new show-cases and encouraging the use of only those materials which appear to be safe. Acid free tissue should be used to cover and protect objects in storage. Cloth covers of unbleached and unfinished linen or cotton should cause no problems. It is unlikely that total prevention of tarnish can ever be contemplated. Most museums have to store textiles in wooden cupboards and drawers in areas where there is little hope of completely removing air pollutants but we should be aware of these problems and endeavour to reduce them to a minimum.

Source of Samples: Copper based filament

4.	Italian,theatre costume, 18c	S.806-1982
25-26	German (?) metal lace, 18c, on an English mantua,	T.88-1978
28	Indian,Kalaga,spangle, mid-late 19c	I.S.219-1960
35	Italian, cut velvet, mid-17c. or modern	
65	English, Warner Pattern Book, 1760	T.373-1972
71-72	German (?) metal lace, 18c	VAM
74-75	Italian, Marionette Theatre curtains, c.1734	W.31-1924
88-89	Indian, Kalaga, mid-late 19c	I.S.219-1960
92-95	English dress, lace and fringe c.1730-40	T.114-1873
104	English, Vardy Bed hangings, selvedge,mid 18c	
	Hardwick Hall	National Trust.
136	English, wall-hangings c.1670	H.H.-1948
139	French (?) braid, Porchester Castle	Institute of Archaeology, Oxford.
176	French "empire" dress, net with gold spots, c1810-20	Private Collection
191	Chinese robe, spangle, mid 19c	T.230-1948
200	European theatre costume, 1920's	C.D. 23
205-206	Indian, gauze dress with embroidery, 18-19c	I.S.5842
287-288	English, Warner Pattern Book, 1819	T.373-1972
306	English, silk textile, 18-19c	VAM Conservation Dept.
310-311	English, silk textile, 18-19c	VAM Conservation Dept.
328	English, heavy upholstery cord, 18-19c	VAM Conservation Dept.
336-337	European, braid 18-19c	VAM Conservation Dept.
342-343	Bangladesh, embroidered silk 1988	Private Collection

Source of Samples: Far-Eastern Membrane

6	Japanese scroll mount, 19c	1645-1888
34	Japanese scroll mount, 19c	Private Collection
85	Japanese scroll mount, 19c	F.E. 139-1983
86	Chinese scroll mount, 19c	Private Collection
100-103	Chinese bed hangings, Calne, 18c	National Trust
121	Central Asian lampas from a chasuble,	
	Regensberg Cathedral 14-15c	8288-1863
153-154	Chinese, Chester Beatty lampas, ?	Chester Beatty Collection
155	Chinese lampas, ?	Chester Beatty Collection
156-157	Screen, borders, 19c	H.H.435-1948
158-160	Screen, borders, 19c	H.H.436-1948

Source of samples: Far-Eastern Membrane (continued)

161-163	Screen, borders, 19c	H.H.435A-1948
177	Chinese lampas early 19c	C. Devitt
178	Chinese taffeta broché, 16 or 17 c	C. Devitt
179	Chinese lampas late, 18c	C. Devitt
180	Chinese robe, 17c	F.E.D. 41-1985
185-188	Chinese, Hong Kong, filaments and two fabrics, 1988	V&A Coinservation
190	Chinese robe mid 19c	T.230-1948
192-3	Chinese robe, late 18c	T.43-1952
194	Chinese lady's jacket, late 19c	T.125-1966
195-196	Chinese lady's robe, early 18c	T.184-1948
197	Chinese lady's robe, late 18c	T.32-1950
302-304	Tibetan thanka borders, 17,18c	
345-346	Chinese Kossu 18c with 17c border, 17,18c	Private Collection

Acknowledgements

I wish to thank Graham Martin P.S.O. for EDXRF analyses of Nos 4-120; David Scoullar S.O. and Helen Ashcroft S.O. for preparing cross-sections; Rachel Swan, Ph.D. student, for allowing me to use A.A. and EDXRF analyses of silver standards; Dr J. Ashley-Smith, Keeper of Conservation, for his encouragement and suggestions during the course of the project.

References

1. **Barker** A.D.: Gold Lace and Embroidery, Northern Society of Costume and textiles Publication I, 1980, 1-14

2. **Braun Ronsdorf**, Margarete.: Gold and Silver Textiles from Medieval to Modern Times, Ciba Review, 1961 (3) 2-16

3. **Darrah Josephine** A.: Metal Threads and Filaments, in: Institute of Archaeology Jubilee Conference London, 1987, Preprints 211-221

4. **Hoke**,E.: Mikroanalytische Untersuchungen von Metallfäden aus Grab I in der Pfarrkirche Traismauer, Fundberichte aus Oesterreich, 1978 (16), 255-259

5. **Hoke**,E. and **Petrascheck-Heim** I.: Microprobe Analysis of Gilded Silver Threads, Studies in Cons., 1977 (22), 49-62

6. **Indictor**, N., **Koestler**, R.J., **Blair**, C. and **Wardwell**, A.E. : The Evaluation of Metal Wrappings from Mediaeval Textiles using Scanning Electron Microscopy Energy Dispersive X-ray Spectrometry, Textile History 1988, 19 (1), 3-22

7. **Járó** Márta: The technological and analytical examination of metal threads on old textiles, in: Proc. of 4th International Restorer Seminar Vol. 2, National Centre of Museums, Hungary 1984, 253-264

8. **Járó** Márta: The investigation of the metal embroidery threads of the Hungarian coronation mantle by SEM and physical methods of analysis, in: Icom Comm. for Cons., 7th Triennial Meeting, Copenhagen 10-14 September 1984. Preprints 1984, 84.1.22-84.1.24.

9. **Járó** M. and **Gondár** E.: Mediaeval membrane threads used for weaving and embroidery, in: Archaeometrical Research in Hungary, Budapest, 1988, 255-266

10. **Stodulski**, L.R., **Mailand**, H.F., **Nauman**, D. and **Kennedy**, M.: Atomic emission spectrographic and SEM. EDXR studies of European, Middle Eastern and Oriental Metallic Threads, in: Application of Science in Examination of Works of Art, Museum of Fine Arts, Boston 1986, 76-91

11. **Stodulski**, L.P., **Nauman**, D. and **Kennedy** M.: Analysis of precious metal threads, Technical appendix to 'Fabrics in Celebration' from the collection of Peggy Stratt Gilpy, Indianapolis Museum of Art, 1963, 361-375

Author's address:

Conservation Department
Victoria and Albert Museum
GB-London
SW7 2RL

GONDÁR Erzsébet – JANKOVICH Katalin

Die Untersuchung der Textilreste an einem Bronzearmband aus der Spätbronzezeit

In Jánosháza (in Westtransdanubien) wurde in den 1940-en Jahren ein spätbronzezeitliches Hügelgrab aufgedeckt. Dieses Hügelgrab gehört zur Späthügelgräberkultur im Bakony-Gebirge, zur Farkasgyepü-Jánosháza Gruppe. In diesem Hügel wurden 6 verbrannte Gräber gefunden. Im ersten Grab waren die folgenden Beigaben: zwei bronzene verzierte Armbänder, eine schlangelförmige Nadel und profilierte Schüsseln. Nach diesen Beigaben wird das Grab in die Reineckesche BD Perioden um 1200 v. u. Zeit datiert. In diesem Grab kann man an einem Bronzearmband Textilreste beobachten. Bei Bestattungsritus wurden die kalzinirten Knochen und Bronzebeigaben vom Scheiterhaufen gesammelt und dann wurden die verbrannten Reste in Textilien eingewickelt. An dem Bronzearmband kann mann diese Textilreste beobachten, es ist unmöglich, dass die Textilreste aus dem Kleid des Toten stammen, weil das Feuer so stark war, dass auch die Bronzebeigaben im Feuer des Sheiterhaufens zerschmelzt wurden. Bei dieser Hitze wurde das Todeskleid vollständig verbrannt. Die Textilreste aus der Spätbronzezeit können wir sehr selten in Ungarn finden. Deshalb war für uns die Bestimmung des Textilgrundstoffes sehr wichtig. Es bedeutete eine besondere Schwierigkeit, dass der gefundene Stoff zu wenig war, und in Korrosion gebettet war. Deshalb haben wir mit lichtmikroskopischer Untersuchung kein Ergebnis erreichen können. Man konnte nur sehen, dass irgendwelche Textilien sich am Bronzearmband befinden. Bei der Bestimmung der Textilreste hat der japanische Elektronmikroskop (RASTER) vom Typ JSM 35 grosse Rolle gespielt. Von dem Armband haben wir ein ganz kleines Muster abgenommen (Abb. 1), so haben wir zum Glück einen Brechungsquerschnitt bekommen, wo man die elementaren Wollfäden in Korrosion im Querschnitt sehen kann. Das Präparieren des Musters wurde von Kohle und Gold gedämpft. Das Muster haben wir in verschiedenen Vergrösserungen untersucht, und wir haben von diesen verschiedene Aufnahmen gemacht (Abb. 2).

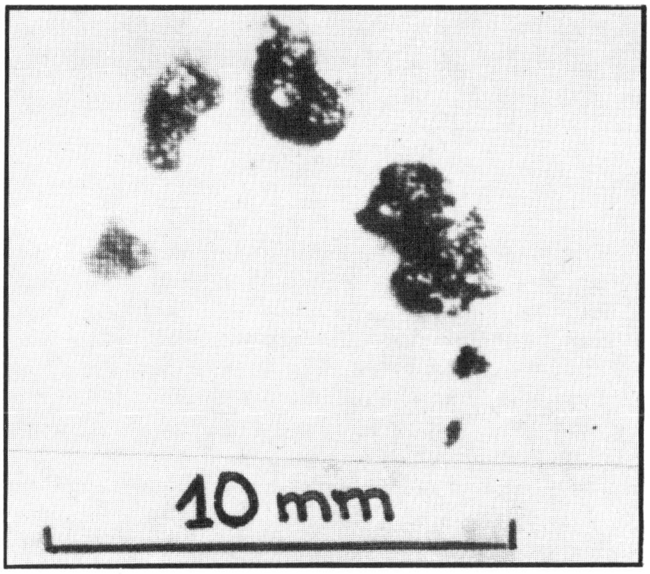

Abb. 1. *Muster von spätbronzezeitlichen Textilresten in Korrosion. Linie bedeutet 10 mm*

In der 60 und 200 fachen Vergrösserung kann man nur die Textilfäden bemerken. In der 600 fachen Vergrösserung kann man schuppige, für die Wolle charakteristische elementare Fäden sehen. Wenn wir den Querschnitt von bronzezeitlichen Wollfäden vergleichen, können wir bestimmen, dass beide verschiedene zeitliche Muster gleiche Morphologie gezeigt haben (Abb. 3). Die erfolgreiche Untersuchung wurde möglich dadurch, dass wir sowohl in der Fläche als auch im Querschnitt für die Wolle charakteristische morphologische Kennzeichen gefunden haben, so konnten wir die Wollfäden eindeutig identifizieren.

Abb. 2. *Elektronmikroskopische Aufnahmen des spätbronzezeitlichen Textilmusters in verschiedenen Vergrösserungen (x60; x200; x600)*

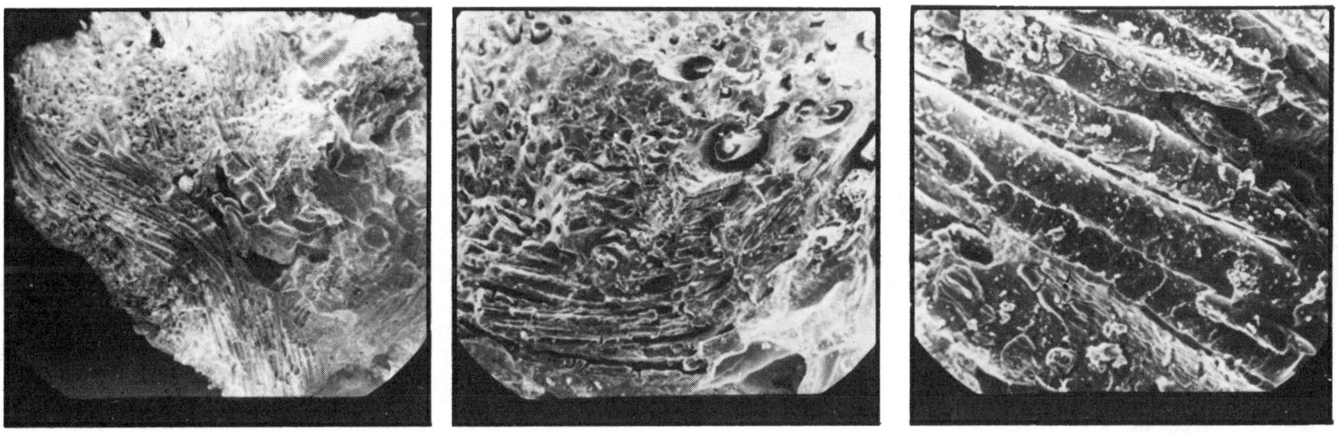

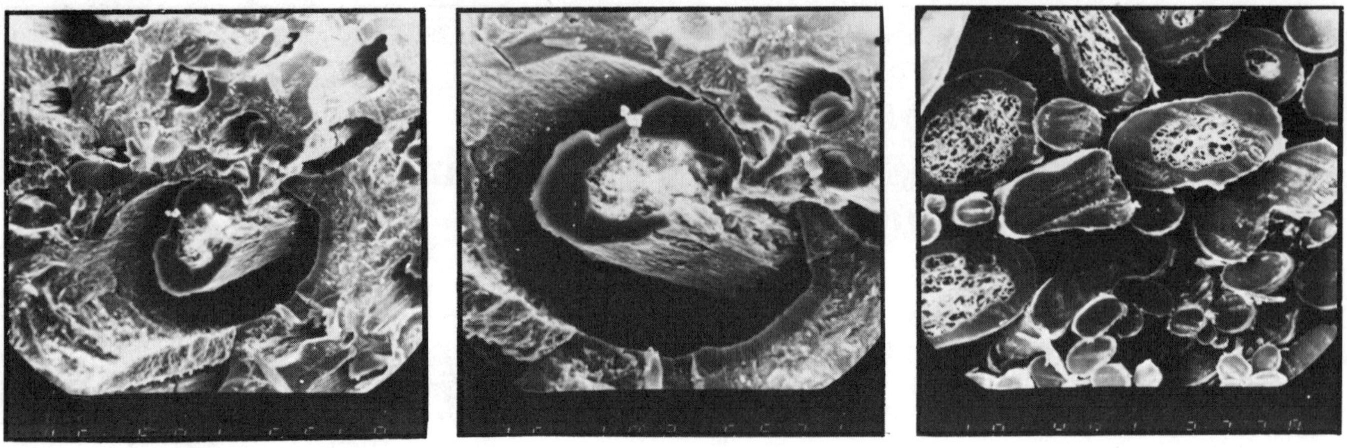

Abb. 3. *Querschnitt von bronzezeitlichen Wollfäden (a, b, Vergrösserung: x600; x1000) und von heutigen Wollfäden (c, Vergrösserung: x1000)*

Der Vorteil der elektronmikroskopischen Untersuchung ist der folgende: Bei der Untersuchung, wo man Lichtmikroskop oder chemischen Untersuchungen kein Ergebnis erreichen kann, können wir mit Elektronmikroskop ein Resultat bekommen. Wir brauchen zur dieser Untersuchung nur sehr wenig Muster. Früher haben wir nur voraussetzen können, dass die im Bakony - Gebirge lebende Späthügelgräbergruppe sich mit Tierzucht beschäftigt hat, weil dieses Gebiet für Ackerbau ungeeignet ist. Das Ergebnis der Textiluntersuchung hat unsere Vorstellung bestätigt, diese Späthügelgräbergruppe hat beim Bestattungsritus Wollstoff tierischen Ursprungs benützt.

Adresse der Autoren:

Gondár Erzsébet
Innovatext Institut für Textilindustrie
Gyömrői u. 86.
H-1103 Budapest

Jankovits Katalin
Institut für Denkmäler,
Wissenschaftliche Abteilung
I. Táncsics M. u. 1.
H-1014 Budapest

INDICTOR, Norman – BALLARD, Mary

The effects of aging on textiles that contain metal: implications for analyses

Abstract

Factors which affect the general condition of historic textiles containing metal are listed and discussed. Specific elements of structure for textiles which contain metal (metal, organic substrate, adhesive, fiber) are discussed according to physical and chemical deterioration processes with emphasis on what a careful observer might expect to see on an historic textile containing metal. The effects of common textile conservation practices on textiles containing metal are described.

Introduction

Recently five categories describing the inclusion of metal in textiles have been defined [41-43, 46, 47].

Table 1 *The Five Categories*

I. Metal applied (with adhesive) to already woven fabrics.
II. Metal wire or flattened strips.
III. Metal wire or strips wound around fiber core.
IV. Metallic surface applied (with adhesive) to organic wrapping wound around fiber core.
 a. Organic = cellulosic.
 b. Organic = proteinaceous.
V. Metallic surface applied (with adhesive) to organic strips.
 a. Organic = cellulosic.
 b. Organic = proteinaceous.

These categories exclude metallic salts or oxides used as mordants or pigments. Reasons for assigning these categories are: 1) to provide a convenient way of describing textiles containing metal; 2) to survey the historic and geographic implications of such categorization; 3) to provide a pedagogic introduction to this aspect of textile technology. Other studies defining similar categories have been noted [8, 14, 25, 35, 54, 55, 90, 97, 109].

Textiles that contain metal stand out from other textiles. In surveying examples of textiles with metal some observations concerning the condition of historic textiles reflect both on their art historical significance and their fragility:

a) Textiles containing metal are generally in poorer condition than those which do not contain metal if comparable conditions of material quality and survival have been provided. Special stresses from the high mass of material and the multiplicity of degradation processes tend to undermine the integrity of the fabric containing metal on aging. Sometimes, because of their artistic, symbolic, or economic value, textiles with metal have been so treasured that special care and treatment has been provided while comparably old unadorned textiles have been treated as disposable.

b) In a textile that contains metal the areas that contain metal will tend to be more degraded than other areas.Not only are the fabric and thread systems which have metal more sensitive to natural aging processes, but they are more susceptible to degradation from the general practises such as cleaning and washing wich are intended to maintain the integtity and beauty of the material.

c) Historic textiles that contain metal and appear to be in good condition (superficially) invariably show considerable evidence of degradation and loss specifically associated with the metallic inclusion upon close (microscopic) inspection.

d) Archaeological and lexicographic evidence point to significant loss of textiles containing metal. Some excavations have unearthed beaten metal strips that appear to have been associated with textiles although no trace of attendant textile remains [33, 73, 79, 85, 109]. It is interesting that several languages retain words for the act of, or the person, reclaiming metal from textiles that are remnants of the tailoring process, have become worn, or lost favor [45, 48, 78, 111].

Aging

The condition of historic textiles may be seen as depending on two large factors: the quality and the peculiarities of the manufacturing processes; and the circumstances and uses to which the textile has been subjected.

a) Manufacture

i. The longevity of a textile material will depend heavily on the skill, patience, and intentions of the craftsmen involved. For example the tightness of weave, the ply of the thread, the fineness of the fibers have a profound effect on the aging behavior of any textile.

ii. The aging characteristics of a textile will also depend heavily on the quality of the materials employed. For example the quality of fibers available, seasonal variations in raw materials, alloy composition of metallic materials, may profoundly affect the aging behavior of the textile.

iii. Regional or workshop practices which may have little to do with skill or material quality but more to do with tradition, aesthetics, or individual experimentation may profoundly affect the aging of textiles. For one example iron mordants, used in a number of dying recipes, seem to be particularly corrosive to fibers. Other examples are the extremely fine Moghul silks of the seventeenth century, so fine because the raw silk may have been subjected to exhaustive degumming, producing a fine filament, initially of very high tensile strength, but which experiences serious embrittlement sooner than courser materials of a comparable age (e.g. Turkish, Persian) owing to the loss of sericin during the degumming.

b) History

Those textiles which were adorned with metal were generally devised for sumptuous or symbolic effect. Usually, but not always, the textile served a function in addition to its decorative value. Beyond its era of functionality the textile may be stored, displayed, or studied in a collection. Because of their portability and sumptuousness many textiles containing metal are known to have travelled great distances (gifts of emissaries, explorers' hoards, war booty, trade goods, heirlooms) undergoing climatic and environmental changes that would tend to place considerable strain on the textile. The following list attempts to take into account the many possibilities that a textile may experience in its history.

i. Wear, abrasion, use, affects the aging of the textile. This kind of deterioration consists of a set of physical phenomena [32] describable in terms of frictional forces interacting, stresses, strains, folding, tearing, bending, compression, twisting, etc.

ii. If the textile is clothing, in addition to the above set of physical phenomena, the cloth may have come into contact with body fluids. The interaction of aqueous material with textiles may produce a wide array of chemical (hydrolytic) and physical (fluid absorption, salt deposition) changes, profoundly affecting the aging of the textile.

iii. The interaction of atmospheric gases - oxygen, ozone, oxides of nitrogen and sulfur, particulate matter, may affect the condition of any textile. Chemical reactions may degrade fibers, organic substrates, adhesives, and metals to form corrosion products.

iv. Archaeological textiles may have had direct contact with earth, decaying animal matter, moisture, ground water and minerals, as well as conflagration.

v. Biological attack - fungal, insect, rodent, ect. may come at any time in the life of a textile, archaeological or not. In addition to producing physical damage owing to loss of material, sometimes biological attack weakens the fabric chemically, so that further biological attack is facilitated. Generally the mechanisms of biological degradation are both oxidative and hydrolytic[84]. Although metallic elements are not susceptible to biological attack the textile system that supports the metal (adhesive, organic substrate, fiber) is not immune.

vi. Display and storage are benign for textiles compared to wear, use or burial, but prolonged exposure to high light levels, cycles of changing temperature and relative humidity, biological attack as well as exposure to atmospheric pollution may have profound effects on the aging behavior of textiles. Some of these hazzards are likely to be visited upon textiles (especially for transported collections of foreign origin) unless storage conditions accomodate these changes. Environmental changes can exascerbate problematic interactions among fibers, metal, adhesive, substrate. Adhesive, for example, used in a particular climate can alter substantially when placed in a different environment.

vii. Conservation treatments have experienced extensive revision and local idiocyncracies at various times. (The germ theory of disease must have given license to wash textiles that were made with the clear recognition that water would never be used on them: cf. paper, membrane, parchment, substrates for metal threads.) Technical training and technological advances enhance consciousness of the structure of matter and deterioration mechanisms as well as our ability to control environments. Connoisseurship and historical acumen tend more and more to enter into the decision making processes of what must be preserved for the cultural heritage. There is a general assumption that treatment, study, maintanance, storage of museum materials constantly improves, and that at least some of the deleterious effects of aging among museum textiles is ameliorated. Conservation treatment requiring any kind of handling is damaging to a textile. Treatment involving liquid agents, aqueous or non-aqueous, will produce irreversible changes in the textile. Of course no treatment or alternative treatments may be worse. Various forms of aqueous treatment - bleaching, soaping, lubricating, wetting, swabbing, humidifying, especially accompanied by agitation may be unsafe. In the first three cases chemicals are used which may be difficult to remove suggesting long term implications for the aging of the textile. In all cases, especially for old or embrittled examples material will be lost. Nonaqueous solvents can produce even greater loss, weakening of metal interfaces, and visual effects that may be unacceptable. Most treatments are performed to remove soils and stains (sometimes consolidants {glues, tapes, etc.} from earlier conservation treatments), to free the textile from undesirable wrinkles and folds, and to reveal and clarify design elements - color, weave patterns, metallic luster, etc. For this reason localized treatments specific to the area of damage and tailored to the type of damage are often the most appropriate. Sometimes the curatorial impulse to display a readable specimen imposes a costly conservatorial intervention, which the textile carries along as its legacy.

Deterioration mechanisms and methods of observation of the individual parts of metal-containing textiles

a) Metals

i. Gold, silver, and combinations of gold and silver, [8, 11, 17, 24, 34, 37, 41-43, 54-57, 73-77, 87, 97, 101, 109, 112] have been used most frequently as textile adornment, although examples of alloys of baser metals have also been observed. Table 1 indicates the various kinds of metal inclusions in textiles. Categories **I**, **IV**, and **V** involve adhesives; Categories **II** and **III** do not.

ii. Table 2 emphasizes the various physical forms by which metal is incorporated into textiles, along with

Table 2 *Incorporation of metal in textiles*

Cate-gory	Metal	Cloth	Examples
	Application of Metal with Adhesive to Textile		
I	Leaf	Silk, cotton, linen	Chinese, Japanese, Indian, Indonesian, Medieval Banners
	Dust	Silk, cotton,	Indian, Indonesian
	Paint/Leaf (with other woven embroidered, or printed features)	Silk	Banners, [Fortuny gowns], Chinese
	Paint/Leaf (as part of a painted cloth)	Cotton ,linen	Tibetan Tangkas, Indian Pichawais, Easel Paintings.
Metal [Thread] without Core		Cloth	
II.	Wire (Woven as weft. Laid and couched in embroidery, sometimes sewn)	Silk Silk, cotton	Turkish velvets Turkish embroideries
	Flattened Wire or Strip (Woven as weft. Laid and couched in embroidery, sometimes sewn.)	Silk Silk	Persian brocades Eur. Stumpwork
	Sequins, Mirrors, Ornaments (embroidered appliqued)	Silk, cotton, linen wool	Eur., No. Afr., Indian, Turkish, Indonesian embroideries.
Metal Thread with Core*		Core	
III.	Wire (Woven usually as weft. Laid and couched in embroidery, sometimes sewn.)	Silk, cotton Silk, cotton,linen	Indian, Persian Brocades Ecclesiastical Embroideries
	Flattened Wire or Strip (Woven usually as weft. Laid and couched in embroidery, sometimes sewn.)	Silk, cotton Silk, cotton, linen Silk, cotton ,linen	Indian, Persian Brocades Eur. tapestries, Bizarre Silks Ecclesiastical Embroideries
*Reindeer sinew in some Scandinav.textiles [24,25]; animal hair in core of few early exmpls.[9, 91, 109]			
Metal Applied to Organic Substrates[a,b]; Wrapped on Core		Core	
IVa	Metal leaf applied to one side.Woven into silk cotton or bast fiber fabrics. Embroidery, Applique	Silk	Chinese, Japanese, Indonesian, Central Asian
IVb	Metal leaf applied to one side. Woven into silk fabrics	Silk, cotton, linen	Chinese, Medieval Eur. Western Asiatic
Metal applied to Organic Substrates[a,b]; No Core		Cloth	
Va.	Metal leaf applied to one side. Woven Embroidered, Appliqued	Silk, cotton, bast	Chinese, Japanese Indonesian, Central Asian
Vb.	Metal leaf applied to one side. Woven, Embroidered, Appliqued	Silk	Chinese

a = Cellulosic substrate; **b** = Proteinaceous substrate typical illustrative examples. See Table 4 for appropriate references.

The identification of metal composition and attendant corrosion products is best achieved by X-ray probe techniques on microsamples which provide elemental analyses and/or identification of inorganic materials. Some instrumentation [15, 35, 38, 41-44, 46, 47, 54-57, 65, 67, 97, 101, 110] which has been used: X-Ray Fluorescence (XRF), Proton Induced X-Ray Emmision (PIXE), Energy Dispersive X-Ray Spectrometry (EDS), which provide elemental analyses; and X-Ray Diffraction (XRD) which identifies inorganic compounds. Sometimes analyses have indicated nearly pure gold or silver, other examples have been found with baser metals, or thinly gilded or silvered copper. Some brass samples have also been found. Later metallic threads are known to contain aluminium and other metallic elements incorporated in synthetic polymers [8, 111].

iii. Deterioration mechanisms for metals surfaces may be both physical (friction, abrasion, stress, etc.) and chemical (reactions with oxygen, water, oxides of sulfur and nitrogen, minerals and salts, etc.) processes. While gold is the metal least susceptible to corrosion, traces of baser metal corrosion can cause severe pitting and displacement of gold surfaces. Gold is most often found as a thin layer over silver; the microscopic appearance of gilded silver often shows a surface on which it is possible to see both gold and silver (corresponding to beating or drawing together a small amount of gold with silver, rather than alloying gold and silver with heat). All the usual corrosion products (oxides, chlorides, sulfides) may be found on silver surfaces, "tarnished" or blackened on most historic samples. Metal threads rich in copper have the potential for producing all the corrosion products generally associated with copper, including bronze disease. The nature of the corrosion products may be heavily influenced by the attendant substrate, adhesive, or fiber.

b) Substrates (paper, parchment, vellum, membrane, leather)

i. Categories **IV [a,b]** and **V [a,b]**, Table 1. Some metallic threads are made by adhesion of a metallic surface (leaf, powder, paint) to an organic substrate [18, 24, 25, 42, 43, 46, 54, 55, 67, 81, 83, 97, 101, 104, 105, 107, 109]. The organic substrate is cut into narrow strips and either wound about a core fiber **(IV)** or incorporated into a textile as a woven, embroidered, or appliqued feature as a flat strip without a fiber core **(V)**. The identification of the substrate is best achieved by microscopy at ca. 50-250x magnification. Distinguishing characteristics [46, 83] of these materials are presented in Table 3.

A major difficulty arises in distinguishing these materials in a deteriorated state since identifying characteristics are often confounded by decomposition products.

The organic wrappings or strips used as substrates are gravely subject to the ravages of time. Paper, which

Table 3 *Appearance (Magnified) of Organic Wrapping Materials Found on Historic Textile Threads. Category IV and V.*

Wrapping Material	Chemical Composition	Appearance	Comments
Paper **IVa,Va**	Cellulose	White yellow brown. Opaque; no translucence No hair follicle pits.	Adhesive material usually visible between metal surface and paper. Silk core.
Membrane **IVb,Vb**	Collagen	Colorless offwhite pale brown. Translucent areas No hair follicle pits.	Adhesive material not uniformly visible. Dark areas of degraded adhesive. Silk, cotton, or linen core.
Parchment **IVb,Vb**	Collagen	White yellow. Generally opaque; sometimes nearly translucent areas. Hair follicle pits.	Adhesive material visible. Dark areas of degraded adhesive Silk, cotton, or linen core.
Vellum **IVb,Vb**	Collagen	Same as parchment. Hair follicle pits may be sparser and smaller than in parchment	Same as parchment
Leather **IVb,Vb**	Collagen	Brown. Hair follice pits larger than in parchment.	Same as parchment

is cellulosic, and the other organic substrates, which are collagenous are all known to deteriorate through a variety of mechanisms. [32, 84].

ii. Physical deterioration[32]. Tension, stress, abrasion, compression, and folding are prominently experienced by the textile during its period of use, all of which may have important effects on the substrate, in the form of losses, tears, unravelling, etc.

iii. Chemical deterioration[84]. Two general types of reaction tend to degrade the polymeric materials which form the basic polymer network: hydrolysis and oxidation. The hydrolytic reactions will tend to break (depolymerize) chains. The oxidative reactions will tend to produce cross-linking of the polymer chains, interact with functional groups that are present, form new functional groups, decompose to species that promote an autocatalytic effect on the oxidative processes. Both oxidative and hydrolytic processes may be catalysed be metallic species, always present in textile system that contain metal. Other types of chemical deterioration that can produce similar effects are: light damage, biological damage, fire damage, reaction with atmospheric pollutants, all of which are extreme examples of oxidative hydrolytic breakdown.

The appearance of aged substrates is usually quite different from new material, often reflecting the use the material has undergone or treatments that have caused both physical and chemical degradation. As might be expected, water damage or treatment may cause degradation and embrittlement of the substrate. Expansions and contractions of substrate from liquid absorption may result in detachment of the metallic surface and adhesive; oxidative processes tend to darken the material generally (cf the yellowing/browning of most organic materials on aging).

c) Adhesives

i. Adhesives are required for the inclusion of metal in textiles for examples of Category **I, IV,** and **V**, Table 1. Treatises and reports have indicated the use of a wide variety of materials used as adhesives [2, 3, 11, 19, 22, 28, 49, 57, 65, 92, 96]: bole, egg white, drying oils, gums, animal and fish glues, etc.

ii. These materials (except for bole) are difficult or impossible to distinguish microscopically. The analyses of these materials is similar to the problem of analysing media and varnishes on painted works [69], organic media used in studio patination of Medieval and later bronzes [97], or furniture finishes [69]. In all these, the original material is a complex organic mixture of high molecular weight substances. Upon aging, the chemical constitution of the material undergoes considerable (oxidative) change. A number of newer analytical procedures have been successfully used to identify organic materials of this kind even after considerable aging. Mass Spectrometry/Gas Chromatography (MS/GC) is an extremely sensitive tool for this analysis. Generally a small sample must be subjected to solvent extraction(s) and chemical derivatization(s) in order to produce material of low enough molecular weight to pass through the instrument. Pyrolysis Gas Chromatography (PyGC) [93, 94] in which samples are subjected to high temperature degradation and chromatographic collection of the fragments has been effective as an analytical fingerprint process especially when similar known samples or closely related materials are available for comparison. In a few studies Pyrolysis Gas Chromatography has been used with Mass Spectrometry (Py/GC/MS) [93, 94] to help identify the pyrolized fragments. Another chromatographic procedure is High Performance Liquid Chromatography (HPLC) [16, 34]. Fourier Transform Infrared Spectroscopy (FTIR) is a sensitive device that provides information on functional groups present in sample [64]. Amino acid analysis [34] should succeed for adhesives that contain egg white or glue. No systematic study attempting to analyse the adhesives specific to the inclusion of metal in textiles has been reported.

iii. The organic adhesives are particularly sensitive to oxidative degradation. The mechanism of degradative processes has received considerable scientific attention [84]. It is similar to the degradative processes described above for the substrates, and similar degradative processes also apply to the fibers, but the adhesive material, among the various organic materials in the system is most sensitive to oxadation. Numerous examples of exposed adhesive on substrates show uneven darkening from yellow to quite black. Water damage or treatment may also produce disasterous effects on the system, weakening either the adhesive-substrate or the adhesive-metal bonding. Fiber, substrate, and adhesive each have different water absorption capacities so that changing cyles of temperature and humidity may produce mechanical stresses in the system.

d) Fibers.

i. The identification of fibers is a well documented subject [89] and is usually achieved with a high degree of success using 10-250x magnification with ordinary microscopy. Occasionally the deteriorated condition of a fiber can confuse an observation, but even severely degraded fibers usually contain enough of their fingerprint characteristics to be identified[89]. Among historic textiles wool has almost never been identified as part of the process of including metal in textiles. Most examples are silk, but cotton and linen are also found. (There are a few examples of gold stamped Cashmere wool; metal threads are also found in embroideries and carpets that also contain wool, but the core for the metal thread has never been identified as wool; a few archaeological ecclesiastical textiles are said to contain metallic threads with animal hair cores.)

ii. A number of general studies have described in detail oxidative, light and hydrolytic degradation of the polymeric materials comprising textile fibers [84].

iii. Occasionally textiles that are heavily laden with metallic threads deteriorate mechanically from the stresses within the system. Among the parts of the system, the fibers seem to resist degradation best. Uniformed treatment, such as washing or dry-cleaning of textiles with threads of Categories **IV** and **V** (Table 1 and 2) may damage metal threads grievously while doing far less damage to the fibers. It is not uncommon to see a historic textile with only a shadow of its original metallic component. Sometimes the cores of metal wrapped threads are almost all that survive of the original.

Conservation

Beyond the need for analysis and characterization of historic textiles comes its treatment in a museum environment which inevitably seeks to minimize the effects of natural aging and to optimize art historic and aesthetic importance. A list of published treatments include:

i. washing (aq/nonaq)
ii. swabbing(aq/nonaq)
iii. humidfying
iv bleaching
v. vacuuming
vi. mechanical cleaning
vii. sewing
viii. mounting
ix. consolidating
x. darning, reweaving
xi. biocides/fungicides
xii. storage
xiii. reuse
xiv. additives (UV screens, softening agts, anti-statics, fire retarders, stain-guards)
xv. no treatment

Some of these treatments (or variants of these procedures) are only historically interesting (consolidating, reuse) reflecting an obsolete sensitivity to the importance of the textile or ignorance of the consequences of treatment. A few of these treatments (bleaching, additives) are generally used in modern textiles in order to forstall the ravages of use or time.

Any treatment is best applied only after identification of the entire textile system. Even after identification is complete and problems clearly appreciated the choice of treatment cannot generally be relied upon by reference to the literature alone; but is best administered by an experienced conservator familiar with the behavior of comparable examples.

References

1. **Atil**. E. The Age of Sultan Suleyman the Magnificent, Nat. Gal. of Art, Abrams, NY 1987

2. **Alexander**, S. M., "Medieval Recipes Describing the Use of Metals in Manuscripts", Marsyas, Studies in the History of Art, Vol XII. 1964–1965, pp. 34–51.

3. **Baumann**, H. P. "History of the Art of Textile Printing" in Proceedings of the Perkins Centennial, H. J. White Jr., Ed., AATCC, Lowell, Mass. 1951. pp. 257–61 (no references)

4. **Bhatnagar**, I. K., "Brocaded Textiles and their Conservation," Conservation of Cultural Property in India, **VIII**, pp. 41–45 (1975) London, 1978.

5. **Bhushan**, J. B., The Costumes and Textiles of India, Taraporavala, Bombay, pp. 49-56. 1958.

6. **Bier,** C. (Ed.) Woven from the Soul, Spun from the Heart, Textile Arts from Safavid and Qajar Iran 16-19th Centuries, Textile Museum, Washington, DC. 1987.

7. **Black,** D. et. al., Islemeler: Ottoman Domestic Embroideries, David Black Oriental Carpets,

8. **Braun-Ronsdorf,** M., "Gold and Textiles, Gold and Silver Fabrics from Medieval to Modern Times" CIBA Review, **3.** pp. 2-16 (1961)

9. **Budney,** M. & D. **Tweddle,** "The Early Medieval Textiles at Maaseik, Belgium, The Antiquaries Journal **1xv** Pt II (1985) {mentions animal tail core of metal thread. p. 361.}

10. **Carlang,** M. & **Salmon,** L. French Textiles from the Middle Ages through the Second Empire (essays), Wadsworth Atheneum, Hartford, Conn, 1985.

11. **Cennini,** Cennino d'Andrea, The Craftsman's Handbook, "Il Libro dell' Arte" Tr. D.V.Thompson Jr., Dover, New York, 1960, pp. 96, 103, 107.

12. Census of India 1961, VOL V PART VII-A Selected Crafts of Gujarat 13. Jari industry of Surat. R.K. Trivedi, Superintendent of Census Operations, Gujerat, Ahmedanad, Delhi, 1968.

13. **Chattopadhyay,** K., Indian Embroidery, New Delhi, 1977, pp. 61-63.

14. **Delacort,** M., A Technical Study of Some Late Medieval Embroideries, M.A. Thesis, Institute of Fine Arts, New York University, 1965.

15. **Darrah,** Josephine A., Metal Threads and Filaments, in Recent Advances in the Conservation and Analysis of Artifacts compiled by James Black, Summer School Press, U. of London, Inst. of Archaeology, London (1987) Abst.: analytical survey of

Table 4 *References for Typical Textiles or Studies by Categories*

I.	3, 20–22, 26, 30, 49, 50, 57, 61, 63, 65, 92, 100,103, 108
II.	6, 7, 31, 66, 82, 97, 112,
III.	1, 4, 5, 6, 7, 10, 11, 13, 14, 15, 23, 27, 31, 35, 38–40, 44, 47, 51, 53, 55, 58, 59, 60, 62, 63, 66, 71, 82, 86–88, 91, 95, 97, 98, 102, 104, 106, 109, 112
IV.	18, 46, 52, 54, 55, 67, 68, 71, 72, 81, 97, 101, 104, 105, 107, 109,
V.	18, 36, 46, 52, 70, 72, 81, 101, 104, 107

metal threads and filaments; a brief history of their manufacture; the methods of examination used and the results and conclusions of the survey. A policy for the treatment of metal threads, at present advocated by the Conserv. Dept. V&A, is given.

16. **Doshi,** R.P., and M.J. **Vaickus,** The Use of High Performance Liquid Chromatography in the Textile Laboratory, in Book. AATCC. Book of Papers, 1982, pp.231-236 (AATA 25-650)

17. Encyclopedia Brittanica, 11th Ed. "Embroidery", by A. F. Kendrick, and A.S. Cole; "Gold and Silver Thread," by A.S. Cole; "Goldbeating," (unsigned), 1911.

18. **Feddersen,** M., Chinese Decorative Art (trans. A. Lane), New York, 1961, pp. 208-229; pp. 268-269 (Bibliog.)

19. **Fischer,** J. et. al. Threads of Tradition: Textiles of Indonesia and Sarawak, University of California, Berkeley, CA, 1979.

20. **Forrer,** Robert, Die Zeugdrucke der byzantinischen, romanischen, gotischen, und spätern Kunstepochen, Strassburg, 1894.

21. **Forrer,** Robert, Die Kunst des Zeugdrucks vom Mittelalter bis Empirezeit, Strassburg, 1898.

22. **Fraser-Lu,** S., Indonesian Batik Processes, Patterns and Places, Oxford Univ. Pr. , Oxford, N.Y., 1986.

23. **Geijer,** A., "Über die bizarren Stoffe, " Festschrift Erich Meyer 1957, Hamburg, 1959.

24. **Geijer,** A,. A History of Textile Art. Pasold Research Fund in Assoc. with Sotheby-Parke Bernet, London, pp. 11-13 et. seq. 1979.

25. **Geijer,** A., Birka III die Textilfunde aus den Gräbern, Uppsala, 1928, pp. 68-74.

26. **Geijer,** A., "Chinese Silks Exported to Russia in the 17th Century", The Museum of Far Eastern Antiquities Bulletin, no. **25** (1953) pp. 1-18, 24 pl.

27. **Geijer,** A. and E. B.-Thomas, "The Viminacium Gold Tapestry," Meddelanden fran Lunds, Universitets Historiska Museum, 1964-65, pp. 223-236.

28. **Gittinger,** M., Master Dyers to the World, The Textile Museum, Washington, D.C. 1982. p. 128.

29. **Gittinger,** M. et al., Eds., Indonesian Textiles, Irene Emery Roundtable on Museum Textiles, 1979. Proceedings, The Textile Museum, Washington, D.C., 1980. See especially: An Introductory Survey to Indonesian Textile Collections in North America.

30. **Golombek,** L., and V. **Gervers,** "Tiraz Fabrics in the Royal Ontario Museum," in Studies in Textile History, V. Gerves, Ed. Royal Ontario Museum, Toronto, 1977. pp. 82-125.

31. **Gönül,** M., "Some Turkish Embroideries in the Collection of the Topkapi Sarayi Museum in Istambul", Kunst des Orients, **VI.** 1, pp.43-76, (1969)

32. **Greathouse,** G.A. and C.J. **Wessel,** Eds., Deterioration of Materials, Causes and Preventive Techniques, Reinhold, NY. 1954.

33. **Greenwalt,** C.H., Jr., and L.J. **Majewski,** Lydian Textiles, Reprinted from University Museum Papers from Athens to Gordion the Papers of a Memorial Symposium for Rodney S. Young, the University Museum, Pennsylvania University, Philadelphia, 1980, pp. 133-147.

34. **Hancock,** W.S. Handbook of HPLC for the Separation of Amino Acids, Peptides, and Proteins, CRC Press, Boca Ratan, 1984.

35. **Harden,** I. and F. Duffield, "Characterization of Metallic Yarns in Historic Persian Textiles by Microanalysis, " CELL 44. Abs. of Papers, 188 th ACS Natl. Mtg. Phila., PA, 8/84. Historic Textiles and Paper Materials, Needles, H.and S. Zeronian, eds. Adv.in Chem. **212**, 231-252, ACS, Wash., D.C. 1986.

36. **Härtel,** H.-M. **Yaldiz,** et.al., Along the Ancient Silk Routes, Central Asian Art from West Berlin State Museum, Metropolitan Museum of Art, N.Y. 1982. *145

37. **Hoffmann,** H.,-P. F. **Davidson,** Greek Gold, Jewelry from the Age of Alexander, Ph von Zabern, Mainz/Rhein, 1965.

38. **Hoke,** E., and I. Petrascheck-Heim, "Mikroprobe Analysis of Gilded Silver Threads from Medieval Textiles", Studies in Conservation, **22.** pp. 49-62 (1977)

39. **Hope,** J., "A Bizarre Twist" Hali, issue 32, **Vol. 8.** no.4, pp.26-34 (1986)

40. **Hughes,** Therle, English Domestic Needlework 1660-1860, London, n.d.(1966); history; no refs.

41. **Indictor,** N. and R.J. **Koestler,** "The Identification and Characterization of Metal Wrapping in Historic Textiles Using Microscopy and Energy Dispersive X-Ray Spectrometry – Problems Associated with Identification and Characterization," Scanning Electron Microscopy (1986), pp. 491-497.

42. **Indictor,** N.-R.J. **Koestler,** C. **Blair,** and A. **Wardwell,**"The Evaluation of Metal Wrappings from Medieval Textiles Using Scanning Electron Microscopy-Energy Dispersive X-Ray Spectrometry" Textile History, **19** (1) 3-22, 1988.

43. **Indictor,** N., "Metal in Historic Textiles" Textile Conservation Group, Meeting of March 9, 1988, Fashion Institute of Design, NYC.(Textile Conservation Group Newsletter, Vol . X, No. 5. Mar 1988)

44. **Indictor,** N. and. C. **Blair,** "Examination of the Use of Metal in Historic Indian Textiles by Scanning Electron Microscopy-Energy Dispersive X-Ray Spectrometry, (in prep.)

45. **Indictor,** N. & N. **Indictor,** "A Glossary of Terms Relating to the Use of Metal in Textiles" (in prep.)

46. **Indictor,** N.,**R.J.Koestler,** M. **Wypysky,** and **A.Wardwell,** "Metall Threads Made of Proteinaceous Substrates Examined by Scanning Electron Microscopy-Energy Dispersive X-Ray Spectrometry" (in press) Studies in Conservation.

47. **Indictor,** N., "The Use of Scanning Electron Microscopy-Energy Dispersive X-Ray Spectrometry for the Examination of Metal Threads on a Karuppur Sari," Studies in Indian Textiles, a Socio-Cultural Survey, Ed.L Varadarajan, Manohar, New Delhi, (in press)

48. **Indictor,** N., N. **Indictor,** and A. **Chakravarty,** "A Glossary of Terms Relating to the Use of Metal in Indian Textiles" (in prep.)

49. **Irwin,** J., and B. **Brett,** Origins of Chintz, London, 1970. (see esp. p. 11)

50. **Irwin,** J., and M. **Hall,** Indian Painted and Printed Fabrics, Ahmedabad, 1971. No. 185, p. 148.

51. **Irwin,** J., and M. **Hall,** Indian Embroidery, Ahmedabad, 1973.

52. **Ito,** Toshiko, Tsujigahana the Flowering of Japanese Textile Art, transl.by M.Bethe, Kodansha Int'l. Ltd., Tokyo, N.Y., 1981.

53. **Jackson,** F. Nevill (=Mrs,) (Old Handmade Lace With a Dictionary of Lace, Dover, NY, 1987, Rpt of 1900; History plus dictionary plus glossary; no refs.

54. **Járó,** M. and Erzsébet **Gondár,** "Mediaeval Membrane Threads Used for Weaving and Embroidery", Achaeometrical Research in Hungary, Eds. M. Járó and L. Költő, National Center of Museums, Budapest, 1988. 255-266.
Forty three membrane samples, 13th-16th C, (Venice, Germany, Toscana, Italy, Regensburg, Lucca, Hungary, Cologne) + four leather samples, 10th-14th C, (Arabia, Byzantine, Sicily, Turkey)
Three categories: A, narrow, 0,25 0,50 mm; medium 0,60–1,00mm; C, wide, 1,00. (A–B=0,40-0,60mm; B–C=0,75-1,00mm)
Five subcategories: 1. Gilding, no silver; 2. Gilding over silver; 3. Gilding traces (visible) over silver; 4. Gilding traces (detectible chemically) over silver; 5. Silver, no gold.

55. **Járó,** M., "The technology and analytical examination of metal threads on old textiles," in Conservation–Restoration of Church Textiles and Painted Flags;Investigation of Museum Objects and Materials Used in Consevation–Restoration, Fourth International Restorer Seminar (2–10 July 1983), Budapest, 1984. pp. 253–264. (An attempt at classification: I. Metal wires (applied as such {Category **II**}or spun round a silk core {Category **III**}; II. Threads made of narrow, thin metal strips (originally applied as such, cf.ephod, {Category **II**} or more commonly spun about silk or linen core {Category **III**}; Threads made of animal membrane, leather or paper covered with very thin leaf gold or silver "...narrow strips made of animal membrane or leather were gilded or silvered then spun round a linen core and thus woven... of oriental origin where the membrane was often replaced by paper... gilded membrane...called 'aurum filatum cyprense' {Category **IV**}.

56. **Járó,** M. "The Investigation of the Metal Embroidery Threads of the Hungarican Coronation Mantle by Scanning Electron Microscope and Physical Methods of Analysis, ICOM 7th Triennial Meeting,Copenhagen,10-14 Sept. 84.1.22-84.1.24, Paris, 1984.

57. **Járó,** M., A. **Timár-Balázsy**, and L. **Kriston**, "X-Ray Diffraction Investigation of the Pigments of Printed Textiles," Scientific Examination of Works of Art, Working Group 1, ICOM Committee for Conservation, 1987.
German, 13/14th C; 6 samples: Red = HgS, cinnabar; Red = Fe_2O_3, hematite+kaolin+$BaSO_4$ (bole+barium white); Green = ?+gypsum+barium white; Silver = Sn; Gold =Cu. Barium white = 19th C

58. **Johnstone**, P., Byzantine Tradition in Church Embroidery, London, 1967, pp. 65-73.

59. **Johnstone**, P., Turkish Embroidery, Viktoria & Albert Museum, London 1985.

60. **Kendrick**, A.F., English Embroidery, London, 1904.

61. **King**, Donald., "Textiles and the Origin of Printing in Europe", Pantheon, 20, 1962.

62. **Krishna**, A. & V. **Krishna** (A.Mookerjee, Ed.) , Banares Brocades, Crafts Mus., N.Delhi, 1966. Parenthetic Romanized Hindi (Sandskrit?) for stylistic and technical terms.

63. **Kuhnel**, E., and L. **Bellinger**, The Textile Museum Catalogue of Dated Tiraz Fabrics, Washinton D.C. (1952)

64. **Lang**, L., J. E. **Katon**, J.F.**O'Keefe**, and D.W. **Schiering**, "The Identification of Fibers by Infrared and Raman Microspectroscopy," Michrochemical Journal, **34** 319-331 (1966) (AATA 25-702)

65. **Lee-Whitman**, L., and M.**Skelton**, "Where Did All the Silver Go?" Textile Museum Journal, **22** pp. 33-52 (1984); Skelton and L.Lee-Whitman, "A Systematic Method for Differentiating Between 18th Century Painted-Printed Chinese and Western Silks," Historic Textiles and Paper Materials, Needles, H.L. & S.H. Zeronian, eds. Adv. in Chem. Series **212**, pp. 131-152, Amer. Chem. Soc., Washington, D.C. 1986.

66. **Mackie**, L., The Splendors of Turkish Weaving, The Textile Museum, Washington, D.C. 1973.

67. **Magagnato**, L., et.al., Le stoffe di Cangrande, Alinari, Florence, pp. 190-193, 1983.

68. **Mailey**, J., Embroidery of Impreial China, New York, 1978.

69. **Mills**, J., and R. **White**, The Organic Chemistry of Museum Objects, Butterworths, NY. 1987.

70. **Minnich**, H.B. (in collab. with Shojiro Nomure) Japanese Costume and the Makers of Its Elegant Tradition, Tuttle, Rutland, VT, Tokyo, 1963.

71. **Nissim**, J. "A Monograph on Wire and Tinsel in The Bombay Presidency," in "Art in Industry through the Ages," Navrang, New Delhi, repr. (1895-97) 1976 pp. 287-302

72. **Nomura**, S., (Collector of Old Brocades and Ancient Costumes) Historical Sketch of Nishiki and Kinran Brocades, Shimmonjen, Kyoto; Branch Nichome Motomachi, Kobe, n.d.

73. **Oddy**, W.A., "Ancient jewellery as a source of technical information – a study of technique for making wire", in Conservation-Restoration of Church Textiles and Painted Flags; Investigation of Museum Objects and Materials Used in Conservation-Restoration, Fourth International Restorer Seminar (2-10 July 1983), Budapest, 1984. pp. 241-252. {Reviews techniques for wire making in No.Europe and Italy, before 1000 AD; specific examples of museum objects, a few may have come from textiles.}

74. **Oddy**, W.A., "The Production of Gold Wire in Antiquity," Gold Bulletin, **10**, (3) 79-87, 1977.

75. **Oddy**, W.A., "Hand-Made Wire in Antiquity: a Correction, "Masca, 1,2 6/79 44-45, 1979

76. **Oddy**,W.A., "Gold Wire in Antiquity, "Aurum, **5**, 8-12, 1981.

77. **Ogden**, J., Jewelry of the Ancient World, Rizzoli, NY, 1982. Ch 4.

78. **Osborne**, H., Ed., The Oxford Companion to the Decorative Arts, Oxford U.P., Oxford and New York, 1985, pp.317, 210, 243, 454, etc.

79. **Pfister**, R., and L. **Bellinger**, The Excavations at Dura-Europas Pt. II The Textiles, New Haven, 1945.

80. **Pope**, A.U., Ed. A Survey of Persian Art, Oxford Univ Press , London, New York, 1939, Vol 3, Articles by P. Ackermann, A.U.Pope, and H. Goetz.

81. **Priest**, A., Japanese Costume an Exhibition of No Robes and Buddhist Vestments, Metropolitan Museum of Art, N.Y. 1935.

82. **Reath**, N.A., and E.B. **Sachs**, Persian Textiles and their Technique from the Sixth to the Eighteenth Centuries. Yale Univ. Press, New Haven, CT, pp. 4-10, 1937.

83. **Reed**, R., Ancient Skins Parchments and Leathers, London 1972.

84. **Reich**, L., and S.S.**Stivala**, Elements of Polymer Degradation, McGraw-Hill, NY, 1971.

85. **Riboud**, K., G. **Vial**, and M. **Hallade**, Tissus de Touen-Houang Conserve au Musee Guimet et la Bibliothèque Nationale, Paris, 1970.

86. **Rothstein**, N., The Elegant Art of Woven Silk. (Chapter in) An Elegant Art, Fashion and Fantasy in the Eighteenth Century. Los Angeles, New York: Los Angeles County Museum of Art, Henry N.Abrams, Inc., pp. 61-87, 1983.

87. **de Saint-Aubin**, C.G., Desinger to the King 1770, Art of the Embroiderer, Trans. and Annotated by N. Scheuer and E. Maeder, Los Angeles County Museum of Art, Godine, Boston & London, 1983.

88. **Sarre**, F., and H.Trenkwald, Old Oriental Carpets, Leipzig, 1926-29.

89. **Sawbridge**, M. and J.E. **Ford**, Textile Fibers under the Microscope, Shirley Institute, Didsbury, 1987. (AATA 25-2035).

90. **Schreier**, Barbara A. & R.R. **Bresee**, History of Decorative Metallic Yarns, Book of Papers, 1979 National Technical Conference, AATCC Oct. 3-5 Hyatt House, Cherry Hill, NJ, Host Delaware Valley Section (Lib.Cat. TP 890.5 N27 1979 MSC: source Mary Ballard)

91. **Schuette**, M. and S. **Müller-Christensen**, A Pictorial History of Embroidery, (trans. D. King) Praeger, NY, 1964. (See p. 297, 9/10 SO-CALLED "CHASUBLE" OF SAINTS HARLINDIS AND RELINDIS, England, ca. 850 for 'gold thread is a metal strip wound on brown transparent animal-hair' Another example of animal-hair core exists in the National Museum of Denmark: personal communication, Irene Skals, Department of Conservation. See also Budny M. and D. Tweddle, The Early Medieval Textiles at Maaseik, Belgium, The Antiquaries Journal, **Vol LXV Pt II** (1985), p.361 ff.)

92. **Shaftel**, A. "Notes on the Technique of Tibetan Thangkas," Journal of the American Institute for Conservation **25** (1986), pp. 97-103.

93. **Shedrinsky**, A., T.P. **Wampler**, N.**Indictor**, and N.S.**Baer**, "Application of Analytical Pyrolysis to Problems in Art and Archaeology: a Review, "Journal of Analytical and Applied Pyrolysis (in press).

94. **Shedrinsky**, A., T.P.**Wampler**, N. **Indictor**, and N.S. **Baer**, "Application of Pyrolysis Gas Chromatography (PyGc) to Problems in Art and Archaeology," Conservation of Cultural Property in India (in press).

95. **Sloman**,V., Bizarre Designs in Silks, Copenhagen, 1953.

96. **Solyom**, B. and G., **Farbric** Traditions of Indonesia, Washington State Univ. Press, Pullman, Washington, 1984.

97. **Stodulski**, L.P., H.F. **Mailand**, D.**Nauman**, and M. **Kennedy**, "Scanning Electron Microscope /Energy Dispersive X-Ray and Atomic Emission Spectrographic Studies of Precious Metal Threads from European, Middle Eastern, and Oriental Textiles," Abstract, Applications of Science in Examination of Works of Art, Museum of Fine Arts, Boston, MA(1983); Ed. P.A. England & L.van Zelst, "Atomic Emission Spectrographic and Scanning Electron Microscope/Energy Dispersive X-Ray Studies of European, Middle Eastern, and Oriental Metallic Threads," pp.76-91, Museum of Fine Arts, Boston MA,1985.

98. **Stone**, C., The Embroideries of North Africa, Longman, Essex, 1985.

99. **Stone**, R., R. **White**, and N. **Indictor**, "Detection of Patination on Italian Renaissance Bronzes," (in prep).

100. **Talwar**, K., and K. **Krishna**, Indian Pigment Paintings on Cloth, Ahmedabad, 1979.

101. Tessuti Preziosi del Giappone Kinran; Shokko-nishiki; Kinsha; secoli xvi/xix; Storia; Techniche; Analisi; Restauro, Museo d'Arte Orientale "Edoardo Chiossone"; Compagnia dei Librai, Genoa, 1985. Failla, Donatella (History&Weaving tech), Romanello, Paolo (Metallic Material). Bensi, Paolo (Dyes), Di Campo, Rosalia (Restoration).

102. **Thornton**, P., Baroque and Rococo Silks, London, 1965.

103. Victoria & Albert Museum, Printed Textiles, London, 1949.

104. **Wang**, Binghua, "Yanhu Gumu," Wenwu, **no.10**, 1973, pp.28-36. (Yuan Dynasty excavation with silk and cotton brocades containing gold threads: nianjin — silk core wrapped with gold foil (category **III**); pianjin — leather strip with gold – no core (category **V**.)].
(Anne Wardwell suggests that nianjin has an organic substrate, i.e.Category **IV**.)

105. **Wardwell**, A.E. a. "Eastern Islamic Silks of the 13th and 14th Centuries Woven with Gold and Silver," (in Prep.); b. "Flight of the Phoenix: Crosscurrents in Late Thirteenth to Fourteenth-Century Silk Patterns and Motifs," The Bulletin of the Cleveland Museum of Art, 74,(1), 1-35, 1987.

106. **Watson**, J.F., The Textile Manufacture and the Costumes of the People of India, Allen, London, 1867, reprint, Varanasi, 1982.

107. **Weng**, W. and B.**Yang**, The Palace Museum: Peking, Abrams, New York, 1982, pp.288-297. ("Golden threads, made either of gold leaf paper strips or of foil strips wound around lacquered paper cores were already in use." (in the Tang Dynasty.) This is probably either a mistranslation or a misunderstanding: The first kind is gold paper strapwork, category **V**; the second is gold leaf applied to paper with red-brown bole wound round a silk core, category **IV**.)

108. **Wilckens**, Leonie von, Der spätmittelalterliche Zeugdruck nördlich Alpen, Anzeiger des Germanischen Nationalmusems, Nürnberg, 1983, pp. 7-18.

109. **Wild**, J.P. Textile Manufacture in the North Roman Provinces, Cambridge University Press, Cambridge, 1970. Suggests 4 categories (3 taken from Geijer): wire alone (**II**); wire spiraled on (silk) core (**III**); flattened metal on a core (**III**); gilded proteinaceous wrapping on core (**IV**).

110. **Williams**, E., and N. **Indictor**, "Determination of Metallic Mordants on Textiles by Particle-Induced X-Ray Emission," Scanning Electron Microscopy (1986) pp.847-850.

111. **Wingate**, I.B., Fairchild's Dictionary of Textiles, 6th Ed. Fairchild Publ., New York,1979.

112. **Yusuf** Ali, A., A Monograph on Silk Fabrics Produced in the North Western Provinces and Oudh, Allahabad, 1900. Reprint, Calico Museum, Ahmedabad, 1974.

Authors' address:

N.Indictor
Chemistry Department Brooklyn College, CUNY
Brooklyn, USA-NY 11 210 Brooklyn and the Graduate Center and Objects Conservation Department Metropolitan Museum of Art
USA-NYC 10028

M.Ballard
Conservation Analytical Laboratory
Smithsonian Institution
USA-D.C. 2056 Washington

TIMÁRNÉ, BALÁZSY Ágnes

Deterioration process of fibers in different chemical actions

The science, of logic distinguishes very clearly between reason and cause. The thoughts which justify our judgement are called logical base = reason (ratio). We have to distinguish from this logical base, the real reason which is the cause (causa fiendi). The first shows whether our thoughts have an adequate logical base, the latter describes what causes the phenomenon, in which circumstances it will occur.

For textile conservators the statement "cellulose-based fibers are sensitive to acids, protein-based fibers are sensitive to bases" is the "ratio" that explains why they do not apply acidic or alkaline media when treating the fibers. The "causa fiendi" has to be sought in much deeper regions, at the chemical structure of these materials and the processes of chemical attack. If we learn the "causa fiendi" of this statement we will see that the "ratio" is not as simple as that.

For that aim I offer a short survey on the chemical structure of the natural fibers, the process of their deterioration by different reagents and draw conclusions on the condition of the textile with metal threads before cleaning, on its sensitivity to different cleaning agents.

1. Chemical structure of natural fibers

On the basis of their chemical composition, natural fibers belong to two different groups: flax, hemp, cotton – plant fibers belong to the cellulose–based fibers, wool, silk amimal fibers, and hairs belong to the protein–based fibers.

Both cellulose and protein are natural polymers. They are composed of small molecules called monomers or units joined together by a special link.

Cellulose

In the case of the cellulose polymer the units are β-D-glucose molecules joined together by oxygen links called glucosidic link. In fibers the polymer chains are said to consist of about 7000-12000 units. This number is called the degree of polymerisation (DP).

Cellulose chain

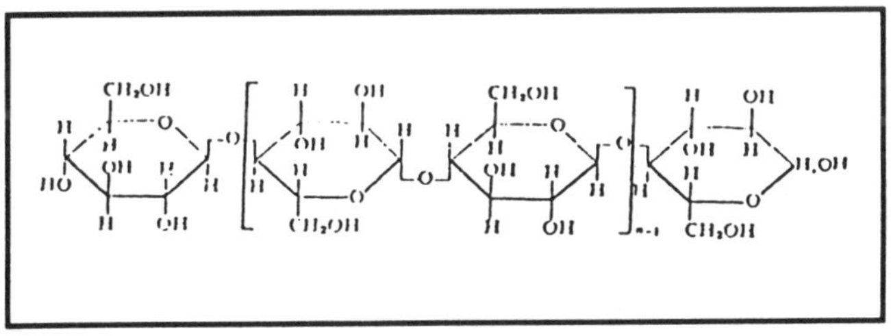

Cellulose chain

Characteristic functional groups are present as side-groups: three -OH (hydroxyl) on each unit. The chemical properties of the cellulose is due to the oxygen links and the hydroxyl functional grops.

Protein

The units of the protein polymers are various amino acids of different sizes, joined together by the so called peptide links. Eighteen different amino acids make up the protein of wool and silk in different proportions. Among them there are different sizes, small and large, containing different functional groups such as hydroxyl, carboxyl, amino groups. The sulphur containing amino acids such as cistine and cisteine provide a resistance to acids for wool but make it attractive to moths. A very small proportion of sulphur containing amino acid is present in the fibroin protein of silk.

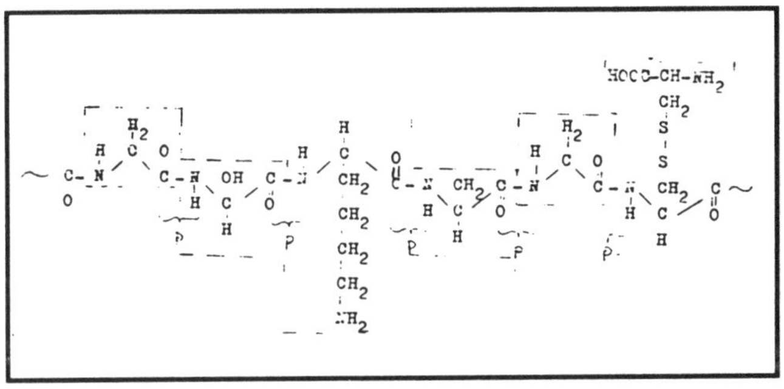

Part of a protein chain (p=peptide link)

Characteristic functional groups are present as side groups on the chain: -OH (hydroxyl), -COOH (carboxyl), $-NH_2$ (amino) and other groups. The chemical properties of protein are due to the listed functional groups and the peptide bond.

2. Primary, secondary and tertiary structure

Primary structure

The atoms and the covalent bonds between the atoms determine the primary structure of the polymer chain. In the case of cellulose the covalent bonds are present as carbon-carbon and carbon-oxygen links in the pyranose ring and between the units.

For proteins mainly carbon-carbon, and carbon-nitrogen bonds create the primary structure. If also the amino acid cistine takes part in the building of the protein (wool and hair), disulphide covalent bonds may be present in the chain, between two parts of the same chain, or between two chains. This also belong to the primary structure of the protein. The primary structure of a protein is determined by its amino acids and their sequences (the order in which they follow each other in the chain).

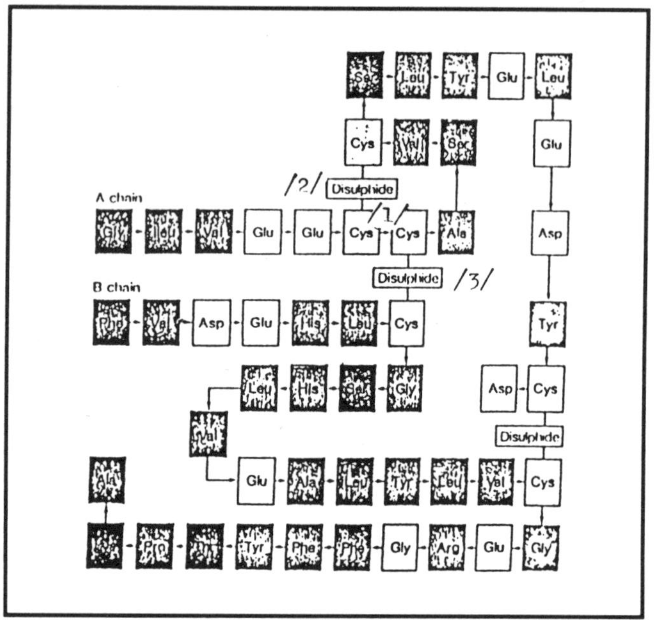

Disulphide covalent bonds in the protein chain beef insulin (1), between two parts of the same chain (2), and between two chains (3).

Secondary and tertiary structure

The secondary bonds (hydrogen bonds, dipole forces, Van der Waals' forces) between two parts of the same polymer chain (intramolecular) or between two or more chains (intermolecular) determine the secondary and the tertiary structure.

Two cellulose chains showing intra-molecular (*) and inter-molecular (Λ) hydrogen bonds between hydroxyl groups

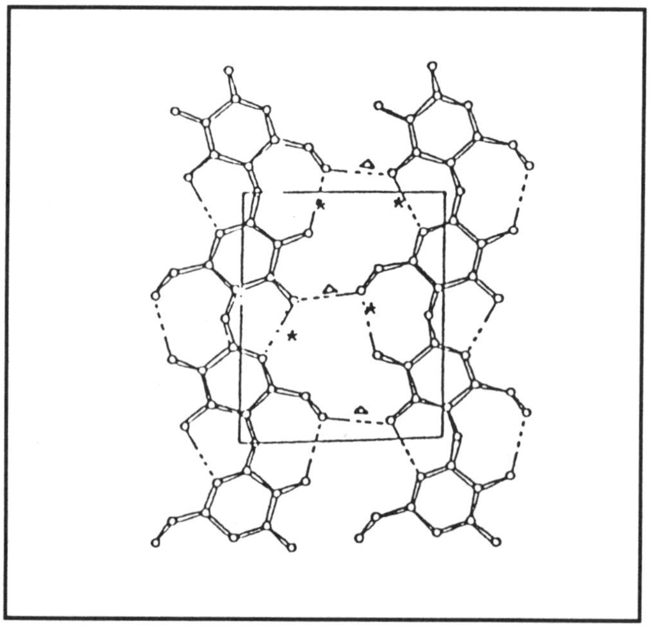

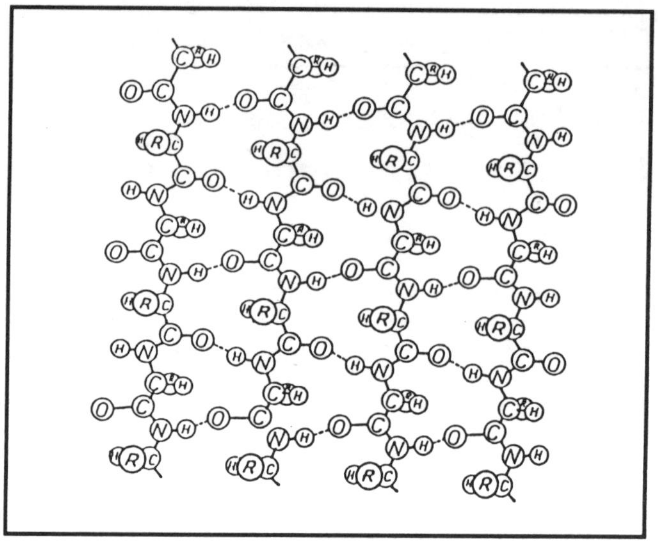

The extended β – structure of the silk fibroin

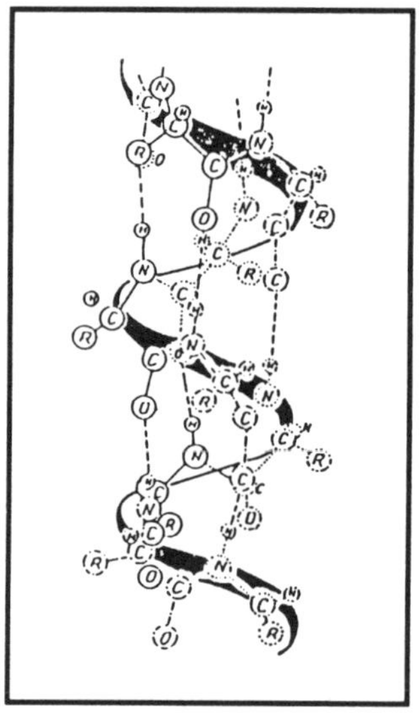

The helical structure (α-helix) of the wool keratin

In the case of cellulose both the intra-, and the the intermolecular secondary bonds belong to the secondary structure.

In the case of proteins the intramolecular secondary bonds determine the secondary, the intermolecular secondary bonds the tertiary structure.

The secondary bonds are created by the hydroxyl groups of the cellulose and the peptide bonds, carboxyl and amino functional groups of the proteins.

The secondary and tertiary structure determines the level of organization between the polymer chains. The more secondary bonds keeping together the chains the

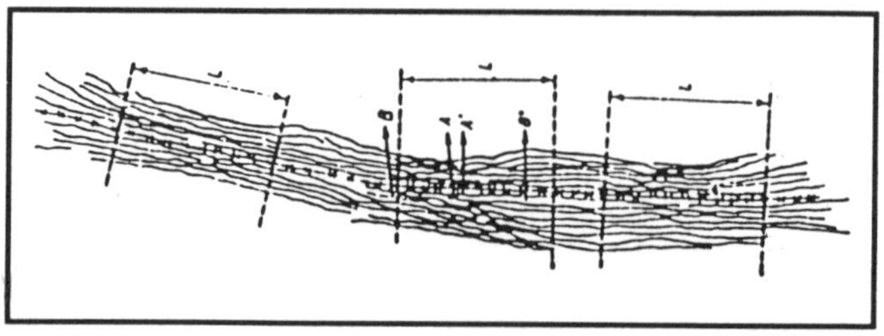

higher the level of organization of the polymer. Highly organized structures are called crystalline. The natural polymers can never be totally crystalline. To create a crystal a material must consist of equal sized molecules. The lenght of polymer chains is always an average lenght (average number of units in the chain). There are always some longer or shorter chains which do not permit a totally organized, i.e. crystalline structure. The result is that the natural polymers have both well-organized (crystalline) and badly-organized (amorphous) parts.

The picture shows the structure of cellulose: one cellulose molecule chain can go through crystalline and amorphous regions. A,A',B' are chain ends in the crystalline region, B is a chain end in the amorphous part. L is the length of the crystallite.

There is a serious difference in the relationship of the crystalline and amorphous parts of different fibers. Linen has about 90%, native cotton and silk about 60%. The crystalline portion of wool constitues a small percentage of the fiber, and as a result wood is a very weak fiber. The conclusion is a different stability against ageing: chemical reagents usually penetrate into the amorphous parts or attack the surface of the crystalline region. The better organized the secondary (tertiary) structure, the better the fiber is protected against chemical attack. Linen, for example, is better able to resist deterioration than cotton unter the same contions.

Result of oxidation of natural fibers

Oxidation of natural fibers occur by light induced oxidation with the oxygen of the air or by treatment with oxidizing agents in acidic, neutral or alkaline solutions.

The oxidation of cellulose is a heterogeneous reaction: this means that attack of the reagent is not neccessarly uniform throughout the fibre. In general, attack occurs most rapidly in the amorphous regions of the fiber followed by a much slower attack on the more ordered ones. The products of oxidation may retain the orginial fibrous form but may, depending on the reagent and the vigour of the attack, be a friable powder.

The anhydroglucose unit in cellulose is subject to attack in at least 3 places. Attack at the carbon 2 and carbon 3 atoms leads to ketone groups, or breaking of the ring leading to two aldehyde groups, which may in turn be oxidized to carboxyl groups. Attack at the carbon 6 may form aldehyde or carboxyl groups. Some of the possibilities are given by formulae 1-7 below.

The yellowing of cellulose by oxidation is due mainly to the appearance of aldehyde groups.

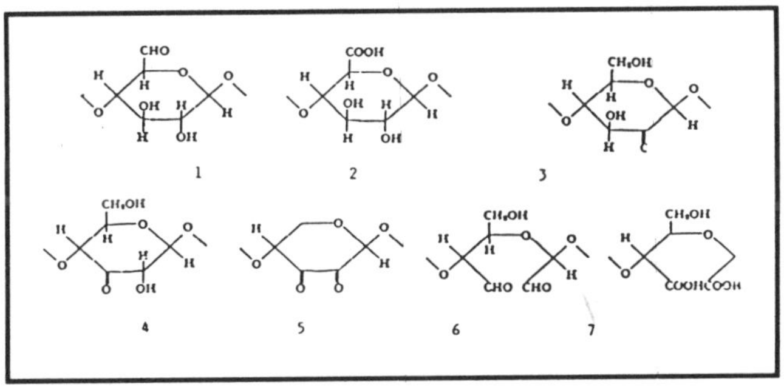

Products of oxidation of cellulose are normally referred to as "oxycelluloses" even though such materials differ widely in their properties. For example, cotton oxidized in alkaline solutions gives products which exhibit reducing properties only to a small extent (low copper number, see acid hydrolysis) but which contain a large content of acid groups(as determined by the absorption of the basic methylene blue). Oxidation in acid conditions leads to the reverse of this – high copper numbers and low quantities of acidic groups.

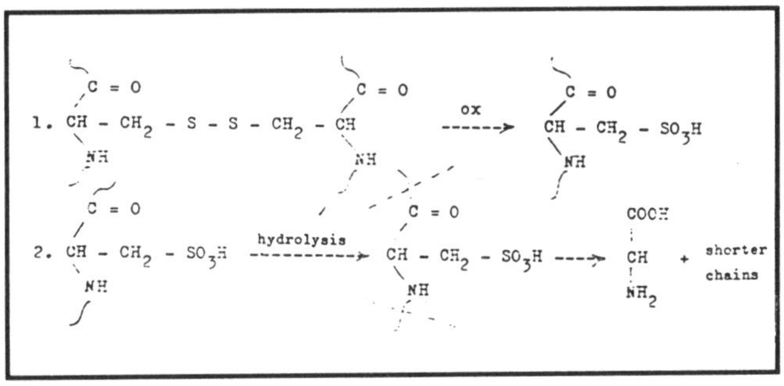

Breaking of chains as a result of hydrolysis

Oxidation of proteins also occurs mainly on the side groups which can be oxidized, like the hydroxyl groups carried by the serine or hydroxy-proline amino acids. Unfortunately the peptide bonds can also undergo oxidation and the disulphide links may break by oxidation reaction with help of later hydrolysis. The oxidized disulphide residue turns into chain scission and through that to a lower degree of polymerization.

By oxidation the peptide bonds of the main chains may also break.In the fibroin silk the presence of the tyrosine amino acid in the polymer causes fast yellowing by oxidation.

4. Result of acid and alkaline hydrolysis of natural fibers

Acid hydrolysis Cellulose is sensitive to acids. Acid hydrolysis of cellulose will break the chain at the oxygen link (glucosidic bond), lowering the DP and releasing reducing and non-reducing groups. The recuding groups originate from the hidden aldehyde groups of the pyranose ring, the non-reducing groups are the original

Products formed by the acid treatment of cellulose are referred to as hydrocelluloses.

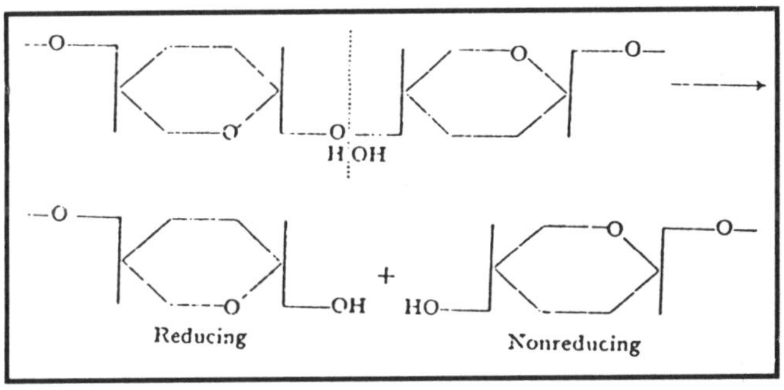

hydroxyl groups of the same ring. Progressive attack causes cellulose to lose strength and ultimately its fiber structure to give a friable powder.

From the practical point of wiew, the extent of acid attack could be determined by the changes in the reducing properties or the viscosity of a solution of the material. While the viscosity decreases, the number of reducing ends increases and can be shown by the increase in the extent of reduction of ammoniacal silver nitrate producing a deposit of metallic silver. A similar phenomenon is the formation of cuprous oxide from Fehling's solution on the fiber. The latter test has been well standardized and gives a figure known as the copper number. Measuring it before and after treatment, the conservator can prove the extent of damage caused by acid hydrolysis or oxidation.

Proteins are less sensitive to acids than cellulose, however this sensitivity increases with increased level of deterioration. Acids break the salt linkages by protonizing the $-NH_2$ (amino) groups and preventing the dissociation of the carboxyl group.

Acids also attack the peptide linkages - the result are shorter chains and a decrease of the DP.

Acid attack cannot lead to the breaking of the disulphide linkages wich are present in the protein of wool or hairs. This explains why wool can resist acid environment and can survive in peat bogs or bloe coffins.

Dissolution of the amorphous regions by acid causes the overall crystalline fraction to increase while the moisture content decreases. It is a believed that chain scission in the amorphous regions creates mobile chain

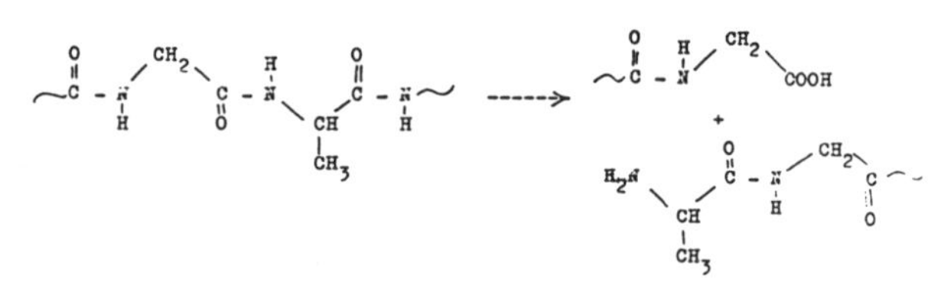

Changing of the ratio of amorphous and cystalline regions during aging

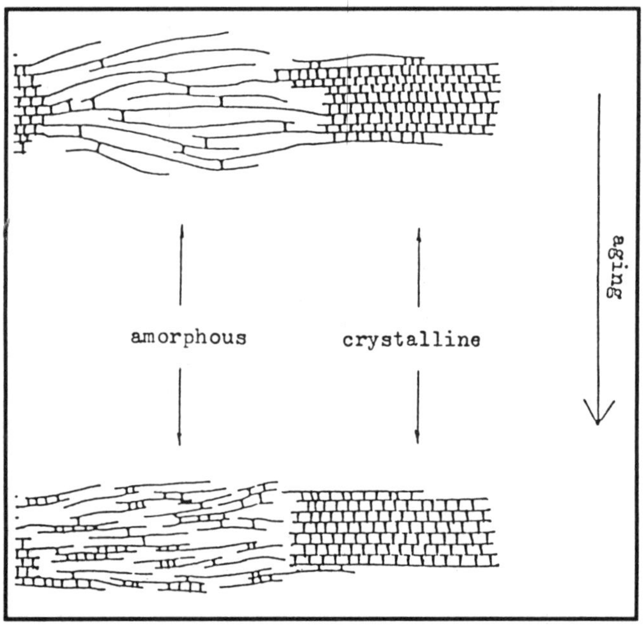

ends which are able to aligh with cyrstallites. This process is called recrystallization and the result can be a friable condition in extreme cases.

Alkaline hydrolysis

Native cellulose undergoes alkaline hydrolysis under severe alkaline conditions. Hydrocellulose (cellulose after acid hydrolysis) and oxycellulose (cellulose after oxidation) are more sensitive to alkaline hydrolysis than native cellulose. It is important for conservators to take into consideration this fact because they usually get the cellulose-based objects at one of the many stages of hydrolysis or oxidation.

The loss of weight of cellulose arising from alkaline treatment is dependent on the number of carbonyl (aldehyde and ketone) groups present on the chain. This suggests that the degradation occurs in stepwise fashion from the reducing end of the chain or from the aldehyde or ketone groups in the chain. The main reaction is called β-alkoxy elimination because the small molecule which is eliminated from the end of the chain must be in a beta position to a carbon possessing both an electon withdrawing group (x) and an acidic hydrogen atom. Because the broken bond occurs near the end of a long chain with the elimination of a small molecule, an alkoxy group, the reaction is called a "peeling" reaction.

The following reaction shows how the reducing end spontaneously changes from a closed ring to an open ring with an aldehyde at carbon 1. The glucose reducing end (I) is supposedly transformed into a fructose end (II) and then β-elimination occurs. This results in the scission of the C-4 oxygen bond and the elimination of a glucose residue (III):

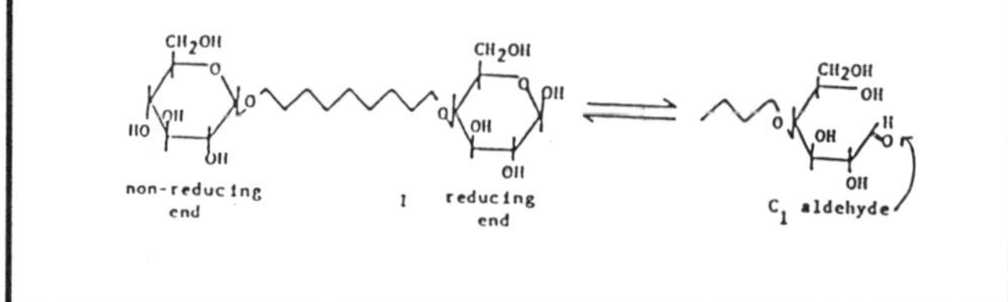

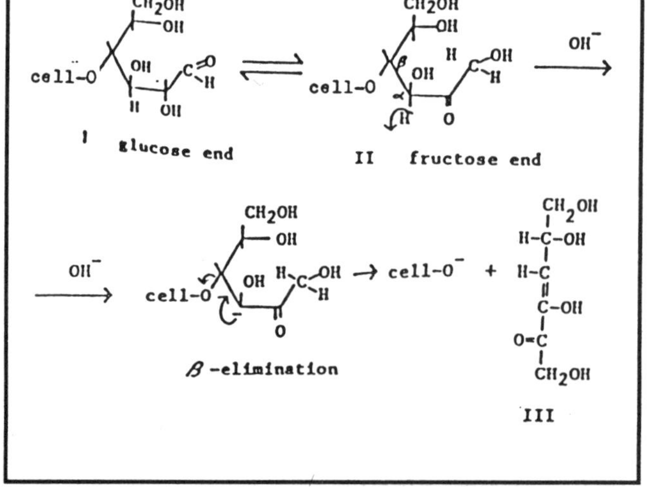

In the presence of an aldehyde group at carbon 6 or at carbon 2 and carbon 3 or ketone groups at carbon 2 or carbon 3, the oxidized glucose unit will break again at the β situation to these carbonyl groups. The result is the breaking of the chain = shorter chains = lower DP.Here is an example with a ketone group at carbon 3:

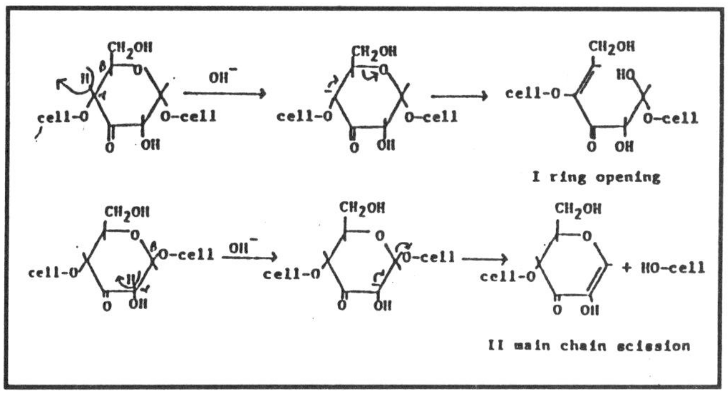

β-elimination in an alkaline medium causes either ring opening (I) or main chain scission (II) and reduction in the D.P. The location of the break along the polymerchain will determine the magnitude of the strength loss.

Bases attack proteins at several points. They break the salt linkages, the peptid bonds, the result is similar to that of acid hydrolysis but they also attack the disulphide linkages. Fifty per cent of the disulphide links break off immediately with weak alkali solution. This explains why not only silk but also wool dissolves in alkaline solutions. The breaking process of the disulphide link leads to the following result:

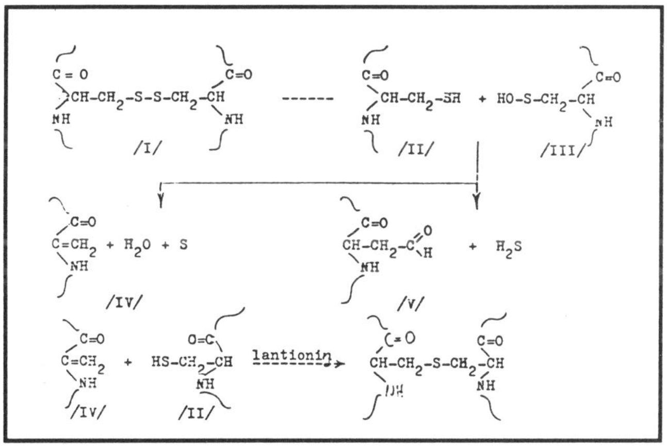

The presence of an aldehyde group in product V and the double bond in product IV causes the yellow colour of deteriorated wool. The hydrogen sulphide developing during the process is able to accumulate in a closed space and tarnish silver or cause other corrosion.

The whole process of alkaline hydrolysis results in shorter chains = lower DP. loss of strength and more crystalline structure = friability of protein fibres. Ultimately it can lead to total degradation.

5. Role of temperature, humidity, microorganisms and other factors in the deterioration of natural fibres

Temperature:

Increasing temperature mainly together with moisture accelerates almost every chemical process, including hydrolysis and oxidation.

Humidity (water):

Water itself can hydrolyse very degraded fibres. If the fibres were previously treated with acid or had absorbed acids from the atmosphere, with the help of the humidity, acid hydrolysis can be carried out. Water also can serve as a solvent for salts which produce an acidic or alkaline pH. These salts may be present as soil on the object. Water can be a source of hydrogen peroxide (H_2O_2) which is an oxidizing agent.

Changes in humidity lead to stretching, swelling and shrinkage of natural fibers.

Microorganisms:

Microbiological degradation can be caused by moulds, fungi or bacteria. Fungi require aerobic environments, but some types of bacteria can grow in anaerobic conditions as well.

In the case of cellulose the microorganisms attack the amorphous areas of the fiber first. The degradation is localized, and the products formed may not affect the degree of polymerization. The reactions which occur are hydrolytic enzymatic reactions, producing glucose which is then metabolized by the microorganism.

The activity of the microorganisms is limited in enviroments of pH less than 5, in environments of very high salt concentration or high concentration of heavy metals.

Wool is susceptible to degradation by moulds and bacteria. Attack by them has been determined as involving an initial rupture of the disulphide bond followed by enzymatic hydrolysis of the peptide links. The result is change in the primary structure, shorter chains, ultimately leading to total degradation.

Proteolytic enzymes do not really attack fibrous fibroin and in this respect silk differs from wool which is attacked by pancreatin.

Metals:

Oxidation is accelerated in the presence of heavy metals like iron, cobalt, manganese or copper. They enhance the decomosition of peroxides which carry out the oxidation. This explains why dusty textiles more easily unergo light induced oxidation. Also this is what conservators take into consideration when applying demineralized water for cleaning.

Hydrolysis of protein is accelerated by the following metals:

Na > K > Ca > Ba in order of efficacy. Heavy metals can serve as catalysts of hydrolysis of cellulose and proteins.

In the copper/textile system copper ions act as localized

biocidic agents. In the iron/textile system the corrosion products form casts around the fibers, which later decay. It is interesting that a higher proportion of wool is found in association with iron artefacts and a higher proportion of bast fibers is preserved in association with copper alloys.

Other factors:

Impurities of the fibers, processing and finishing materials also contribute to the chemical deterioration of natural fibers.

Mechanical damage not only leads to physical ageing but opens the surface of more and more crysstalline parts of the fiber to chemical or biological deterioration.

Summary

It is evident that there is a considerable difference in the condition and appearance of textiles exposed to light during their lifetime, or spending most of their time under the earth or in a crypt.

The level of photo-oxidative deterioration is different in a textile object with metal threads from one without. The metal wire or strip usually prevents the fiber core from light induced oxidation. Where the metal has been damaged this preventive "tub" opens and at that places the fiber core may be oxidized. Around the metal thread the fabric may be damaged much more by photo-oxidation than in the core. The conclusion: possibly the core fibers could survive the cleaning treatment with less damage than the surrounding fabric.

We have seen that oxidation leads to new functional groups on the polymer chain of the fibers which make them more sensitive to any kind of hydrolysis. Because oxidation does not necessarily shorten the chains the fibers retain their fiber structure and seem to be in quite good condition before cleaning. If the oxidized fiber is put into water it may show such a sensitivity to hydrolysis that also in neutral pH the wet treatment can conclude in a much more degraded condition of the object.

The circumstances of the burial, the composition of the soil in the ground and other factors determine the condition of archaeological textile. The principal agents of decomposition are biological. If the burial conditions are suitable for the range of microorganisms that will decomose a certain type of fiber, then it will eventually disappear. In anaerobic conditions there are no microorganisms that can easily decompose keratin and fibroin provided that they are in neutral to basic conditions. Cellulose will only survive in anaerobic conditions where the pH is high. Solid compact clay is very suitable for preserving linen, cotton, or bast fibers.

Fibers coming from loosely packed, aerated soil can be expected to be at some stage of oxidation. Ground with much surface water can greatly contribute to the oxidation of the fibers because of the presence of hydrogen peroxide in different concentrations. Transition and heavy metals are catalysts of this oxidation. This oxidized condition again has to be taken into consideration when choosing a wet treatment which can cause further degradation.

The alkalinity of the ground without oxygen causes deterioration on protein fibers over a long period of time this may explain why wool and silk have been found in the alkaline water of Swiss lakes. With oxygen the alkali in the ground destroys all fibers. Silk resists oxidation better in alkaline conditions.

The acidity of the ground causes hydrolysis of cellulose and silk but does not harm wool or hair. Cellulose is less likely to oxidize in acid or neutral environments. This permits us to choose a mildly alkaline treatment for the purpose of neutralizing cellulose-based fibers which are acidic.

Because hydrolysis occurs mainly in the amorphous region, when washing out the products of hydrolysis the more fibrous crystalline part results in an increase of tensile strenght. If the loss of weight of the fabric does not cause considerable difference in the structure, wet cleaning impoves the condition of the object.

Hydrocellulose is rather sensitive to oxidation. It is suggested that such objects be kept in the dark and that the cleaning and conservation be carried out with very limited lighting.

In the absence of acid or alkali, high salinity decreases microbiological degradation.

High salt concentration of the groud will result in more acid degradation of wool because it narrows the pH region in which acide attack is resisted (originally it is pH 4 to 8).

Naturally all the above described conditions of the object are influenced by many different factors.

This and may other aspects have to be investigated when choosing a cleaning agent for the metal threads on valuable fabrics. My aim was not to recommend different chemicals which can be used for this purposes but to show that we have to dig very deep when investigating the condition of the object before cleaning. We have to find the "causa fiendi" of our decisions and not only the reason.

References

Bogle, M.: Technical data on cotton, wool silk, Textile Conservation Workshop Notes, Merrimack Valley Textile Museum, 1979, 3p each.

Csűrös Z. Rusznák I.: Textilkémia, Tankönyvkiadó, Budapest, 1964. 628 p

Jakes, K.A. Sibley, L.R.: Survival of cellulosic fibers in arcaeological contexts, Science an Archaeology n. 25. (1983) pp. 31-38

Janaway, R.C.: Dust to dust: The preservation of textile materials in metal artefact corrosion products with reference, to inhumation graves, Science and Archaeology No. 27. (1985) pp. 29-34

Kerr, N.: The degradation of cellulose: causes and prevention, A literature review, School of Textiles, North Carolina State University, 1980

Peters, R.H.: Textile Chemistry Volume I-III Elsevier Publishing Company, 1963

Sibley, L.R. Jakes K.A.: Survival of protein fibers in archaeological contexts, Science and Archaeology no 26. (1984) pp 17-27.

Taylor, R.L.: The chemistry of proteins. A Unilever Educational Booklet, 1972

Weaver, J.W.: Analytical methods for a textile laboratory, American Association of Textile Chemists and Colorists, 1984

Author's address:

National Centre of Museums
Könyves K. krt. 40. H-1087 Budapest

SIPOS Enikő – GONDÁR Erzsébet

Effect of different treatments on textiles with metal threads

Every restorer has met with difficulties in hadling objects of composite materials. This problem is especially true for fabrics ornamented with metal threads. Namely, an effective cleaning agent for metals may do harm to core yarns, to ground fabric, or to needlework pads. No solution is possible other than by compromise.

Since in restoring practice, the possibility of separating different materials requiring separate treatment is exceptional, a correct decision is rather difficult even after careful material tests and experiments.

This fact induced us to test all the suggested cleaning methods for the sake of determining the cases where they are effective, and where their application may be inadvisable because of their inconvenience.

The suggested cleaning methods which could be found in the special literature are:

a) Mechanical treatments such as cleaning with:
— sodium hydrogen-carbonate
— magnesium carbonate
— glass pencil
— ultrasonic means

b) Chemically active, relatively neutral acid or basic solutions

Gold

— non-ionic detergent in distilled water
— formic acid, against the coloured copper compounds, if the core yarn is silk
— diluted ammonia, against the $CuCO_3$, if the core yarn is cellulose
— emulsion

Silver or gilded silver

— non-ionic detergent in distilled water
— thiourea in acidic conditions, against silver sulphide, silver cloride and copper compounds
— formic acid (see above)
— emulsion

Copper

— formicacid (see above)
— EDTA[Na] against chloride, oxide and sulphide
— reduced plasters
— emulsion

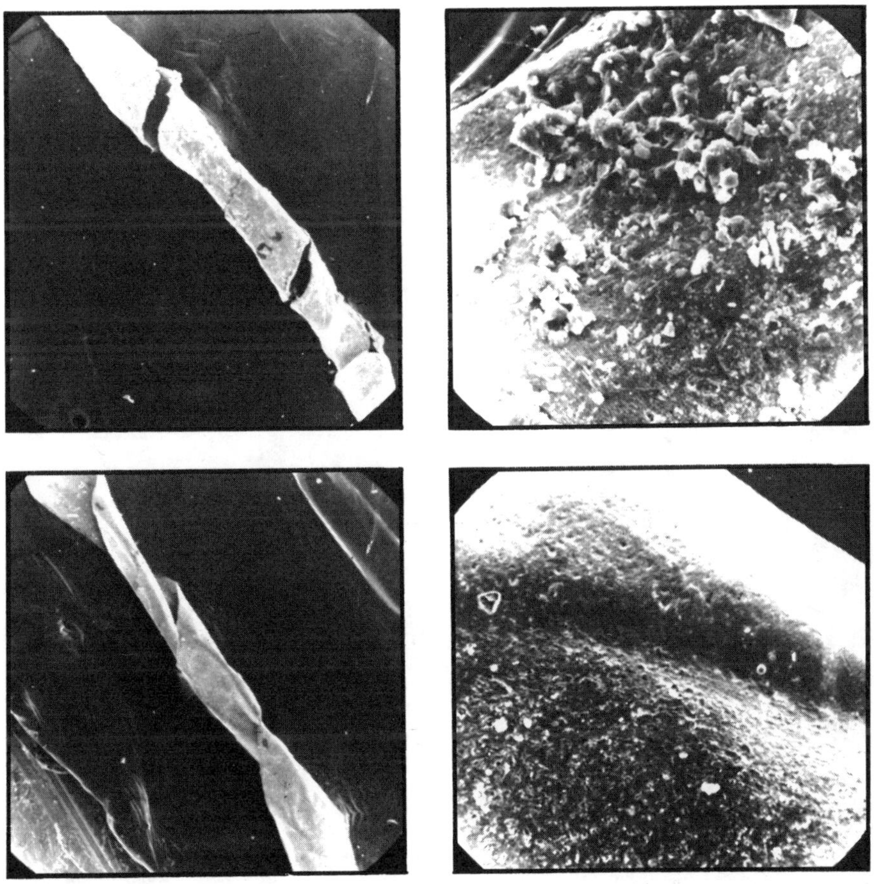

Fig. 1 a) *Gold thread before cleaning*

b *Detail of the uncleaned gold surface*

c *The cleaned thread (detergent + distilled water)*

d *Detail of the cleaned gold surface*

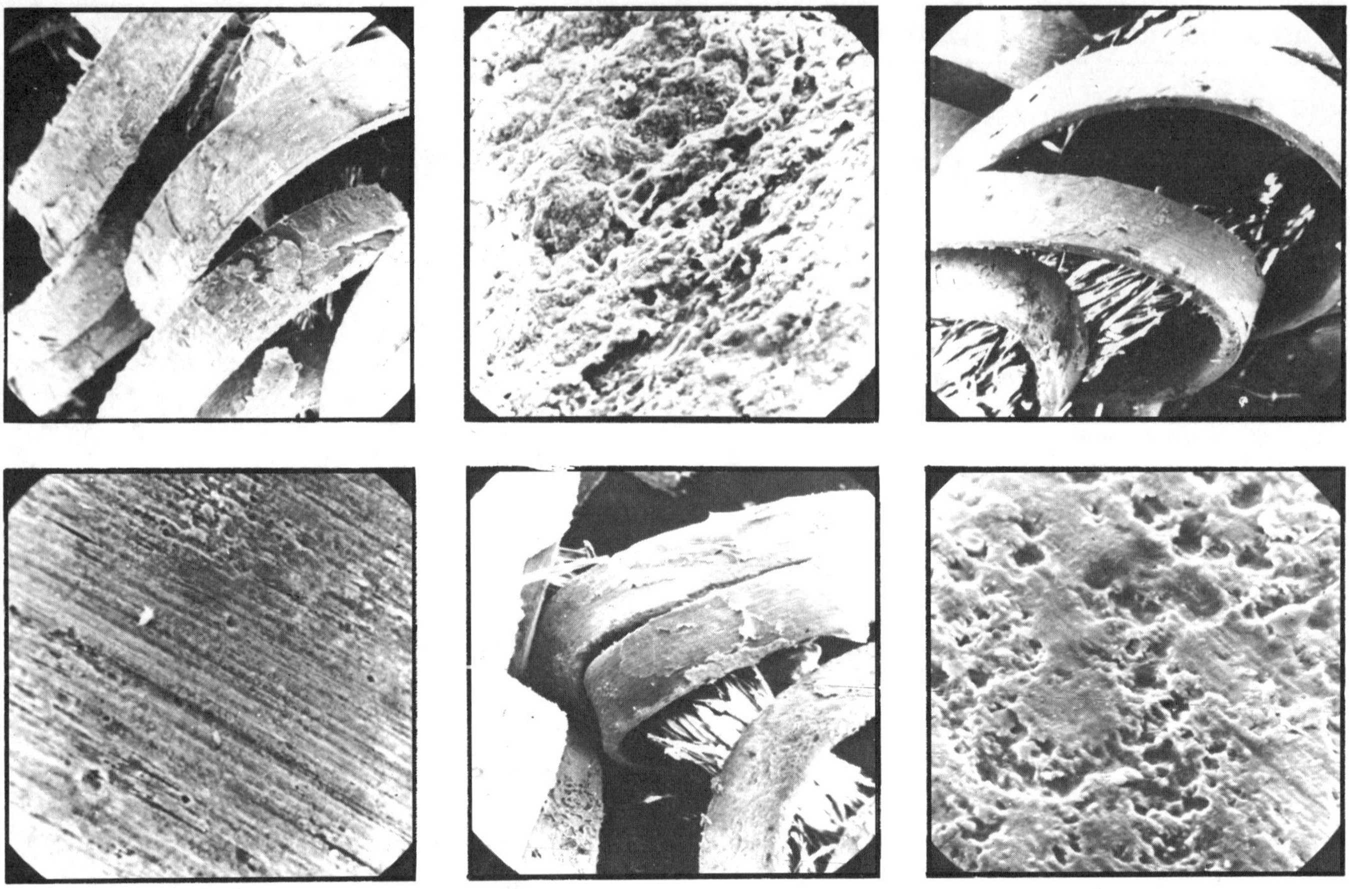

Fig. 2 a *Silver thread before cleaning* **b** *Detail of the uncleaned silver surface* **c)** *The thread after cleaning with thiourea* **d** *Detail of the cleaned silver surface* **e***The thread after cleaning with "Silver Dip"* **f** *Detail of the cleaned silver surface*

Fig. 3 a *Gilded silver thread before cleaning* **b** *Detail of the uncleaned metal surface* **c** *The thread after cleaning with glass pencil* **d** *Detail of the cleaned metal surface* **e** *The thread after cleaning with thiourea* **f** *Detail of the cleaned metal surface*

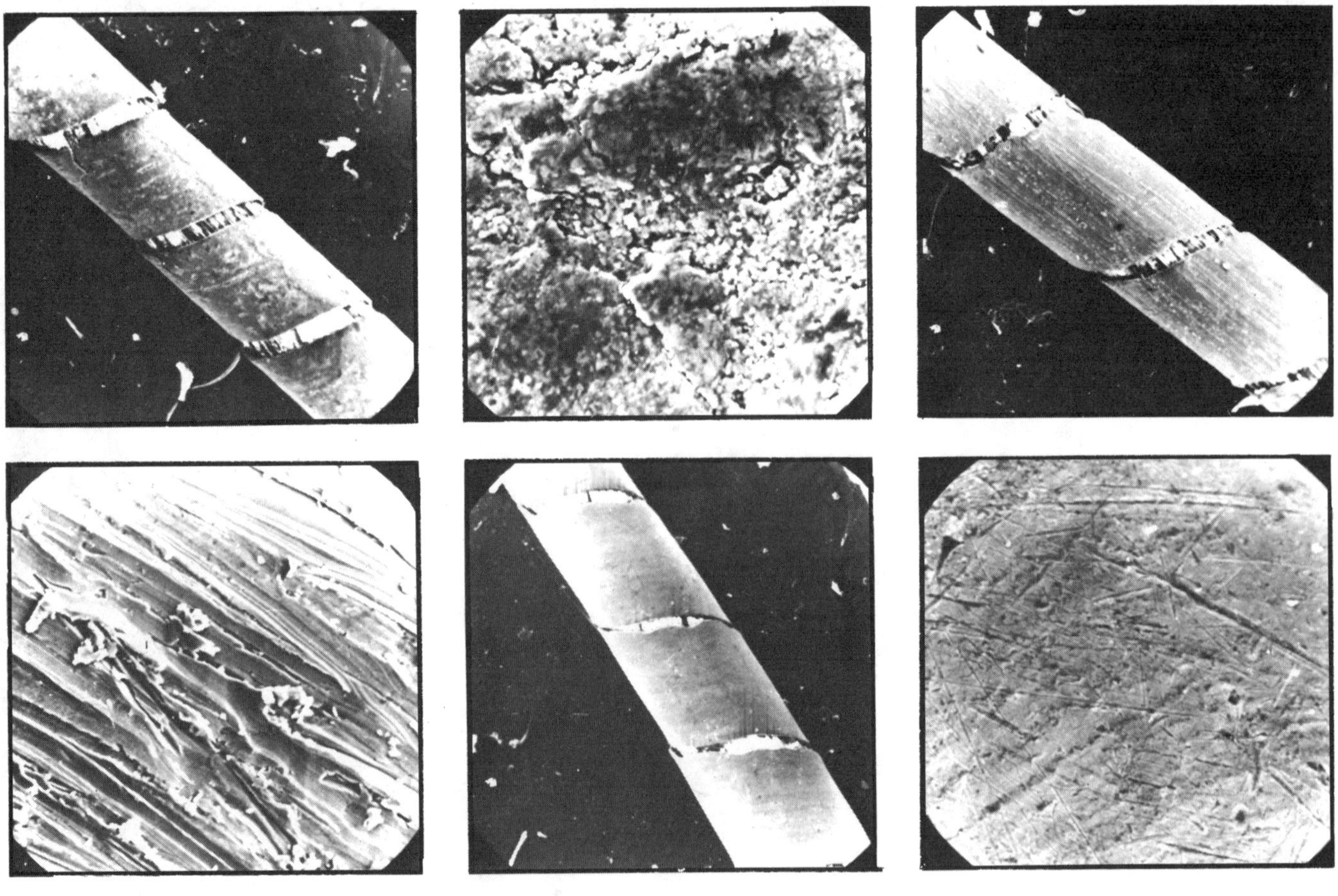

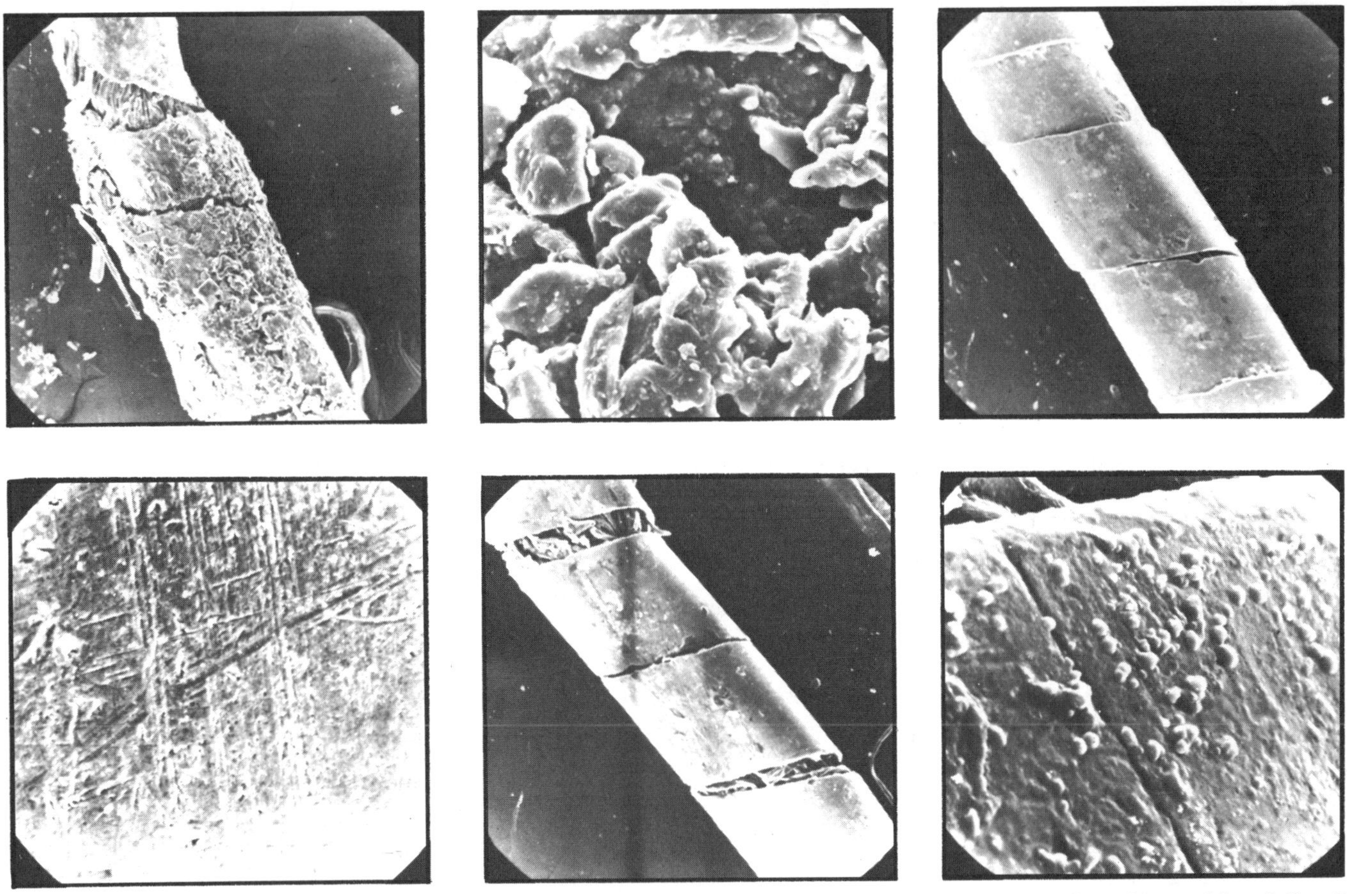

Fig. 4 a *Copper thread before cleaning* **b** *Detail of the uncleaned copper surface* **c** *The thread after cleaning with emulsion* **d** *Detail of the cleaned copper surface* **e** *The thread after cleaning with EDTA[Na]* **f** *Detail of the cleaned copper surface*

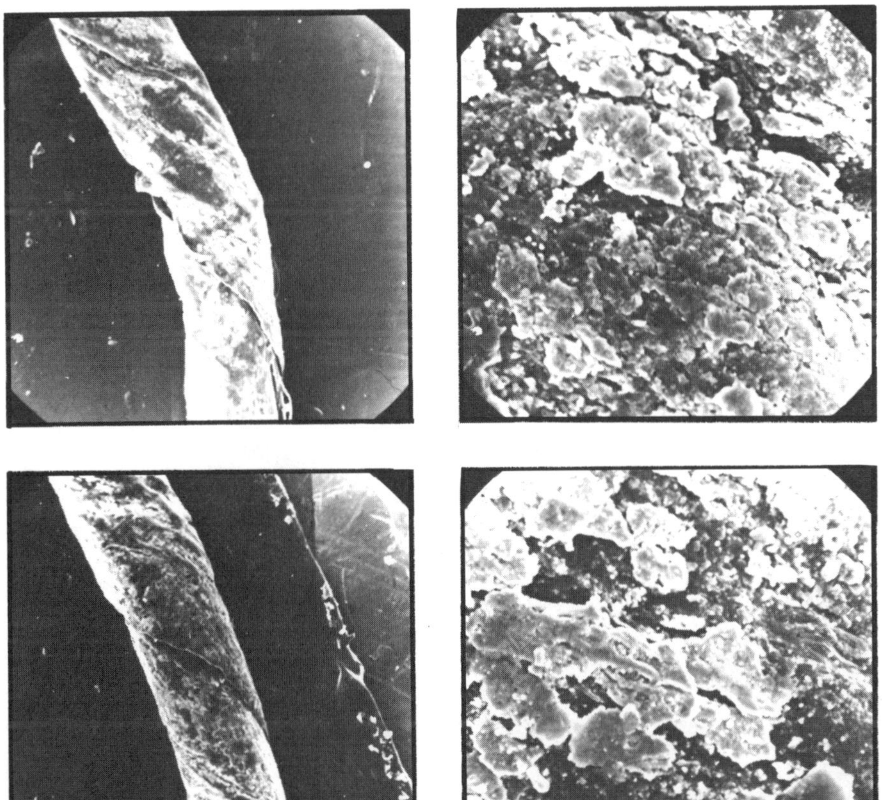

Fig. 5 a *Gilded membrane thread before cleaning* **b** *Detail of the uncleaned membrane surface* **c** *The thread after cleaning with detergent in distilled water* **d** *Detail of the cleaned membrane surface*

For threads from excavation

— ion exchange resins in distilled water
— EDTA[Na]
— non-ionic detergent in distilled water

Our aim was to investigate the effect of some of the cleaning methods used in the treatment of metal threads.

The chemicals used:

— non-ionic detergent in distilled water 1 g/l
— 3 % EDTA[Na] and non-ionic detergent
— 2 % thiourea and 3 % formic acid
— 12 % thiourea, 30 % citric-acid and non-ionic detergent (Silver Dip)
— emulsion: 9 parts of 0,5 % detergent in distilled water and 1 part of organic solvent

Effects of the different cleaning methods have been examined by JSM-35 Scanning electron microscopy. The tests were made on 5-10 mm long samples.

The results have been documented with photos. (See Figs. 1-5)

Conclusion

Before cleaning, in every case the type of metal thread and

the type of corrosion have to be determined, and the surrounding fabrics analysed.

Any chemical treatment is justified only by serious motivation because the re-corrosion of the metal cannot be avoided and repeated cleaning remove yet another layer.

Generally, mechanical treatment and commercial products containing strong inorganic acid do harm to old fabrics.

Authors' address:

Sípos Enikő
Hungarian National Museum
Múzeum krt. 14-16.
H-1088 Budapest

Gondár Erzsébet
INNOVATEXT, Textile Engineering
Gyömrői út 86.
H-1103 Budapest

HOWELL, David

Experiments with chemical cleaning for metal threads

At the Textile Conservation Studios, at Hampton Court Palace, there is a general policy of not cleaning metal threads. However, in 1986, circumstances forced us to look into the possibilities of cleaning one particular object, and I present our findings here as a type of "Case Study".

The object in question is the Canopy of State from The King's Audience Chamber at Hampton Court Palace. It is English, and was bought for the Palace at the end of the seventeenth century for King William III. It consists of crimson silk damask, which is sewn onto a buckram backing with a linen thread, decorated with metal braid and metal fringing. Up to four years ago the canopy was in fairly good condition, and in normal circumstances it would still be on display now.

However, in March 1986, there was a tremendous fire at Hampton Court Palace, which caused the canopy to be removed from display for us! The main parts of the canopy were buried under the rubble, and had to be excavated by Archaeologists from English Heritage. Because most of the pieces were wet from the fire hose water, and burst pipes, it was thought sensible to remove as much soiling as possible by immediately washing them in de-ionized water.

Nearly all the pieces were recovered, mostly in reasonable condition, and it was considered possible to conserve the object and return it to display.

One problem was that lead had melted from the roof and had solidified on both the silk and the metal braid. I have been able to think of no way that this lead could be removed without damaging the metal or fabric.

The Curatorial staff at the Palace decided that because this room had been so devastated, it was possible to rebuild it and furnish it so it would appear as it was in the late seventeenth century. This would mean new wall coverings, new flooring, new curtains, and generally the room would look new. They had doubts about returning the canopy into this setting in its present condition because it may look inappropriate. They made two suggestions. A replica could be made, or we could try to clean the metal thread to make the original canopy look more impressive.

Initially, the idea of a replica seemed very attractive. But to make a convincing copy of the canopy would cost a very large amount of money, and having spent so much it is unlikely that there would be any money left for the conservation and proper storage of the original canopy. What would probably happen is that it would be shut away in a storage room in non-ideal conditions, and be forgotten about until the replica started to fall to pieces. By then it is likely to be impossible to conserve the original canopy.

So it was decided to try to investigate the cleaning of the metal thread. One of the main problems was that the metal was extremely brittle, and any attempt to remove it from the main fabric of the Canopy would have resulted in a great deal of damage. It was important, therefore, to be able to clean the metal 'in situ' on the object. Several small samples had been broken from the canopy in the fire, and these were used for experimentation. Several methods were tried, including electrochemical reduction, electrolytic reduction, and ultra-sonic cleaning. None of these methods were successful; in fact, the ultra-sonic method resulted in a "Metal Thread Soup"!

At this point it was decided that to pursue the enquiry a more sophisticated analysis would help. Samples were taken to Exeter University Engineering department to be analysed using their Scanning Electron Microscope and Energy Dispersive X-ray Microprobe. This particular apparatus was chosen as the microprobe has a windowless detector, so that low atomic weight elemets such as carbon and oxygen can be detected. The threads were of a type previously reported by Darrah (1), with a flat metal strip being wrapped around a silk core.

Looking at the surface of the metal, one can see that the surface is very cracked, and that corrosion is presumably still taking place under the corrosion layer (See Fig. 1).

Fig. 1

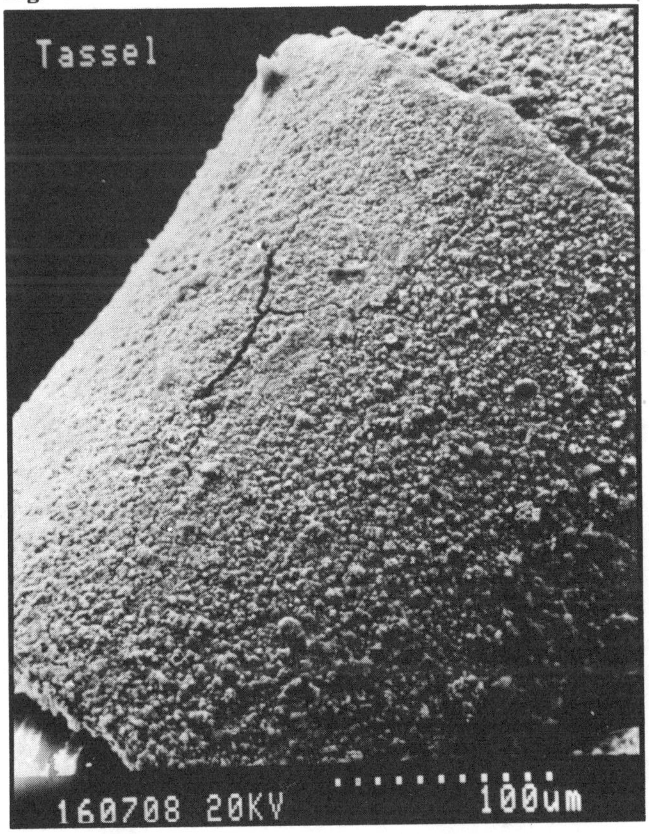

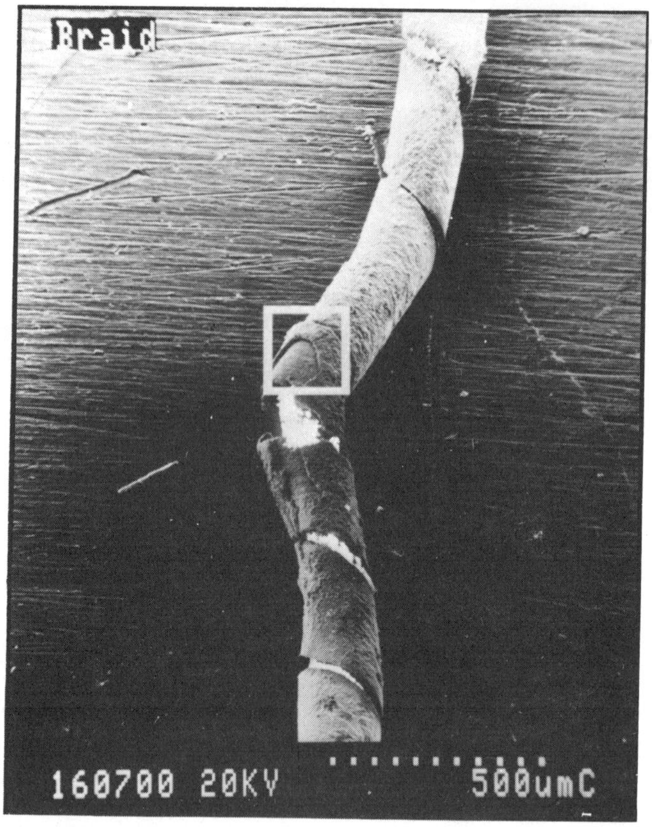

Fig. 2

The analysis was rather disappointing, because although carbon and oxygen were actively looked for, the only elements found were silver and sulphur, with some chlorine.

Another sample was found where some of the corrosion had fallen off (See Figs. 2 and 3).

The analysis of this piece showed that under the corrosion was a layer containing gold, and when just a small area of the metal underneath was analysed it was found to be a gold and silver surface with no residual corrosion.

Cross sections were taken of the threads, and these were analysed. One can see clearly the corrosion layer on top of the sound metal layer, and in the analysis one can see the thin gold layer.

From this information it seemed that there were only two real possibilities for removing the corrosion, mechanical action, or chemical cleaning. Mechanical cleaning was ruled out because brushing with a glass bristle brush would inevitably scratch the surface and leave it prone to more corrosion, and any more gentle technique would take too long.

As silver sulphide is not soluble in any solvents, the most obvious cleaning agents were acids, and the most common acidic metal cleaning agent for silver is Silver Dip. This has a high mineral acid content, and so there are fears for its effect on textile material, but it was tried, and found to be very effective. The other main components of Silver Dip are thiourea and a wetting agent.

It was decided to change this formulation by replacing the mineral acid with an organic acid, methanoic (formic) acid. The thiourea was felt to be relatively safe as it has in fact been used as an after-treatment for silk to prevent photo-degradation. An aqueous solution of 3% formic acid, and 2% thiourea was tried, and over a period of one hour, with gentle brushing with a soft brush, the results were as good as those obtained with Silver Dip.

Fig. 3

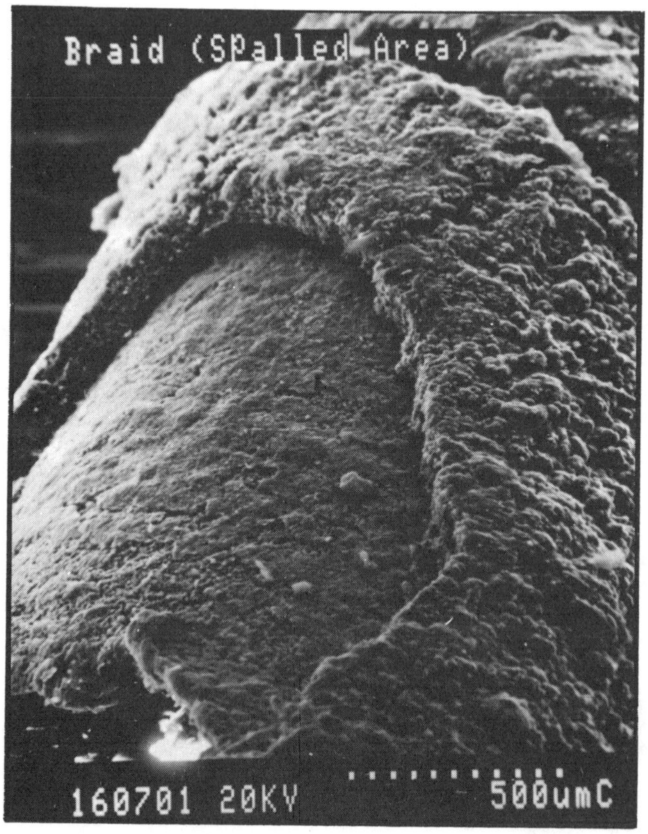

Fig. 4

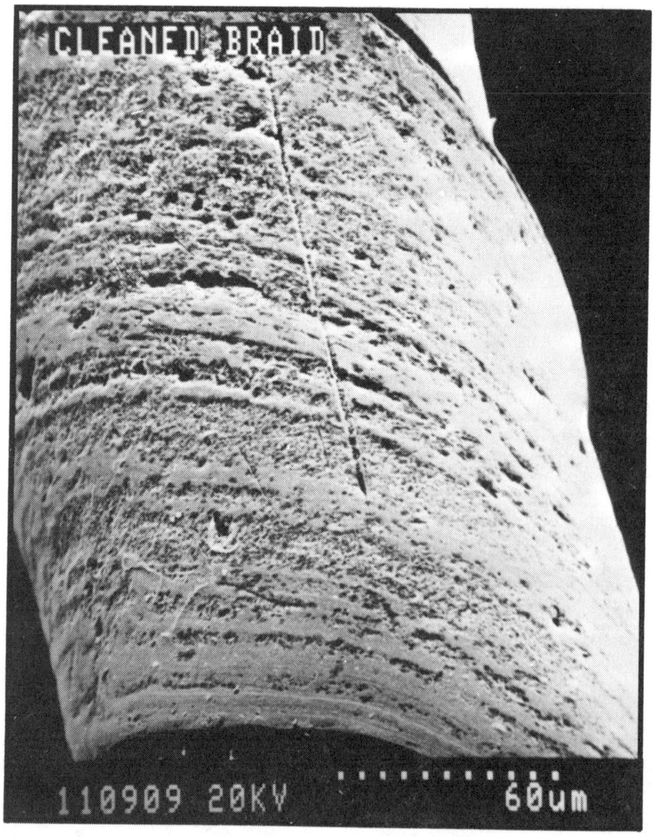

Following a suggestion from Mr. Andrew Oddy of the British Museum that a non-aqueous treatment may be less disruptive to the silk, a solution of 3% formic acid and 2% thiourea in propanone (acetone) was tried, and this again gave very good results.

Samples of the cleaned metal were then taken to Exeter University for inspection and analysis. The analysis of a cross-section of cleaned metal showed that the corrosion had been removed, but the gold layer was still intact.

Looking at the surface of the metal and its analysis one can see that the metal surface is pitted, but not too disrupted (See Fig. 4).

The analysis shows that the corrosion is no longer present. With the surface now exposed it was possible to look for two elements that one might expect in this type of thread, iron and copper. Although there is a small trace of copper, there did not seem to be any iron. From an aesthetic point of view, the cleaned metal has a much more attractive appearance than the uncleaned.

It should be pointed out that this sample was cleaned nearly two years ago, and has been treated very badly since then. It has been handled by countless visitors, and stored in a chemical laboratory in a cardboard box, but no sign of re-corrosion can be observed.

The effect of cleaning on the silk component is the most difficult area to evaluate.

I have soaked samples of new silk, old silk, and old buckram in the cleaning solution for three hours, heat aged one sample for twenty eight days at sixty degrees centigrade, and light aged another sample in a 'Microscal' light fastness tester for forty-eight days. These samples were then tensile strength tested, and there was no significant loss in strength. This cannot be accepted as a conclusive confirmation that the silk is not affected, as textiles are notoriusly variable in their strength. I also believe that quite severe damage can occur to textiles before their strength is affected.

What is required is a more reliable procedure for evaluating degradation, such as viscosity measurement, or some other more sophisticated method. I am about to start work on this at the moment.

The last factor in metal cleaning is whether to lacquer or not. I have carried out some experiments using methacrylate resins, and electrostatic air-brush application in the hope of coating the metal, but not impregnating the silk. These have been totally unsuccessful: subsequently I do not believe that this object should be lacquered.

We have not yet decided whether to clean this object or not, and any comment's are welcome.

Acknowledgements

I would like to thank Dr. Bill Hooper of Exeter University for his help and advice with the use of the Scanning Electron Microscope, and the staff at the British Museum Conservation Research Department for their comments and help.

Reference

Darrah, Josephine: Metal Threads and Filaments. Recent Advances in the Conservation and Analysis of Artefacts. Ed. Black, James (University of London: Institute of Archaeology: Summer Shools Press, London, 1987) pp 211-222.

Author's address:

Textile Conservation Studios
Apt 11A, Hampton Court Palace
GB-KT8 9AU East Molesey, Surrey

CHOCHA, Crystyna

Conservation of the XVIIIth century embroideries and tassels from the royal castle collection. Cleaning of metal threads

In the year 1982 the National Museum in Warsaw handed over to the rebuilt Royal Castle the haberdashery and embroidery constituting mainly the elements of decoration of the back and canopies from kings' old thrones.

The objects handed over were as follows:

1. Tassels from the throne of the last Polish king named Stanislaus Augustus Poniatowski – 18 small pieces and 17 big ones – made in France, in the eighties of the 18th century. These tassels were then used to decorate the throne of Tzar Nicolaus I., during his coronation as the King of Poland in the year 1829 and, subsequently, together with the throne carried out of Poland to Moscow after the November Uprising. They were regained by the Poles on the basis of the Riga Treaty in the year 1921 and until the outbreak of the Second World War they were exhibited in the Royal Castle rooms.

From the year 1939 to the year 1982 they were stored in the National Museum in Warsaw.

2. Tassels from the Ballroom, with the strings or cords on which the chandelier was hung – 12 pieces – also made in France in the eighties of the 18th century.

3. Embroidered rosette (embroidery for the application), made in France in the second half of the 18th century, part of the decoration of the king's throne from the Throne Hall of the Castle.

4. Embroidered palmettes (embroidery for the application), made in Poland in the first quarter of the 19th century, from former king's thrones from the Castle (14 types, in total number 62).

All the above mentioned objects were coated with patina, were dirty, and were damaged as a result of improper preservation and transport in the past. In order to make it possible for them to be returned to their proper historical spot, i.e. to the thrones – being under reconstruction at present, they had to undergo a thorough process of conservation in the Atelier of Textile Conservation in the Royal Castle.

In the years 1985-1988 47 such tassels, 12 palmettes and one rosette underwent a process of conservation. The conservation of these objects in our atelier differed much from the usual work done here.

Palmettes and rosette are embroidered with various metal threads on linen cloth. Paper and cardboard were the basis for the convex embroidery. Metal threads were supported by silk thread. In order to make the embroidery, various metal threads were used along with metal plates and sequins. Palmettes are reinforced with paper in blue or white colour.

With regard to the tassels, all of them that have undergone the process of conservation in our atelier are characterized by identical structure. The base consists of a wooden mould, with a small head, stamped as one whole piece. In the case of the tassels from King Stanislaus Augustus' throne the mould is in the shape of an elongated bell; in the case of the tassels from the Ballroom they are in the shape of a dumpy bell. In both cases the wooden mould is wrapped in metal thread. A net with quadrate eyes made from a strand of 8 metal threads is put over a mould in the tassels from King Stanislaus Augustus' throne.

The skirt of a tassel is formed out of a fringe made from metal and silk threads; these threads are decorated with canetilles made from metal wire and with flowers of two types: flowers made of thick parchment which is wrapped in metal plate; flowers made from various metal threads. The flowers are decorated with sequins.

Tassels from the Ballroom are on a wooden mould wrapped in a metal thread. A net which forms irregular quadrilaterals is put on that. The net is made from parchment tape and string.(The parchment tape is coiled round with metal plates; the string consists of 3 twisted metal threads.) The bottom part of the mould is covered with a linen cloth reinforced by cotton threads. A plait consisting of triple twisted metal thread and plate is laid on the head of that tassel.The skirt is made from the fringes which consist of double twisted metal thread; the outer part of the skirt is decorated with canetilles consisting of metal wire. The string used for hanging the chandeliers is made from twisted linen threads in a cover of metal threads.

All the metal threads used for the production of tassels have silk core; all the threads, wires, plates, and sequins are gold in colour.

Before all the conservation work was undertaken all the metal threads from the inverstigated objects had been examined in the physical and chemical laboratories of the Ateliers of Conservation in the Royal Castle. The purpose of this examination was to identify the metal threads of the above mentioned objects.

In each of these samples it has been proved that:

sample 1

a thread from a fringe belonging to the central part – the existence of a large amount of gold, copper and tin, and small amount of silver.

sample 2

a plate from a strand of a flower – the existence of a large amount of copper and silver, a small amount of gold and a trace of tin.

sample 3

a thread from a head – the existence of a large amount of copper and silver, a small amount of gold and trace of tin.

(The above samples were designated by M.Zawartko-Laskowska).

From the above presented description of the objects it is easy to learn what different materials had been used and what a complicated technique had been utilized to make them. Along with metal we have wood, silk, parchment, cotton, linen, paper and cardboard.

The main problem was the cleaning of those objects. When choosing the cleaning method we assumed that it had to be an adequate method for all the various kinds of materials used, and it should not be harmful to any of them. We decided not to use drastic chemical preparations. We executed some tests. Among others, for cleaning our objects we used distilled water with detergent. The result of this test was very positive.

I should like to present here a detailed description of the conservation work carried out in our atelier on the objects under discussion exemplified by the tassels from the Ballroom and some of the palmettes at present decorating the back of the kings's throne from the Senators' Hall.

The tassels were covered with dust and were dirty, and the metal threads had lost their gloss and had become black. The metal threads in which the mould was wrapped were broken and tangled. The net put on the mould was drawn aside. The plait of the parchment tape was damaged in many places, and was torn and frayed. The spirals covering the plait supporting the skirt were torn and unseamed. The fringes were tangled, the metal thread of the fringe was frayed in many places. The canetilles had flattened and bent wires. At first, the tassels were dusted with a soft brush, the strings fastened to the tassels' heads with hand hammered nails were taken off. Then every tassel was dismantled, every process was noted, and every part of the tassel was numbered.

The decision to dismantle was undertaken for two reasons:

- it was feared that the wooden mould of a tassel could seriously be deformed having dried after being cleaned with distilled water with detergent.
- arranging metal thread (in which wooden mould was wrapped) under the net placed over this mould would mean repeated action of plaiting the thread, the metal thread under the parchment tapes and would result in new damage to the threads and plates.

Simple parts of tassels, disassembled and designated, were put subsequently into distilled water, with a temperature ranging from 30-35°C, and non-ionic detergent (0.2% solution) and then left in them for 1 hour.Next, they were taken out, rinsed several times in distilled water and dried in flannel. Detergent foam was spread over some dirty fragments in order to soften the dirt and these were then cleaned mechanically with soft brushes and tampons of cotton-wool. They were rinsed several times, repeating the spreading of the foam detergent as many times as was necessary. In order to dry the tassels, single fragments were stretched on a plate covered with melinex; the net made from parchment tape was dried on a special form made of wood and wrapped in a flannel. Metalized thread in gold colour of the same width as the metal plate covering the parchment tape was applied. Missing metal coil fragmenis of the parchment tape were reconstructed. Clean elements were fixed back on a wooden mould in the opposite sequence to that used when they were disassembled, maintaining the original order of fastening. The canetilles were sewn back to their previous place. (Some of the canetilles were missing. They were reconstructed in the workshop of Mr. K. Wojciechowski in Warsaw.) The tassel strings were cleaned with detergent foam. They were hung for drying, but they were not fastened to the tassels (Figs. 1-4).

Fig. 1 *One of the tassels from the Ballroom.*
a *Before cleaning and consemvation;*
b *After cleaning and conservation*

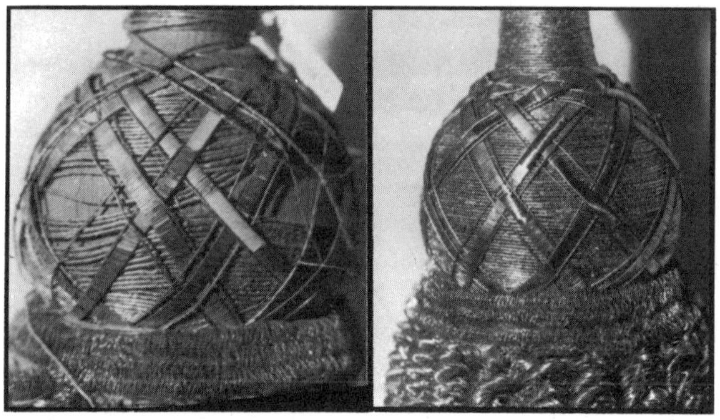

Fig. 2 *Fragment of the tassel.*
a *Before cleaning and conservetion;*
b *After cleaning and conservation*

Fig. 3 *The canetilles (one of the tassel's parts) during cleaning. Washing in distilled water with non-ionic detergent*

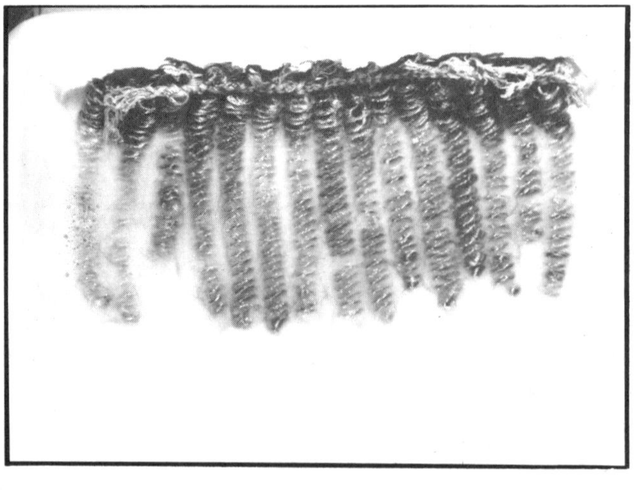

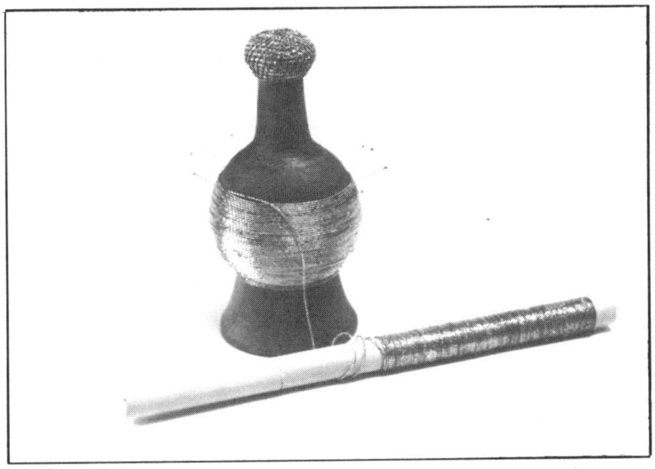

Fig. 4 *Fragment of the tassel – during conservation*

Fig. 6 *Fragment of the palmette – during cleaning. A dirty place is covered with detergent foam*

Palmettes were generally dirty, covered with dust, certain parts of the plates and sequins had gone black. The embroidery was less damaged. In some places silk threads supporting metal thread and plates were broken. Plates were bent. It was stated that single sequins and some metal threads were missing. They were cleaned in the same way as were the tassels except for the washing in the solution of distilled water and non-ionic detergent for fear of damage to the paper and cardboard which had been used as the base for the embroidery.

Dirty places were covered with detergent foam and cleaned mechanically with soft brushes and tampons of cotton-wool. Foam was washed away with distilled water preventing the base of the embroidery from being soaked. It was dried in the flannel cloth. If necessary this activity

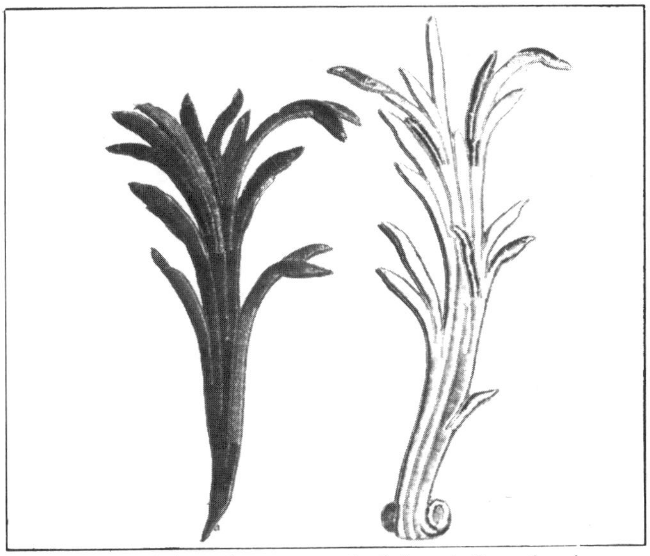

Fig. 5 *Two of the palmettes: on the left – before cleaning; on the right – after cleaning*

Fig. 7 *Rosette during cleaning*

was repeated several times. Very dirty spots were cleaned with ethyl alcohol. The embroidery which had been cleaned was secured, by fastening all the useamed elements of the embroidery with specially chosen thread. Some of the missing parts were not reconstructed (Figs. 5-7).

The objects which have been dealt with here were returned to their historical spot. They can be admired in the reconstruced rooms of the Royal Castle in Warsaw.

Author's address:

Zamek Krolewski w Warszawe
Pl.Zamkowy 4
Pl-00277 Warszawa

KALININA, Luydmila Andreevna

The treatment of metal therads in Russian north-European textiles

The north-European part of the USSR occupies a specific place in the history of Russian culture. It is this part that became the depository of Russian folk art and other works of art. The principal central museums of Moscow, Leningrad, and Arkhangelsk are proud of those masterpieces that remain from long ago in northern dense forests, in small Arkhangelsk villages.

The branch of the All-Union Arts Centre of Restoration named after academician Grabar recently started its work in Arkhangelsk. This branch is due to concentrate its work on restoring different types of old relics from the collections of numerous Arkhangelsk regional museums. Among the collections from these regional museums there are numerous valuable textile relics, but a great number of them can no longer be exhibited because of their poor condition. The methods of different centres are now being used by our textile restorers in the cleaning and conservation of textiles. We follow the main principles of restoration: minimum artistic interference and a minimum of chemical treatment. This concept is followed in order to retain the authenticity of the textiles and to help prolong the life of the textile.

The treatment of metal threaded textiles

Those textiles which have metal threads are usually of religious origin or are from ornate parts of folk dress. Where metal threads have not become coated with metallic salts but have their own natural protective layer, we only clean the textiles. In such cases we use distilled water and soft liquid "Kashtan" or "Progress" soap made in the USSR.

It is in this way that the aer showing the Entombment, dated to the middle of the 18th century, has been cleaned. It is fulfilled in Old-Russian embroidery in silk and metal threads and decorated with beads and metallic hand-made lace. When we started the work the aer was dirty, the centre-piece had at some later stage become covered with a dark-blue dye, like ink, which had been poured into the embroidery in metal threads. The cleaning and the removal of the dye was done following the recommendation of the All-Union Arts Centre of Restoration. The method of tamponing on filter paper with a solution of water and liquid "Kashtan" soap was used, followed by thorough washing in distilled water with the same method. Each part taken separately was washed several times uninterruptedly, because of various physical effects upon the threads during drying. To enable quick drying the embroidery in metal threads was treated with alcohol. As the embroidery is in relief, small sandbags were used to smooth it out. These bags prevented deformation during the full drying. This kind of smoothing out proved to be more effective than the small electric irons that are sometimes used.

In ethnography, namely in the ornate folk dress of the people of this region, metal threads were used for decorating various types of head-dresses. In the easten part of the Arkhangelsk region women put on pearl caps (named "kokoshniks") with white calico shawls with splendid embroidery in metal threads. Almost all of these shawls need cleaning and washing before they can be exhibited. Some of them have dark metal threads which need to have the oxides removed. We use the methods of the Vilnius Restoration Centre for cleaning such shawls. First of all the shawl is treated in an aqueous solution of chloramin, then in an aqueous solution of acetic acid. Thus all spots of calico come out and the textile becomes white. In order to neutralize it, the shawl is treated in an aqueous solution of hyposulphite, after which it is washed thoroughly in water. In order to soften the textile of the shawl it is sponged with a water-alcohol solution with glycerine. It is then first dried with filter paper, then with a hair-drier and smoothed out on the inner side with a hot electric iron on top of a thick layer of material in order to retain the texture of the embroidery.

To remove silver sulphide from the metal threads we soaked a head-dress in an aqueous solution of thiourea and then cleaned the threads with a brush. After such treatment the metal threads should be covered with polymeric lacquer to avoid any further corrosion. Such treatment can be applied only to silver threads. The copper content of the metal threads gives undesirable results after such treatment.

From what has been said it is obvious why it is necessary to carry out laboratory analyses to attribute the material of the relics, not relying solely on visual perception. In the near future our department will have at its disposal such a laboratory too.

Author's address:

Arkhangelsk State Museum of Wooden Architecture
Naverejnaja Lenina 86
SU-Arkhangelsk 163061

JÁRÓ Márta

Re-corrosion of silver and gilt silver threads on museum textiles after treatment

Abstract

Tarnishing or formation of greyish corrosion products can often be observed on the surface of silver or gilt silver threads after the treatment of the textile decorated with. The author gives a possible explanation for this phenomenon.

Textile restorers can often observe that metal threaded embroideries or weavings which had been cleaned for 10-20 years or sometimes for a shorter time are in very bad condition. Tarnishing or formation of greyish products can be observed on the surface of the metal. At the same time pieces which were not treated are better preserved under the same conditions.

A possible explanation is given for this phenomenon.

Tarnishing of silver and gilt silver

From the moment that silver metal is "born" – meaning that it has been reduced from one of its ores – its reaction with the elements of its environment begins in spite of the fact that silver is a noble, nonreactive element.

The three dimensional structure of the metal lattice is broken on the surface. Atoms of the surface have unsaturated valencies which attract ions or atoms from the environment. If these particles are almost the same size as metal atoms, saturation of the free valencies is reached without the violation of the manner of crystal expansion. The thin corrosion layer can be considered as the continuation of the metal. The broken metal lattice on the surface is completed, the corrosion layer protects the underlaying metal [1]. This phenomenon can be observed by the oxidation of silver, the thin oxide layer cannot be observed with the naked eye, the silver seems not to be corroded.

Table 1 *Solubility of some silver compounds [2]*

Name of the compound	Solubility in water mol/liter, 20 °C	Solubility in other reagents
Silver oxide	$1{,}08 \cdot 10^{-4}$	ammonium hidroxide potassium cyanide
Silversulphide	$1{,}8 \cdot 10^{-17}$	nitric acid
Silverchloride	$1{,}31 \cdot 10^{-5}$	ammonium hydroxide potassium cyanide

Despite the protective properties of the silver oxide layer the silver and also the gilt silver in some cases becomes tarnished. Tarnishing is the consequence of silver sulphide formation on the surface of the metal. Silver sulphide is one of the most stable silver compounds (See

Fig. 1 a *18th century silver thread, x78, scale bar= 0,1 mm*

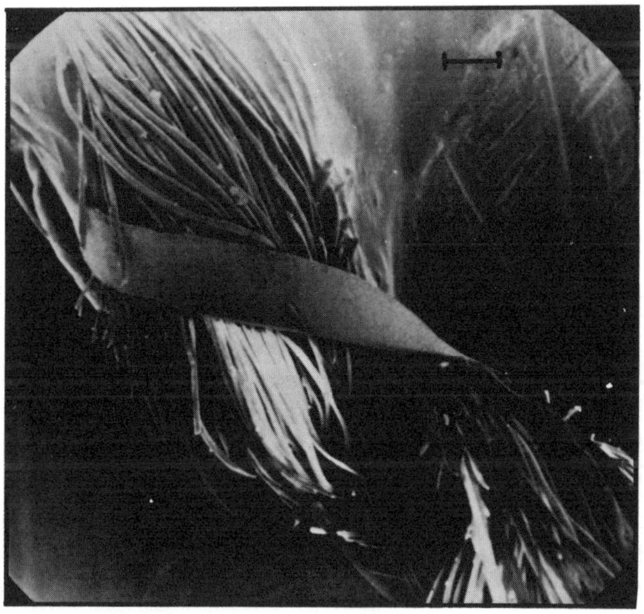

Fig. 1 b *Detail of the surface of the metal strip with "growing" silver sulphide, x6000, scale bar = 0,001 mm [3]*

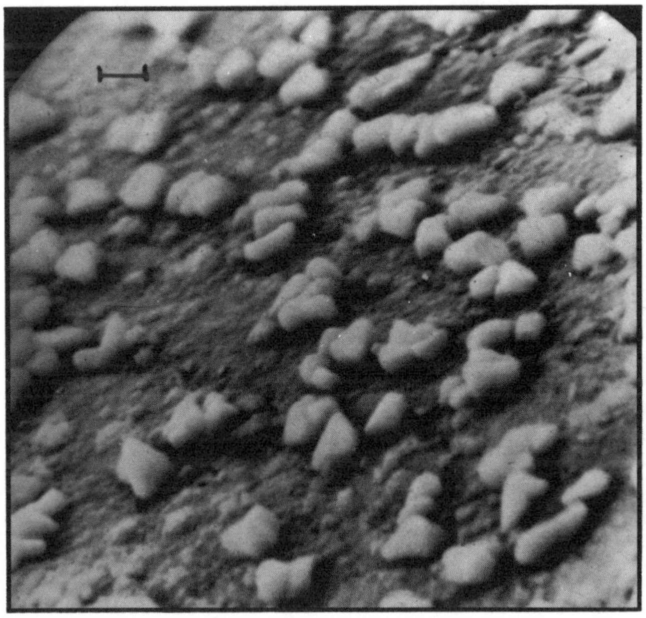

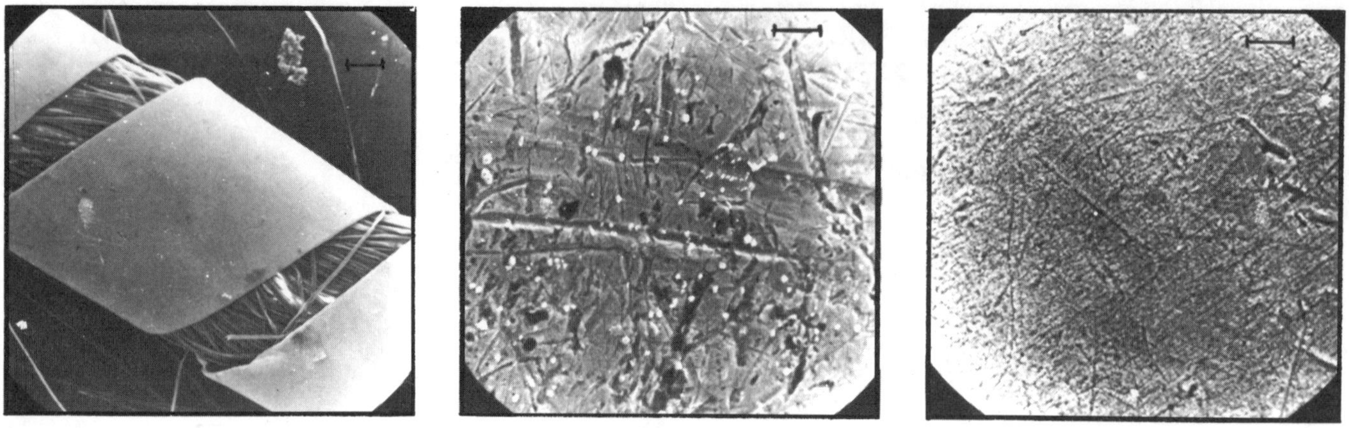

Fig. 2 a *17th century silver thread, x78, scale bar = 0.1 mm* **b** *Surface of untreated part of the thread, x1000, scale bar= 0.01 mm* **c** *Surface of treated part of the thread, x1000, scale bar = 0.01 mm*

Table 1, where the solubility of some silver compounds in water and other solutions is given).

The hydrogen sulphide content of the polluted air attacts silver and gives rise to silver sulphide (Fig. 1a and 1b).

The formation of this black, water insoluble compound occurs only in the presence of both oxygen and moisture; dry hydrogen sulphide does not attack silver [4]. There are more explanations for the steps of this reaction [4, 5], but the decomposition of the protective oxide layer prior to the reaction is surely indispensable for the formation of silver sulphide.

In the case of gilt silver, silver sulphide grows on places where gold does not form a continuous film, usually when it is too thin or damaged. It can often cover the whole gold layer.

Re-tarnishing of silver or gilt silver threads after cleaning

Nowadays air cannot be said to be pure any more, in most cases it is very polluted, it nearly always contains – besides other gases – hydrogen sulphide as well. If a tarnished metal threaded textile is treated with chemicals to remove the silver sulphide layer, the bare metal surface (with the unfinished metal lattice structure mentioned above) will be uncovered. The silver atoms on the surface have no time to develop the protective oxide layer, they react not only with oxygen but also with hydrogen sulphide, tarnishing can be observed soon after cleaning. Figure 2a shows a silver thread originating from a 17th century embroidery. The surface of the metal strip was slightly tarnished. Half of the sample was immersed in acid-thiourea mixture, rinsed in distilled water and dried. Then the treated and not treated parts were exposed to hydrogen sulphide gas for five minutes. The result can be seen in Fig. 2b and 2c.

The surface of the untreated silver was attacked only a little bit, it is protected by the oxide layer. The surface of the treated sample is badly corroded. The protective oxide layer was cleaned off during the treatment, silver atoms on the bare metal surface could easily react with hydrogen sulphide, the silver tarnished.

In every case we remove the corrosion products

Fig. 3 a *19th century silver strip used for weaving, x78, scale bar = 0.1 mm*

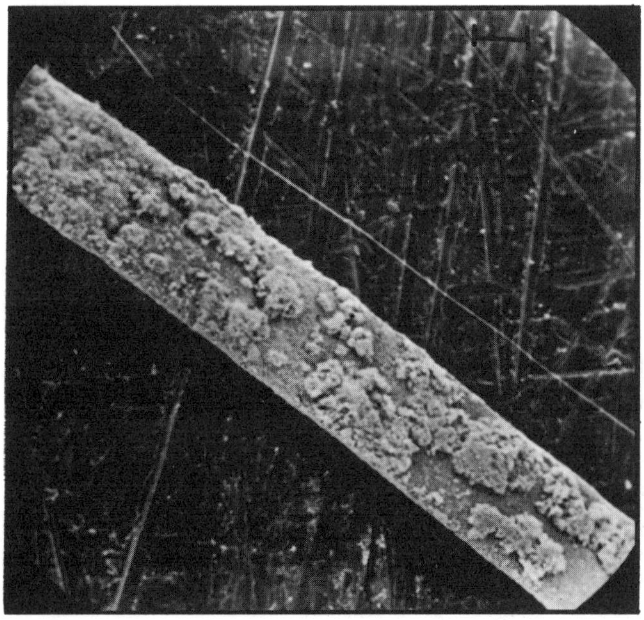

Fig. 3 b *Silver chloride on the surface of the strip, x3000, scale bar = 0.002 mm*

from the surface of silver with chemicals, we can count on the quick formation of silver sulphide in polluted air containing moisture.

Formation of silver chloride on the surface of silver or gilt silver

The other corrosion product of silver which can be observed on silver objects (first of all on archaeological ones) is the greyish silver chloride.

In normal circumstances silver does not react with chloride ions, nor does it dissolve in hydrogen chloride solution. If the protective silver oxide layer is destroyed, silver reacts with chloride ions forming silver chloride. Under burial conditions (e.g. in crypts, tombs) there is the possibility of the formation of ammonia which dissolves silver oxide and gives place to the reaction between silver and chloride ions of the environment. Figure 3a shows a metal thread sample discovered in a 19th century coffin. The surface of the sample is covered with silver chloride [6] (Fig. 3a and 3b).

The silver chloride does not create a protective layer on the surface of the silver, the metal can completely transform to silver chloride. The transformation of silver to silver chloride can be catalysed by copper(II) chloride present [7]. In the case of gilt silver, the setting up of galvanic cells between silver and gold can also play a role in the reaction.

Re-covering of silver or gilt silver threads with silver chloride after celaning

If the silver oxide layer is removed by any chemicals during cleaning, in the presence of chloride ions silver can

Fig. 4 a *Gilt silver thread from a 17th century textile, x78, scale bar=0.1 mm*
b *Corroded sample from the same textile, x 78, scale bar= 0.1*

Fig 4 c *Surface of the uncorroded sample, x1000, scale bar= 0.01 mm*
d *Surface of the corroded sample, x1000, scale bar= 0.01 mm*

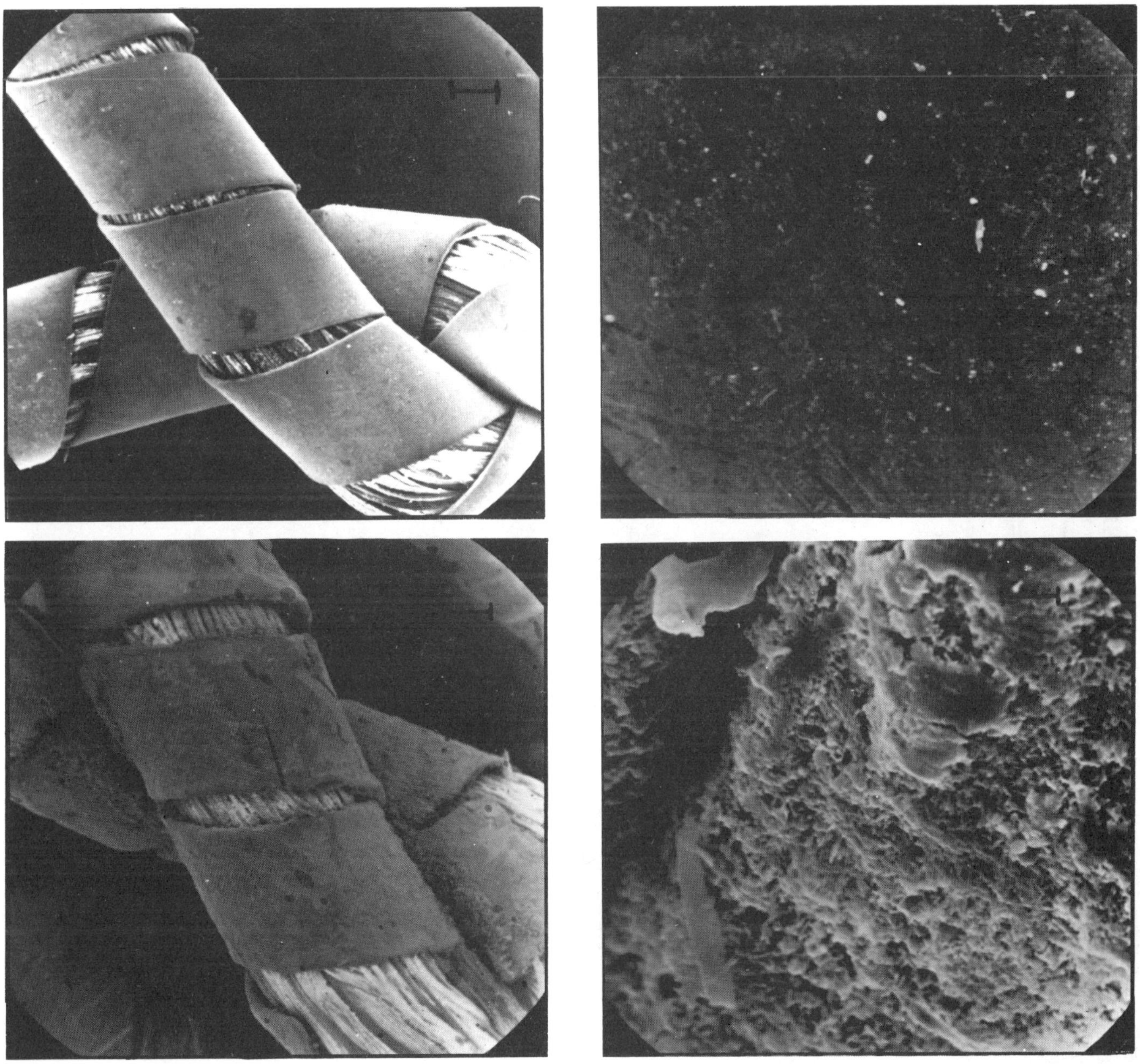

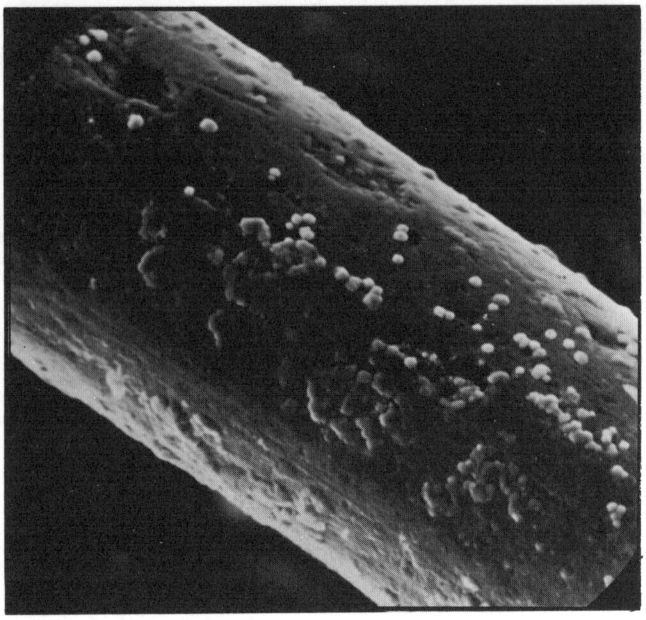

Fig. 5 a *Silver wire from a 15th century weaving, x1000, scale bar= 0,01 mm*

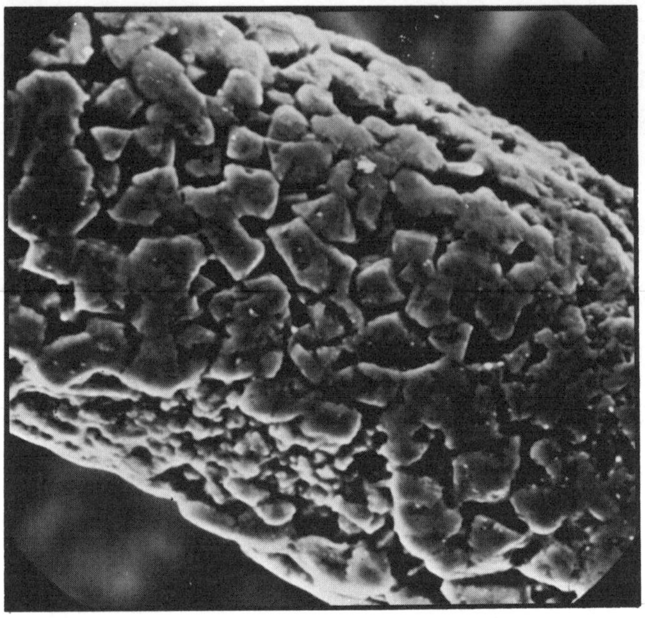

Fig. 5 b *Wire sample from the treated part of the same textile, x1000, scale bar= 0,01 mm*

form not only silver sulphide but also silver chloride. Some years ago ammonium hydroxide was sometimes used for cleaning corroded silver and gilt silver threads. This product solubilizes greasy dirt on the surface and dissolves silver oxide and chloride. It cleans the surface of the silver but at the same time if the bare metal surface is uncovered (and there is the dissolution of a small amount of copper, which, in almost every case, is present in the silver) it creates the possibility of the reaction between silver and chloride ions of the surroundings, or present in the fibrous core of the thread. Some years after cleaning, greyish silver chloride appears on the surface. Figure 4a and 4b shows gilt silver threads taken from different parts of an embroidery cleaned with ammonium hydroxide about 20 years ago. Parts of the textile (presumably where chloride ions were present) are now covered with a thick silver chloride layer [3]. Figure 4c and 4d shows the difference between uncorroded and corroded surfaces of the samples presented in Fig. 4a and 4b.

Something similar happened in the case of a textile interwoven with silver wires. The restorer used oxalic acid solution for removing an iron corrosion stain. In some days, on the cleaned part of the textile the silver wires were covered with a greyish black corrosion product (Fig. 5a and 5b). Silver chloride could be identified on the surface [6] We can suppose that oxalic acid transformed silver oxide to silver oxalate which reacted with chloride ions and gave rise to silver chloride [8].

Conclusion

Different chemicals are sometimes used for the treatment (cleaning) of silver or gilt silver threaded museum textiles. If the chemical used uncovers the bare metal surface (dissolves not only the corrosion products but also the protective oxide layer) or transforms the oxide layer, we can count – in the presence of hydrogen sulphide or chloride ions – on the re-corrosion of the metal. This fact has to be taken into consideration when the cleaning method is chosen.

Acknowledgement

The author is grateful to Ms Erzsébet Gondár, INNOVA-TEXT, Budapest, for the scanning electron micrographs.

Notes and References

1 **Stambolov**, T.: The Corrosion and Conservation of Metallic Antiquities and Works of Art, Central Research Laboratory for Objects of Art and Science, Amsterdam, 1976, p. 7

2 Data taken from Stambolov's publication (see Ref. 1, p. 155) and from the Lange's Handbook of Chemistry (Handbook Publishers, Inc., Sandusky, Ohio, 1956).

3 Mr Attila **Tóth**, Research Institute of Technical Physics of the Hungarian Academy of Sciences, Budapest is kindly thanked for the identification of the corrosion product by microprobe analysis.

4 **Lilienfeld**, S.– **White**, C. E.: A study of the reaction between H_2S and Ag, Journal of Amer. Chem. Soc., 52/1930, pp. 885-92

5 **Stambolov**, T.: op. cit. p. 156

6 Mr László **Kriston**, Institute of Criminology, Budapest, is kindly thanked for the identification of the corrosion products by X-ray diffraction method.

7 **Stambolov**, T.: op. cit. p. 156

8 **Gmelin**: Handbuch der Anorganischen Chemie, Silber, Teil B5, Springer Verlag, Berlin-Heidelberg-New York, 1975, p. 149

Author's address:

National Centre of Museums, Dept. for Conservation
Könyves K. krt. 40.
H-1087 Budapest

HEATH, Diana

The conservation of painted and polychromed metal work, two case studies from the V&A Museum

Introduction

The use of paint and polychromy on metal surfaces is obviously widespread and common throughout history and in most geographical locations.

Because of the intrinsic qualities and natural beauty of polished and patinated metals many are left exposed to the atmosphere where physical and chemical alteration of surface takes place almost immediately in the case of the majority metals, with a few notable exceptions.

One way of preventing the alteration and in some cases ultimate breakdown of the metal structure, is to apply a protective coating which acts as an impervious barrier between metal and atmosphere. This is often found as a coating of another less reactive metal, ie., gilding or tin or a layer of paint or varnish or a combination of these.

The secondary reason is to decorate and embellish, to emphasise certain features, to add variety of colour and texture and even to make the metal appear to be made ot another material altogether.

Two objects in the V&A 's collection illustrate well these uses of coatings for protection and decoration in a variety of ways.

They originate from very different parts of the world, of different periods and were created for quite different purposes. One is the plateau from the Sèvres Egyptian Centre Piece and the second is a sculpture of the Buddhist Goddess, Tara.

In looking at the conservation requirements and approach to the treatment of these objects it is possible to make some interesting comparisons and explain our current approach to the problem of conserving polychromed metal artifacts.

Plateau from the Egyptian Centre Piece

The Centre Piece is part of the Egyptian Service made for the Empress Josephine in 1812 by Sèvres originally inspired by engravings by Denon and to commemorate Napoleon's Egyptian Campaign this was then subsequently presented to the Duke of Wellington in 1818.

The Centre Piece is almost 7 metres long, comprising seventeen separate units of biscuit porcelain, depicting temples, obelisks, monumental Egyptian gateways, seated figures and avenues of sacred rams on plinths. They are mounted on a plateau made up of seventeen bases which in turn rest on seven much larger bases rectangular in shape, made from rolled, tinned black iron-plate. The bases were constructed from 3mm thick sheets cut to size and soldered along the seams.

Separate wooden blocks of laminated oak were bolted to the inside of each base to give additional rigidity and a firm fixing for bolts securing the porcelain pieces in place. Handling holes were incorporated in the upper bases. The painted finish on the bases is an interesting example of the 'tôle peinte' technique which basically involves applying successive coats of paint and pigmented varnish, each being 'stoved' at a low temperature and polished, prior to the next coat. The number of coats and use of materials depended very much on the desired effect. When complete the surface was given a final, clear coat of varnish, so producing a tough, resistant finish.

The seventeen upper bases were painted to imitate porphyry, a pinky red colour. The large lower bases were painted to resemble grey granite.

Paint samples taken from the bases revealed the layer structure. First, a gesso ground (which was found to be calcium carbonate) was applied to the metal, then a base coat of dark red for porphyry and dark grey for granite followed by applications of tiny speckles in light grey, black and pink, and light grey, black and white respectively. This appears to have been achieved with a "spattering" technique which produces circular and oval spots, regular in distribution. The whole surface was completed with 2 coats of colourless varnish.

Condition prior to treatment

The entire Centre Piece and its plateau was acquired by the V&A for Apsley House in 1979, until that time it had been the property of the Duke of Wellington. It had inadvertently spent the 30 years prior to 1979 in store in the basement of Apsley House in straw filled, packing crates which periodically became soaked when flooding occurred and were thus exposed to a considerable level of damp and fluctuating environmental conditions.

The varnish layer had obviously deteriorated, it is likely that the effects of ultraviolet light over a period of time, plus the regular handling and flexing of the bases, resulted in crazing to the varnish. This was compounded by the effects of moisture on the surface. Natural varnishes are not impervious if soaked for any length of time. Once cracks had occurred in the film and moisture and other products penetrated beneath, the deterioration would accelerate.

The varnish was found to contain linseed dryers and a copal resin which accounts for the general cracking and crazing that had occurred.

Once the film shrinks this can pull the paint layer from the gesso which can deform and so this in turn exposes the underlying tinned iron to potentially harmful elements.

The paint had an irregular craquelure overall. The coating of tin on iron normally offers considerable protection, as tin is non-reactive to most substances. The problems arise, if there are any defects in the tin coating even tiny 'pin-pricks' or fine cracks. These may be as a result of the way the coating was applied, because of irregularities in the underlying iron substrate or because of stresses that have occurred to the metals during the course of time due to physical movement. Weighty structures such as these are likely to flex when been lifted. Expansion and contraction of the metal caused by fluctuating temperatures, even the effect of direct sunlight on the surface, can result in disintegration of coatings, particularly as the gesso ground and paint layers may have become increasingly brittle with age.

Given the presence of an electrolyte even in minute cavities, corrosion is all too likely to occur. Deterioration is accelerated because of the galvanic action between the tin and iron. The tin acts as a cathode to the iron and increases the corrosion where the iron is exposed. Once the corrosion reaction had begun the expansion is rapid causing pitting of the iron and rust to spread between the tin and paint layers. Resulting in eventual break up of all coatings because the volume of iron oxide as it expands is much greater than that of the base metal.

The effects of severe deterioration were evident from the state of the bases. Those most badly affected were the larger granite-finish bases. The varnish had altered colour dramatically and had turned yellowish brown giving the bases an olive green appearance. The crazing of the varnish was widespread and rust patches were visible over many areas, with 15 to 20% of the paint and gesso missing on five of the larger bases, and much of the remaining surface bulging and blistering where corrosion was active beneath; rendering remaining paint very insecure and liable to crumble at the slightest pressure.

Conservation considerations

When deciding on the correct treatment method for these bases it was necessary to consider the following points: the objects were for continuous use and viewable on all facets ie., they would continue to support the porcelain components. Therefore they must be durable, useable and as intact as possible. Given the symmetrical pristine nature of the whole Centre Piece, it was important visually that the surface was restored and matched as closely as possible with the original.

Because there was so much of the original surface remaining it was not necessary or feasable to retain and consolidate all the broken paint fragments.

The rust must be stabilized and any future risk of deterioration minimized, given that it was not thought likely at that point to be able to create a completely controlled environment for the display area.

Method

The decision was taken to remove all the flaking, blistering areas of paint as it was obviously out of the question to treat the rust without exposing it. This would allow for the extent of the corrosion to be checked.

This was simply done with scalpels, and metal scrapers, sizeable areas came away where ever the rust had encroached, revealing a thin film of red rust under the gesso though certain eruptions were caused by nodules. Care was taken not to disturb any sound paint layers. The debris was brushed from the surface.

The nodular rust reposits though not extensive were hardest to remove requiring mechanical removal using scrapers and 'blunted' scalpel blades as well as fine '0000 grade' steel wool wrapped around bamboo skewers. A mild abrasive paste was used to remove traces and the residue was washed off with swabs soaked in white spirit.

There remained many exposed areas of iron with a reddish film of rust. The use of proprietary brand rust removers was not deemed suitable, as many contain powerful acids and require thorough rinsing, with water, which would have been difficult to control, plus the risk of leaving harmful residues which could cause recurring corrosion or weakening of the remaining paint structure.

After the tests the product found to be suitable was a rust convertor 'Trustan 23' a complexed tannic acid solution containing isopropanol and a small percentage of weak tannic concentrate which acts readily on active iron oxide goethite FeO(OH) converting it to magnetite Fe_3O_4, – a stable black oxide. The reaction is complete in twenty four hours and no harmful residue is left. The application was with fine, sable hair brushes which allow accurate control of the liquid.

A red lead oxide primer 'Chemiclene R099' was then applied carefully to all treated areas of exposed iron. It was important to paint over the tinned areas too and 'seal' the surface completely.

Removal of varnish

The original varnish coating, which was yellowed and unstable, was removed using I.M.S. soaked cottenwool pads appled to the surface and left for ten to fifteen minutes. Once softened the varnish was wiped with swabs of I.M.S. until all residue was removed.

Refilling and retouching

The thickness of the original gesso ground varied considerably so that is was necessary to fill up the treated 'patches' with a gesso substitute. This had to be a flexible, durable filler with minimum shrinkage on drying, which allowed for sanding or polishing to obtain an even smooth surface.

The filler chosen was fine surface Polyfilla in which powder pigment was mixed to match the coloured ground, this required sanding down to the correct level to achieve an even finish. The sanding was done using waterproof silicon carbide paper (320 grit) with a small amount of white spirit, water was not used as this could have initiated further outbreaks of corrosion if it had seeped through to the iron.

The retouching was a particularly painstaking part off the operation. It was done using acrylic paint 'Rowney Cryla' colours. The ground was applied first over the filler and then in three stages the speckles and spots applied. It became evident that the effect could not be produces by 'flicking' with a paint laden brush or spraying on, as the areas were too restricted and there was not enough control of where the paint was deposited. The only satisfactory way was to apply each spot deliberately with a paint laden brush ensuring that each was a circular or oval spot, as opposed to an irregular shape as would be produced by a stippling brush.

The underside of the bases had not been originally painted they were treated with 'Trustan 23' where rust had formed, coated with red oxide primer and then given too coats of 'Ebony Matt' black metal paint (Manders).

Protective coating

The finishing of the upper surface of bases required a tough inert varnish.

After testing of several nitrocellulose and acrylic based lacquers Paraloid B72 was selected, dissolved in IMS 40% and toluene 30% this was successful giving a gloss finish. Two coats were applied using a De Vilbiss spray gun.

Figure of the Goddess Tara

In marked contrast the examination and conservation of a standing figure of the goddess Tara (a Shakti of the fourth Dhyari Bodhisattva Avalokitésvara, the Buddhist God of Compassion).

From the stylisation this figure is thought to originate from the Katmandu Valley, Nepal made in the 15th to 16th century A.D. and exported to Tibet as a monastic image. The Sculpture is constructed from sections of beaten and repousséd copper with body adornments, jewellery, crown and lotus wand made separately in copper. The sections were then joined by soldering along interlocking seams, to form the main part of the upper torso and the lower body. The head, the hands arms and feet were added in several separate units and riveted in place, the rivets run right through from back to front.

What is particularly striking and unusual about this piece are the contrasting surface treatments used.

The lower half of the body was gilded back and front with a rich, yellow gold and no other colour apart from the decorative use of semi-precious stones set in the anklets. The copper surface was finely tooled with a herring bone texture to help key the gilding onto the surface, this is visible where the gilding is worn.

The upper body and head were treated very differently polychromed, and set with semi-precious and imitation stones including turquoise, lapis lazuli and garnet on the jewellery and the detachable crown.

The face and flesh areas were covered with what appears to be gold flake paint on a kaolin ground. Paint sections revealed a complex layer structure on certain areas of the figure. The gold flake paint could be seen in two separate layers of paint and ground. The lips, inside of eyes and further lines below and above the eye are coloured with red pigment; vermilion on a gesso ground and red glazes have been used to produce this effect.

There is a finely drawn 'third eye' in the centre of the forehead a feature of this particular goddess, there as with the other two, the pupil and centre of the eye as well as the upper lid, are picked out with black pigment (as yet unidentified) and the eyebrows are done in the same way, over gold paint and gesso.

This leads to the conclusion that the figure has been repainted and retouched several times in its history though using essentially the same colour scheme and pigments.

The treatment of the hair is particularly notable. The 'hair' is indicated on the copper surface by narrow parallel lines running over the back of the head to the nape of neck with little curls above the browline and ringlets extending to the shoulders, all deeply 'drawn' in the metal with chasing punches. This was originally a vivid blue which was produced using ground azurite pigment applied on a gesso ground. At least two layers of this have been detected. Azurite can chemically alter over the course of time resulting in a colour change to grey, the additional layer of dirt over the blue pigment imparts a dull blackened appearance with glimpses of bright blue where the top layer has been recently worn away.

Azurite pigment used in this way was coarsely ground to obtain the greatest intensity of colour, for this reason dirt, old varnish and oil deposits become ingrained between pigment particles, part of the reason for a darkened, appearance. The use of blue pigment on the hair of Buddhist deities and temple figures is characteristic of those found in Tibetan monasteries and shrines. This would seem to indicate that the figure although made in Nepal was exported to Tibet and polychromed and painted there, in the traditional way.

Condition prior to treatment

The figure of Tara has been in the collection of the Indian Department, V&A, since 1927, undisplayed for most of that time. The condition is exceptionally good considering the age of the object, the only structural damage is the loss of the lower section of the lotus stem and the rivets which secure the upper and lower halves of the figure.

Of the gold painted surface two thirds remains, though there are worn and missing areas overall. The most obvious damage is on the front of the torso and abdomen, where only traces of paint remain. These are rather obscured by deposits of a brown residue possibly from smoke produced by burning incense.

The exposed gesso is generally discoloured and dirty in appearance. Where this has worn very thin the copper base metal shows through, so darkening the overall effect. The pigment remaining is for the most part stable and well adhered to the metal. There is limited crizzling and flaking usually on the edges of paint and gesso deposits or where there are changes of surface direction.

The hands and arms are particularly worn with areas of exposed copper visible, this is most likely where worshippers through the centuries have stroked the surface as an act of devotion. Other areas of exposed copper occur mainly on the front of the upper torso. On the right shoulder and inside and upper arm and on several unsightly patches on the face namely the right cheek, lower nose and around both eyebrows. The flower of the lotus wand has lost almost half of the polychrome. All the exposed copper has tarnished to a dark brown and has a waxy, dirt film overall, the coloured stones have survived intact though many have similar sooty dirt deposits hiding the surface.

The lower half of the figure was very heavily blackened. It is apparent that the gilding is worn particularly on high points, on some areas there is very little gilding left. The copper has oxidised and reacted with pollutants in the atmosphere to form an even film of dark brown/black copper oxide all over the gilded surface, combined with carboniferous accretions from the incense smoke and airborn dirt.

Treatment method

The stages were as follows:

- To check the conditions of all surfaces prior to cleaning and consolidate any friable polychromy as necessary with a suitable consolidant.
- To test clean all the different types of surface and select suitable cleaning solutions where appropriate. The aim being to clean the polychromy, gilding and copper of harmful dirt deposits and generally improve the appearance without overcleaning or destabilising any of the surviving painted surfaces.
- Once clean to apply a suitable protective coating to prevent discolouration, retarnishing or further loss of paint, as appropriate. Bearing in mind the aesthetic requirements and subtleties of the object.

It is not possible to go into details of all the solvents tested for this purpose, though many were found to have little or no effect on the heavy dirt layers of the polychromed surfaces.

The successful cleaning of most of the gesso and gold paint was completed done with saliva applied on swab sticks and gently rolled over the surface, acetone was then wiped over the cleaned area.

Any flaking or loosened paint flakes were consolidated with Paraloid B72 (30%) in Toluene (40%) and IMS (30%) applied with a pipette and 00 sable paint brush to ensure that consolidant had penetrated beneath the flake. Excess Paraloid B72 was removed with a swab. As there was no curling or distortion of most paint flakes the exercise was straight forward if time consuming.

Where there was light dirt and grease on the surface particularly on the facial features and copper metal, Cellosolve (2-ethoxyethanol) was used to good effect, with salvia where needed. The heaviest black film was very resistant to removal with most solvents tried. Finally, where the gold paint was discoloured and most complete this was cleaned with a solution of Linament of soap (20 mls) thinned with white spirit (5 mls) to which was added, 4 drops of ammonia this was applied briefly to the gold paint surface and neutralised with a mixture of 1:1 white spirit and acetone. This removed the dirt layer revealing a fresh gold paint surface beneath. White spirit and acetone in equal parts proved to be useful for the cleaning of the hair where the blue pigment remained not all the surface responded certain areas had to be left showing little improvement because the surface here was so granular and prone to catch on cottonwool swabs, the cleaning was largely carried out through a layer of Japanese mulberry paper which allows dirt to be absorbed while minimising the abrasion the pigment, solvent is simply applied on top of the paper, which acts rather like blotting paper.

Retouching

On completion of the cleaning and consolidation a decision was made to retouch the largest areas of missing pigment, on the shoulder and arm and patches on the right cheek and nose, because it was felt they were particularly distracting when viewing the object. The retouching was not an attempt to completely hide the damage but to unify the appearance as a whole.

The retouching was carried out using acrylic paints and thinner ("Rowneys Cryla colours"). The missing ground was made up using acrylic base coat followed by the application of several thin layers of colour as appropriate. Ochres, umbers and creams with a little gold and bronze powder dusted on were enough to suggest the correct texture and coloration.

The gilt copper lower section was successfully cleaned using a proprietary brand cleaner 'Horolene' which is ammonia based. This is applied on swabs soaked on the metal for half a minute while being gently agitated, the area is then thoroughly rinsed with deionised water before being gently dried and degreased. It was necessary to work systematically over the surface working on no more than $30mm^2$ — at a time.

The cleaning process removed all the oxide layer and revealed a rich gold surface with considerably more gilding than anticipated. The lower torso was completely coated with 'Incralac' lacquer which contains an effective corrosion inhibitor.

Protective coating

A decision on coating the upper section was more difficult as most lacquers designed for use on metals give a high gloss finish microcrystalline wax would also have altered the overall appearance giving too much of a sheen. It was used, however on the remaining areas of exposed copper.

The painted surfaces had a most subtle sheen with the exposed gesso appearing almost matt. For this reason an acrylic matt varnish was applied which does not interfere with the appearance of the existing finish and is compatible with all materials.

The varnish was to enhance the appearance, protect from any further alteration and act as a barrier against pollutants and dust.

Summary

The main differences between the two objects can be summarised thus. The plateau was to be used as a supporting structure and required a surface that could bear certain stresses and wear. The appearance had to be uniform and retouching was imperative as large patches of missing or non-matching paint would be unsligthly.

The complication, as is commonly the case, was the severe corrosion of the iron base metal which can only be successfully treated if the paint layer is removed. Adhesion of paint and polychromy on metal is only strong if the interfaces remain undisturbed.

The Tara was intended exclusively for display.

The copper surface was not actively corroding beneath the polychromy. The complexity of the construction of the object and the layer structure of the polychromy contributed to difficulties of treatment, especially where the pigment was unstable. Retouching of this object was open to discussion but because of the distracting nature of the damage it was felt appropriate. It also acted as a stable coating for the exposed copper.

In both cases it is clear that thorough examination prior to conservation including the taking of paint sections and analysis of varnish and grounds present, supplies valuable information regarding the original appearance of the object and its history.

References

Truman C.: "The Sèvres Egyptian Service 1810-12", Victoria and Albert Museum, 1982, ISBN 905209.249

John W.D.: " Pontypool and Usk Japanned Wares", The Ceramic Book Company, Newport England, 1953

Harris R. and **Service** S.: "Conservation of the Centrepiece of the Sèvres Egyptian Table Service", 'The Conservator', UKIC No. 6, 1982

Lauf D.I.: "Tibetan Sacred Art --The Heritage of Tantra", Shambhala, Berkeley and London, 1976

Author's address:

Victoria and Albert Museum
Metalwork Conservation Department
Cromwell Road, South Kensington
GB-SW7 2RL London

TAARNSKOV, Bodil

Examples of conservation of painted metal objects

At the National Museum's Department of Conservation in Copenhagen we have tried a number of different methods of cleaning painted metal surfaces, ranging from simple cleaning with distilled water and detergent to more complicated methods involving chemical rinsing combined with fixing of the paint layer during the cleaning process. Often a combination of various methods is employed. This article describes the conservation of iron and copper alloy objects – two materials with many different problems.

Copper coffin from 1681

An exceptional coffin from the Church of Our Lady in Roskilde, near Copenhagen, together with two other coffins, were brought to the National Museum's Department of Conservation (See Fig. 1).

Fig. 1. *Detail from copper coffin. Flowers and fruits painted with translucent enamel in gold green colours*

It was a beautiful lustrous copper coffin from 1681, plated with gilt and silver brass on which were fixed gilt and silver plated fittings in relief. The entire coffin was in green, blue and red. Individual fittings were originally picked out in bright fire gilding. The colours, originally very strong, had become obscured by corrosion which had penetrated the layers of silver, gold and colouring. The lid was the worst affected because dust and dirt had settled on top, thus holding moisture and accelerating corrosion. In addition, the coffin had a number of lime stains because it had not been covered during a restoration of the church.

Before deciding on methods of treatment, we had our chemical laboratory carry out some analyses of the materials. The colour layers' binders were identified together with some of the pigments, and the presence of gold leaf and fire gilding was established. This information was mapped out on a sketch. Some of the yellow, pigments were vegetable extracts, the preservation of wich is generally poor; others were green metal salts which are inorganic salts, which in their make-up often resemble the corrosion products we have on the objects. The colours would therefore change or disappear if the metal was cleaned.

The colour layer's binder was a lacquer, for which neutral solutions would be best suited. This can be explained by looking at the lacquer's chemical composition. The lacquer consists of resins mixed together in a drying oil. The oil consists of three fatty acids which are bound to glycerine with ester links. Together these form a triglyceride. The long fatty acid chains undergo a number of changes during the process of hardening, so that some form a permanent network with cross-linking, while others are broken and form short, free fatty acid molecules. One can imagine that these small molecules act as a kind of softener in the lacquer. The resin molecules are held in the network. If the lacquer comes into contact with, amongst other things, acid or alkali, the ester links will be broken, after which the whole "spine" of the oil molecules will be broken down. Use of a chemical (it can be acidic, alkaline or neutral) will get the lacquer to swell and will result in a rinsing out of the smaller molecules. This can result, if not in the immediate destruction of the lacquer, then in a brittle and lustreless lacquer. Consequently, we decided simply to clean the coffin with distilled water and a little added detergent. The very disfiguring lime stains, which had destroyed the paint layer, were removed with a scalpel. Finally the coffin was lacquered with a very thin layer of "Incralac" so that the colours became saturated and lustrous while the coffin remained matt.

Functional artefacts

Objects for everyday use are another interesting group. Metal objects were often painted to give the impression that the metal was more precious than it actually was. For example, an oil lamp made of brass on an iron base was coated with gold lacquer (spirit lacquer) and bronze gilding with black "shadows". Likewise, a warming-pan of sheet iron was painted a bronze colour. Due to dust the paint has almost completely disappeared from the lid (the same phenomenon seen with the coffin). This kind of painting is often overlooked, so that the deceptive appearance of some objects cause them to be incorrectly catalogued, with regard to the nature of the metal, in museums' registers. Conservation is a compromise between the painting and the metal, therefore we cannot guarantee a truly durable result, and we set strict standards for subsequent storage with regard to a dry atmosphere, often in conjunction with DIM lighting.

We have experimented with various corrosion removing chemicals in cleaning sheet iron (coated with tin) and copper alloys. From our experience with the coffin, we knew that the paintwork could tolerate only a more or less neutral solution, and we therefore concentrated on chelating agents. For the time being we have concluded that the most gentle treatment for painted copper alloys is disodium salt of EDTA (Komplexon III) with a pH value of 6, and for painted iron a 5% solution of diammonium citrate, pH value 5.6. In the following will be described a couple of examples of the use of the latter.

Sheet iron box

A sheet iron box with colour printing was, after preliminary tests, immersed in a 5% solution of diammonium citrate. During the course of the seven hours which the treatment lasted, the box was frequently taken up and brushed with a stiff brush. The treatment was stopped when the cleaning process appeared to have ceased, although there was still a little rust left. For the sake of the paint we worked with an unheated solution instead of heating it, as the literature on unpainted iron (Madsen, H.B.) recommends. Similarly, in order to prevent the reprecipitation of oxides on the surface of the metal, cleaning could not be completed by raising the solution's pH value to 9. Finally the box was lacquered with "Paraloid B44" which put a gloss on the colours.

Fig. 2 *Detail from the fire screen. Flowers painted in oil paint. The lower half part is cleaned with diammonium citrate*

Fire screen

Crackled and flaking paint layers presented another problem, upon which we had to work. A fire screen from the Open-air Museum, north of Copenhagen, provides an example of just such a problem (See Fig. 2).

The screen consists of a framed sheet of iron, the front of which is painted black with oil-painted floral decorations in rose-pink and green. The black point was secure although small spots of rust were breaking through. As for the flowers, the paint had been applied more thickly and was flaking and crackled. This can be explained by the repeated heating and cooling during use which has made the paint brittle. For test purposes we immersed the lowest quarter of the sheet, which was the best preserved, in a 5% solution of diammonium citrate for six hours, after which it was possible to remove the rust with a soft paint brush. As a result of this test we decided that the most damaged areas would require more gentle treatment.

Areas of about 10-15 cm^2 were covered with a fine polyester net which held the loose paint in place. Over this was placed cellulose paper moistened with the solution of diammonium citrate, which was covered in foil (plastic wrap) to reduce evaporation (See Fig. 3).

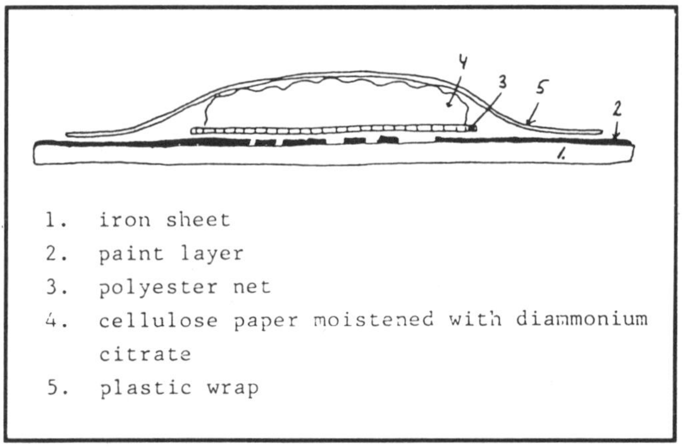

Fig. 3 *The removal of corrosion on areas with loose paint*

After cleaning, the loose paint was secured with "Paraloid B44" solution in toluene after which the whole sheet could be immersed in the solution of diammonium citrate and brushed clean. Finally a coating of "Paraloid B44" was applied to the sheet to protect against corrosion and to bring out the colours.

Lacquering of cleaned objects

Since we are not in a position to achieve optimal results, as far as cleaning corrosion products from our painted metal objects, it is extremely important that we provide the surface with an effective protective covering to prevent further degradation.

We have chosen to use acrylic lacquers: "Paraloid B44" and "Incralac". The latter consists of "Paraloid B44" with the addition of benzotriazole, which inhibits the corrosion of copper. It does not yellow and compared to other types of lacquer remains relatively soluble. An important characteristic required in the lacquer we use is a low permeability to vapour. According to experiments carried out by E. de Witte "Incralac" was found to possess such a characteristic. Furthermore, "Incralac" has good adherence properties.

It is a matter for discussion how far one can actually remove lacquer from an object after a number of years without damaging paintwork. As a rule it is not possible, but what is the alternative to lacquering? It is to allow the objects to be destroyed, since, in practice, we cannot always store them in ideal climatic conditions.

Literature

Gettens, R.J., **Stout,** G. L.: Painting materials - A short Encyclopedia, Dover Publications Inc. (1966)

Hald, P.: Maleriets teknik,
Nyt Nordisk Forlag, København (1978)

Horie, C. V.: Materials for Conservation, Organic Consolidants, Adhesives and Coatings.
Butterworth & Co., Ltd., London (1987)

Madsen, H. B.: Die Reinigung von Eisen mit ammonikalischer Citronen-säure.
Arbeitsblätter, Heft 2, Gruppe 1 (1974)

de Witte, E.: The protection of Silverware with Varnishes, Bulletin, Institute Royal du Patrimonie Artistique, 14 (1973)

Author's addres

Nationalmuseet
Bevaringssektionen
Brede
DK-2800 Lyngby

SLESINSKI, Wladyslaw

Entwicklung der Reinigungsmethoden von Silbererzeugnissen

Der Zweck dieses Referates ist die Lenkung der Aufmerksamkeit auf die Notwendigkeit der Reinigung von Silberoberflächen, und deren Ausführung in verschiedenen Epochen, weiterhin eine Darstellung zur Entwicklung dieser Methoden.

Silber ist ein sehr schönes Metall. Es läuft aber schnell in der Luft an und verliert dadurch schnell seinen metallischen Glanz. Salz- oder Schwefelhaltige Luft besuchleunigt das Anlaufen von Silber. Es wird von den meisten Schwefelverbindungen angegriffen und überzieht sich dann mit einem gelblich-braunen bis schwarzen schwefelhaltigen Film.

Die Ursache für das Anlaufen des Silbers liegt in seiner chemischen Affinität zu niederen Schwefelverbindungen, mit denen sich Silber zu Silbersulfiden verbindet. Optisch ähnliche Veränderungen am Silber erfolgen unter dem Einfluß freier Halogene oder Spuren von Salzsäure. Gelegentlich kann sich diese Chloridbildung bis zur Korrosion steigern.

An diesem Prozeß optischer Veränderungen und nachfolgender Zerstörung von Silbererzeugnissen nimmt auch der Staub teil, der mit fetthaltigen Substanzen zusammengeklebt ist, und Schmutz unterschiedlicher Zusammensetzung und Art enthält.

Grundbedingung für die Entfernung der eben erwähnten zerstörenden Faktoren ist die entsprechende Reinhaltung der Oberfläche von Silbererzeugnissen.

Die Reinigung von Silbererzeugnissen selbst besitzt eine sehr lange Tradition. In der vorliegenden Arbeit beginne ich mit der in Anbetracht des begrenzten Umfangs meiner Ausführungen. Darstellung der Reinigungsmethoden vom Anfang des 19. Jahrhunderts. Sehr populär war damals die Reinigung von Silber und plattierten Gegenständen unter Verwendung von Kaliumhydrogentartarat, Kreide und Alaun. Es wurden 2 Teile Cremor tartari, 2 Teile feine Schlammkreide und ein Teil Alaun genommen, zu feinem Pulver zermahlen, intensiv miteinander vermischt und in einer Flasche mit dichtem Verschlußaufbewahrt. Zum Gebrauch wurde ein wenig davon mit Wasser angefeuchtet und dann mit einem Leinenlappen der betreffende Gegenstand abgerieben, danach in Seifenwasser gewaschen und in Sägespänen getrocknet. Wenn die Gegenstände von Sulfiden stark angegriffen waren, sollten sie einen Augenblick in Chlorwasserstoffsäure getaucht werden. Diese Rezeptur entstammt u.a. von der polnischen Zeitschrift "Pamietnik Rolniczo-Technologiczny" [1] von 1834 und von den deutschen "Neueste Erfindungen und Erfahrungen" [2] von 1899. Eine andere Variante der Verwendung von Kaliumhydrogentartarat zur Reinigung von Silber wurde unter anderem in 1866 so beschrieben: "Man mische zusammen 4 Theile auf das feinste gepulverten Pfeifenthon und ein Theil gereinigten, sehr feinen Weinstein. Braun angelaufene Gegenstände werden erst mit Essig und einem leinenen Lappen gesäubert."[3]

In der 60er Jahren des vergangenen Jahrhunderts war die Reinigung mit einer Kaliumzyanidlösung sehr beliebt. "In ein Trinkglas wird eine kleine Quantität mäßig concentrierter Cyankaliumlösung gebracht und daneben zwei Gläser mit destilliertem Wasser aufgestellt. Mittels eines Zängchens taucht man die Gegenstände in die Lösung, und zauberhaft schnell schwindet der gelbliche oder schwärzliche Überzug. Im zweiten und dritten Glas wird rasch die Abspülung und dann mit einem leinenen Tuch die Abtrocknung vorgenommen. Da Cyankalium bekanntlich zu den stärksten Giften gehört so ist bei diesem Verfahren äußerst Vorsicht notwendig"[4]

"Schwarz gewordenen silberne Gegenstände können in sehr schönen Weise wieder blank gemacht werden durch Kochen in einer wässerigen Lösung von Cyankalium. Es ist dies jedoch ein Verfahren, das man wohl für die Werkstatt, nicht aber für den Haushalt empfehlen darf". [5]

Nach der Zeitschrift "Gewerbehalle" von 1893 kann man auf folgende Weise gelbgewordenes Silber wieder rein weiß machen: "Man bereitet eine gesättigte Lösung von Ätzkali in Wasser, bringt die Flüssigkeit zum Sieden und taucht die betreffenden Gegenstände in einem Zinksiebe, welches mit möglichst großen Löchernversehen ist in die Flüssigkeit ein. An Stelle des Zinksiebes kann man auch jedes andere Sieb benutzen, nur müssen die Gegenstände dann, um eine elektrische Erregung hervorzurufen mit einem Zinkstabe berührt werden". [6]

"Feste noch ganz metallische Stücke sind in dünner Ammoniaklösung zu waschen, dann in lauwarmen Wasser abzuspülen und vorsichtig zu erwärmen, um das Ammoniak wieder zu entfernen" [7]

Das Reinigen von Silbererzeugnissen geschieht ebenfalls leicht durch Abreiben mit einer Lösung des unterschwefligsäuren Natron. [8] Man kann zu diesem Zweck auch einen dünnen Teig aus Schlammkreide, unterschwefelsäurem Natron und destilliertem Wasser zubereiten, der mit einer Brüste auf die Gegenstände gerieben wird, die danach in reinem Wasser gespült und in Sägespänen getrocknet werden.

Ein gutes mechanisch wirkendes Putzmittel für angelaufenes und schmutzige Silbergegenstände erhält man durch Mischen von 10 Teilen Schlammkreide, 5 Teilen geschlammten Pfeifenton, 3 Teilen Bleiweiß, einem Teil Polierrot und einem Teil Magnesia. [9]

In der Zeit zwischen den Weltkriegen (1919-1939) wurde schon festgestellt, daß die Reinigung von Silber als notwendiges Übel angesehen werden muß und nur dann vorzunehmen ist, wenn ein solcher Eingriff unentbehrlich ist. [10]

Gegenstände aus Silber (besonders die mit einem geringeren Feingehalt und die gelötete) sind am besten mechanisch, aber mur mäßig, zu reinigen. In vielen Fällen reicht ein einfaches Überreiben mit einem Ledertuch oder weichem japanischen Löschpapier. Es wurde sogar erkannt, daß bestimmte Unsauberkeiten, insbesondere in den Ornamentvertiefungen und Gravierlinien, dem Gegenstand Räumlichkeit und eine deutliche Zeichnung verleihen. Es sind ausschließlich die mildesten Mittel anzuwenden, die z.B. oxidierte oder vergoldete Teile nicht beschädigen können. Es sollten also weder Schmirgel, Karborund, Kreide noch Pasten zur Metallreinigung verwendet werden, die Säuren enthalten, die das Metall angreifen. Man spricht auch davon, daß es ausreicht, Silbergegenstände mit warmen destillierten Wasser und milder Seife abzuwaschen. [12] Für die Reinigung des Silbers von ihrer Oxidationsschicht (Anlaufen) kann eine Lösung aus 1g Kampfer in 0,25 l Wasser angewendet werden. Danach müssen die Gegenstände abgetrocknet werden. Sehr populär war die Verwendung von Ammoniak zur Silberreinigung, insbesondere von größeren Gegenständen. [13] Das mit Ammoniak gereinigte Silber muß anschließend gut unter fließendem Wasser abgespült und in Sägespänen getrocknet werden.[14]

Gegenwärtig werden zur mechanischen Beseitigung von Oberflächenverschmutzungen des Silbers weiche Pinsel, Bürsten mit verschieden hartem Haar und unterschiedliche Gewebe und Tücher verwendet. Es kommen auch milde Schleifmittel, wie Talg, Kölner oder Französische Kreide, Polierrot, Silicagel (Siliziumkoagulat in Spiritus) zum Einsatz. Zur mechanischen Beseitigung hart gewordener Korrosionsschichten wendet man manchmal eine lange, spitze Nadel aus gehärtetem Stahl mit abgeflachter Spitze an, die einen Griff besitzt und in ein Binokular, bzw. Vibrotool (engl. Bezeichnung) eingesetzt wird. Es ist ein Ziselierapparat, in den man eine Nähnadel einsetzen kann. Ein ähnliches Gerät unter der Bezeichnung "Fasserhammer" bot die Fabrik Woeckel in Stuttgart-Bad Cannstatt an. Eine andere Methode der Silberreinigung, insbesondere bei hartgewordenen Korrosionsansätzen, nutzt die Ultraschallwirkung aus. Ultraschallreinigung beruht auf einer Aufhängung des zu reinigenden Silbergegenstandes in einem Glasbehälter, der ein Reinigungsmittel für die Behandlung im Ultraschallbad enthält, das sich aus 8% Thioharnstoff, 5% Phosphorsäure und zu 0,3% aus einem anionischen Mittel besteht, das in Wasser oberflächenaktiv wirkt. Das Glasgefäß mit dem aufgehängten Silbergegenstand wird im Behälter des Ultraschallapparates untergebracht, der mit Wasser gefüllt wurde. Den Ultraschall läßt man sechsmal je eine Minute mit einer Frequenz von ungefähr 24kHz einwirken. Eventuelle zusätzliche Ultraschallbehandlung sollte nicht länger als 2 bis 4 Minuten andauern. Der Vorteil dieser Methode liegt in einer schnellen und wirksamen Reinigung, während zu den Nachteilen Schwierigkeiten bei der Kontrolle des Verfahrensablaufes und die Gefahr zählen, daß es zur Kavitation kommen kann, insbesondere bei einer Frequenz von 20 bis 40 kHz, wenn in der Leitungsflüssigkeit Luftblasen auftreten. Es kommt dann zu starken hydraulischen Stößen, die das Metall beschädigen können.

Vor der eigentlichen Reinigung ist der Gegenstand zu entstauben und entfetten, z.B. durch Zusetzung von oberflächenaktiven Mitteln (anionische Detergenten) oder die Verwendung von alkalischen Lösungen, die hydrolisieren, verseifen oder aber die organischen Bestandteile im Wasser emulgieren.

Nach einer Behandlung mit Laugen oder Säuren muß das Silber gründlich ausgelaugt bzw. ausgekocht werden, um alle Reste von Verbindungen dieser Stoffe zu entfernen. Nach dem Abspülen ist der pH-wert mit Lackmuspapier zu überprüfen. Die Oberfläche der Silbergegenstände muß nach dem Spülen mit Alkohol abgerieben und getrocknet werden. Für altes Silber ist es sehr wichtig, daß man es rasch trocknet. Wassertropfen, die am Silber langsam trocknen, können Flecke hinterlassen. Bei Silber ist es Vorteil, die Gegenstände in einen auf 100 ^{o}C erhitzten Trockenschrank einzulegen. Das Nachpolieren erfolgt mit einer feinen Neusilberbürste.

Werfen wir nun einen Blick auf die gegenwärtig wichtigsten angewendeten chemischen Methoden:

a) Wasser und Seifenwasser: Kleine Korrosionsstellen kann man durch Auflegen von Tampons mit heißer, abgekochter Wasserstoffperoxidlösung beseitigen, die alle 30 Minuten ungefähr zwanzigmal gewechselt werden müssen. Erzeugnisse aus poliertem Silber werden mit Seife oder einer Detergentenlösung in Wasser gereinigt, trockengerieben und mit einem weichen Flanelltuch poliert. Leichter Schmutz wird beseitigt, indem man den Gegenstand mit in Wasser aufgelöstem kohlesäurem Natron abpinselt, dann unter Wasser ordentlich abbürstet und abspült. Man kann den Silbergegenstand auch in warmes Seifenwasser für einige Stunden eintauchen und dabei öfters bewegen. Haftet der Schmutz noch etwas, so reibt man das feuchte Objekt mit feinem Kochsalz ab.

b) Wäßrige Ammoniaklösung: 5 bis 10% ige Ammoniaklösung beseitigt wirkungsvoll dunkle Silberchloridflecken, 8%ige Lösung löst ebenfalls Schichten aus Kupferbindungen auf. Eine schnellere und wirkungsvollere Reaktion wird durch Anwärmen des Bades auf 30 bis 50 ^{o}C erreicht. Ammoniaklösungen sind unter großer Vorsicht einzusetzen, obwohl sie Silber nicht auflösen. Ungenaues Abspülen des Gegenstandes nach der Reinigung kann nach einiger Zeit ein goldfarbenes oder dunkles Anlaufen der Überfläche bewirken.

c) Ammoniakpasten: Zur Reinigung schwarz gewordenen Silbers wird manchmal Kreide verwendet, die mit Methylalkohol und ein wenig Ammoniak angefeuchtet wurde, oder auch Kreide, Ammoniak und einige Tropfen eines Detergents. Gereinigt wird mit einem Flanelltuch.

d) Kaliumnatriumtartarat in einer 10%igen Lösung oder 70g Kaliumnatriumtartarat, 25g Natriumhydroxid in einem Liter Wasser oder aber auch die oft verwendete Mischung von Kaliumnatriumtartarat und Ammoniak.

e) Silbernitrat von 5 bis 20%ige Lösung beseitigt nicht nur Kupferverbindungen sondern auch metallisches Kupfer.

f) Ammoniumthiosulfat wird zum Zwecks der Aufweichung und Auflösung des Silberchlorids als 15%ige Lösung eingesetzt oder 90g Ammoniumsulfat, 10ml Ammoniak (25%ig) ind 0,5ml eines Detergents in einem Liter Wasser oder 1 Teil Ammoniumthiosulfat (20%ig) und 1 Teil Ammoniak (25%ig).

g) Thioharnstoff gibt in der Anwendung gute Ergebnisse bei der Reinigung von Silbersulfiden in Form einer Säurelösung, wie z.B. 80g Thioharnstoff, 4ml Ameisensäure auf 100ml Lösung oder 8g Thioharnstoff, 1,1ml

Salzsäure, 0,5ml anionisches Detergent, 86,4ml destilliertes Wasser.

h) Komplexon III wird für die Entfernung der grünen Kupfersalze und Beseitigung von Korrosionsschichten (nicht für Silberchlorid) verwendet, z.B. 40g Komplexon III, 30g Ammoniumchlorid, 200ml Ammoniaklösung, 1500ml destilliertes Wasser.

Die Oberflächenreinigung von Silbergegenständen mit Säuren gibt in allgemeinen gute Ergebnisse, kann jedoch teilweise auf das Silber und die Bestandteile seiner Legierung einwirken. Es werden also eine genaue Neutralisierung und ein genaues Auswaschen verlangt.

i) Essigsäure wird meistens zur Entfernung von hartgewordenen Anläufen und kalkähnlichen Ansätzen verwendet, z.B. als Mischung von 10 Teilen Essigsäure (40%ig), 10 Teilen Ammoniumchlorid und 10 Teilen destilliertem Wasser mit Aluminiumspänen.

j) Ameisensäure löst Kupferverbindungen auf und zur Entfernung von Patina an Silber wird oft eine 5 bis 25%ige Ameisensäurelösung in einem eventuell bis auf 60 bis 80 ^{o}C erhitztem Bad eingesetzt.

k) Salzsäure in einer 2 bis 4%igen Lösung löst verkalkte Stellen (Kalziumkarbonat) wieder auf und neutralisiert mit einer 5%igen Ammoniaklösung und mit destilliertem Wasser abgewaschen wirkt Salzsäure in kaltem Zustand sehr langsam auf Silber ein.

l) Schwefelsäure löst Oxide und Kupfer-II und -III-Salze auf und lockert Kalkansätze. Es wird in einem Verdünnungsverhältnis von 1:10 bis 1:40 verwendet.

m) Zitronensäure beseitigt Kupferverbindungen, darunter die Salze von der Silberoberfläche. Am häufigsten legt man die Gegenstände in ein Bad mit 5 bis 10%iger Zitronensäure. Neutralisieren mit 5%iger Ammoniaklösung und gründlich Spülen. Dieses Mittel, das sehr oft angewendet wird, verlangt ein sehr vorsichtiges Vorgehen.

n) Natriumzitrat von 5 bis 10% ige Lösung wird zur Beseitigung stark korridierter Stellen auf sehr feinen Silbergegenstände und des schwarzen Kupferoxids, Kupfersalze (bläulich-grün) sowie zur Lockerung von Kalkschichten, meistens in einem Bad, verwendet.

o) Natriumhydroxid wird zur Reinigung sehr stark schwarz gewordener Silbergegenstände in einem Bad aus einer Lösung von Natriumhydroxid und Natriumkarbonat (5%ig) im Verhältnis von 1:1 oder aus 6g Natriumhydroxid und 5g Natriumkarbonat in 100ml Wasser eingesetzt.

Im Handel befinden sich industriemäßig hergestellte Silberreinigungsmittel, wie Blank-Blink der Firma Heienbrock u. Co., Grotsch, Bezirk Leipzig; Duraglit, Renes (Polen); Edelman - Fabrik Hallerstaedt u.Co., Rastatt (Baden); Renovirin - Fabrik P.Fischer, Pforzheim, Silbo - Fabrik Delu, Bad Honnef; Sona - F.Hoffmann u. Söhne KG Chemische Fabrik Neuburg/Donau (BRD).

Neben den chemischen Methoden der Reinigung von Silbergegenständen werden Methoden der elektrochemischen Reduktion und thermische Verfahren der Korrosionsreduktion von Silber angewendet. Diese Methoden sind jedoch nicht Gegenstand des vorliegenden Referats.

Es scheint, daß nach 1945 ziemlich allgemein man sich bewußt wurde, daß die Reinigung von Silbererzeugnissen ein individuelles Herangehen an jedes Objekt und die Kenntnis der Materialzusammensetzung und der Art der Korrosionsprodukte verlangt. Es tauchten eine Reihe neuer Techniken auf und die Palette der zur Reinigung verwendeten Mittel wurde größer. Man erkennt nun an, daß der Zustand eines Objektes und Grad sowie Korrosionsart, also die Anwesenheit von Niello oder Silberverbindungen mit anderen Materialien, der Hauptfaktor bei der Entscheidung für eine der Methoden ist. Die Auswahl der entsprechenden Reinigungsmethode verlangt die Durchführung zahlreicher Versuche und Untersuchungen, da man nicht voraussehen kann, welches Mittel im gegebenen Fall das beste ist. Große Aufmerksamkeit widmet man gegenwärtig der Neutralisierung, dem Auswaschen und dem Trocknen der Objekte.

Zusammenfassend ist es festzustellen, daß die Entwicklung der Reinigung von Silber nicht nur in einer quantitativen Entwicklung der Methoden und Mittel zum Ausdruck kommt, sondern vor allem in ihrer individuellen Anwendung an Objekten und in einer größeren Sorgfalt um ihre Neutralisierung liegt.

Literatur

1. Proszek do czyszczenia naczyn srebrnych."Pamietnik Rolniczo-Technologicznych" 2,1834,S.145-146.

2. Reinigungsmittel für den Goldschmied. "Neueste Erfindungen und Erfahrungen" 26,1899,S.545.

3. Putzpulver zum Reinigung von Silberwaaren. "Gewerbehalle" Stuttgart 1866,8,S.128.

4., G., Reinigung angeläufener Silberner Münzen, Medaillen oder Vergoldeter Silbersachen. "Gewerbehalle", Stuttgart 1865,12,S.188.

5. Schutz von Silberwaaren gegen Oxydierung. "Westdeutschen Gewerbeblatt" 3.1885,1,S.11.

6. Gelbgewordenes Silber rein weiss zu machen. "Gewerbehalle" 31,1893,4,S.2.

7. Zur Erhaltung von Altertümern. "Technische Mitteilungen für die Malerei" 5,1888,51,S.115.

8. Metallputzmittel, "Gewerbehalle" 26.1888,7,S.2.

9. Putzmittel für Silbergegenstände (in) F.G.**Wiecks**, Deutsche Illustrierte Gewerbezeitung" 61,1896,46,S.366.

10. **Grein** J. ,Przyczyny chorób srebra w gablotach muzealnych. "Rocznik Muzeum Narodowego w Warszawie" 1,1938,S.4.

11. **Grein** J., Przyczyny.o.c.,S.5.

12. **Reyman** T., Jak ratowac przed zniszczeniem zabytki archeologiczne w muzeach regionalnych. "Ziemia" 15,1930,S.364.

13. T.P., W jaki sposób uchronimy wyroby ze srebra od oksydowania. "Przeglad Zegarmistrzowski i Zlotniczy" Poznan,2,1926,1,S.10.

14. **Starykoñ-Grodecki** K., Konserwacja broni i uzbrojenia. "Broñ i Barwa" 3,1936,5,S.106.

Die Adresse des Autors:

Senatorska 25/9
PL-30-106 Kraków

LEITNER-SZAPÁRY, Barbara

Konservierung und Restaurierung von barocken Leuchterappliken
um 1730. aus dem Stift St. Florian Oberösterreich (Austria)

Ich habe diese Arbeit ausgewählt, da sie mir als gutes Beispiel erscheint, mit welchen Anforderungen und Schwierigkeiten ein selbständiger, freiberuflicher Restaurator zu kämpfen hat (Koordinierung der verschiedenen Sachgebiete, Zusammenarbeit mit Untersuchungseinrichtungen, Kunsthistorikern, Handwerkern...).

Die barocken Wandappliken auch Leuchterappliken genannt, stammen aus dem Stift St. Florian in OÖ. Ihre Entstehungszeit ist mit ca 1730 anzusetzen. Die "Reflexionsspiegel" um 1780. Die Bezeichnung Reflexionsspiegel ist von mir für den Teil der Applike gewählt, der auch bei Gebrauch das Licht der Kerze reflektierte. Ob dieser Teil ursprünglich ein Spiegel, ev. ein Venezianischer war, ist ungewiß. Bis zum ersten Weltkrieg wurden die Appliken im großen Festsaal des Stiftes verwendet. Aus dem späten 19. Jhd. stammt auch die letzte Überarbeitung der Stücke (hinzufügen von 2 Leuchterarmen in neogotischen Stil). Ungefähr 60–70 Jahre waren sie dann im Keller des Stiftes gelagert. Die schlechten Konditionen haben deutlich ihre Spuren hinterlassen.

Abb. 1. *Eingangsfoto*

Kurze technische Beschreibung:

Die Appliken bestehen aus:

- zwei parabelförmigen, miteinander vernieteten Kupferblechteilen (diese sind handgeschmiedet und mit Band-, Akkanthusblattmotiven, sowie Teufelsfratzen verziert ziselliert, getrieben, ausgemeißelt.
- Die Oberfläche ist an der Schauseite teilweise Feuervergoldet und Feuerversilbert sowie handpoliert.
- zwei ebenfalls aus Kupferblech getriebene Blattbänder mit Eierstabmotiv, verdecken die Stoßkanten der beiden Teile.
- in der Mitte befindet sich, ebenfalls aus zwei Teilen gearbeitet, der Reflexionsspiegel (Messingblech getrieben, ziselliert, ausgemeißelt). Die beiden sich überlappenden Hälften sind auf einen darunter befindlichen Holzkeil genagelt. Der "Spiegel" ist an vier Stellen mit Linsenkopfschrauben (Messing) an den Untergrund befestigt.
- je ein Leuchterarm (Messing gegossen) mit einer Tropftasse (Kupferblech getrieben) und Kerzenhalterung – die beiden seitlichen Leuchterarme sind die Beifügungen des 19. Jhd. (siehe Eingangsfoto):
- die ursprüngliche Aufhängung erfolgte an der Rückseite mittels eines Eisenhakens (direkt mit den Kupferteil vernietet).

Wissenschaftliche Untersuchungen:

Über den Materialaufbau die Oberflächenbehandlung und Zustand wurden vom Bundesdenkmalamt in Wien durchgeführt. Anhand der Ergebnisse und den Unterlagen, die mir teilweise zu Verfügung standen, wurden die Arbeitsmaßnahmen entwickelt.

Übernahmezustand:

Oberflächen stark verunreinigt (schlechte Pflege-Putzmittelreste, Lagerung). Unter der Schmutzschichte großteils an den blanken Stellen des Kupfers, schöne Cu-Patina (vor allem an der Rückseite).

Stellen wo Feuchtigkeit und verschiedene Metalle zusammentrafen stark korrodiert. (Angriff, bezw. Zerstörung des Grundmaterials und Oberflächenveredelung durch chemische oder elektrochemische Reaktion mit der Umgebung – das Metall oxidiert – speziell die Eisenteile. Später zeigte sich nach der Reinigung, wieweit die Korrosion die Versilberung vom Grundmaterial unterwandert und gehoben hatte.

Deformation der Blechteile und Leuchterarme durch unsachgemäße Lagerung und Überlastung des Materials (hinzugefügten Leuchterarme waren zu schwer).

Versprödung des Materials – die Spiegel sind aus sehr dünnem Messingblech getrieben (2 - 3/10 mm) und durch ihre Legierungsbeschaffenheit und schlechte Montage, sehr versprödet. Viele Bruch- und Fehlstellen, Risse...

Die Tropftasseneinsätze waren aus Eisenblech.

Alle Einzelteile waren mit Eisenniete (verrostet) verbunden.

Nach der Druchführung der Zustandsaufnahme erfolgte die *Zerlegung in die Einzelteile.*

Ziel meiner Arbeit war:

Konservierung des Vorhandenen, Lesbarmachen der Oberfläche und des Objektes, k e i n e Wiederherstellung des Originalzustandes!

Fragestellungen:

a) Reinigung der Oberfläche

b) Problem der Aufhängung und Fixierung der Einzelteile-Minimierung der Belastung für das Material

c) Ergänzung der fehlenden Teile, soweit benötigt

d) Behebung der Bruchstellen und Risse

e) Lesbarmachen der Oberfläche/Oberflächenschutz.

Durch die längere Betrachtung der einzelnen Stücke ist mir ein deutlicher Unterschied in der Machart von Einem zu dem Anderen aufgefallen. Nicht nur die Oberfläche, auch das Grundmaterial waren unterschiedlich. Die Ornamente waren weitaus präziser gearbeitet, das Kupferblech viel homogener und weicher, die Vergoldung und Versilberung daher auch besser haftend. Auch war ich auf der Suche nach der ursprünglichen Form der Spiegel. Es war eindeutig, daß die jetzt vorhandenen erst später hinzugefügt wurden. Genau bei diesem einen Stück welches differenziert gearbeitet war, fand ich auf der Rückseite eine feinpunzierte Linie, die die Umrisse des Spiegels andeutet. Wieweit je ein Original vorhanden war, ob die anderen Appliken nach diesem einen Stück kopiert wurden etc., wäre eine Arbeit für einen Kunsthistoriker...

Konservierung:

Vorgangsweise:

A) a) Reinigung der Oberfläche: Entfernen der Schmutzschichte mit Seifenlösung und Ultraschall.

b) Verwendung von verschiedenen Lösungsmitteln.

c) mechanisch: Holzstäbchen, -spachteln, Skalpell Polierpaste (Rouge).

d) Entfetten der Obverfläche und sorgfältiges Trocknen. Nach der Reinigung zeigte sich wieweit die Oberflächenveredelung noch gut haftete. An Stellen wo sehr gut mit dem Blutstein (Hämatit) oder Polierstahl, handpoliert wurde, war sie am Besten erhalten (die Oberfläche wurde gut verdichtet).

B) Aufhängung und Fixierung – Minimierung der Belastung für die einzelnen Teile.

a) vorsichtiges ausrichten, -klopfen der deformierten Einzelteile mit Holzpunzen, Hauthammer und speziell angefertigten Hilfswerkzeugen.

b) Herstellung eines Trägergerüstes aus nichtrostendem Stahl in Kreuzform. Übertragung der Niet- und Bohrlöcher, wie Montage des Wandhakens auf dem Gerüst.

c) Anfertigung von Senkkopfschrauben ohne Schlitz, statt der Eisennieten und Rundkopfschrauben (aus Messing -je nach Bedarf versilbert oder vergoldet), für die Montage aller Einzelteile und zum Schließen der nicht benötigten Löcher an der Schauseite.

Jetzt sind alle Einzelteile jederzeit zerlegbar.

C) Ergänzung der fehlenden Teile: erfolgte mittels Kopien die auf galvanoplastischem Weg hergestellt wurden:

a) Benetzung eines Positivs mit einem Trennmittel und Abformung in Silikonkautschuk Nr. 56 mit Härter T (2 %), der Fa Wacker. Diese Silikonkautschukmasse besitzt eine relativ niedrige Viskosität ist daher gut gießbar. Die Verarbeitungszeit beträgt 35–45 Min. bei ca 20 ^{o}C, die Vulkanisationszeit 3–4 Stunden. Das Vulkanisat ist elastisch.

Diese beweglichen Formen wurden, besonders bei der Abformung der Tropftasse, mit einem Glasfasern verstärkten Poliurethanschaum umgeben (Stützkappe). Die flachen Teile wurden mit Plexiglasplatten verstärkt, mit einem ca 2 mm starken Kupferdraht rundherum leitend gemacht (die Aufhängung im Bad war ebenfals aus Kupfer). Die Negativform mit Spiritus benetzt und mit einem Fehhaarpinsel Cu-Staub (Superdruckschliff 9900) aufgetragen. Der überschüssige Staub mit Druckluft entfernt.

b) Galvanobad – 20% Kupfersulfat $Cu_2SO_4 . 5H_2O$
3% H_2SO_4 Schwefelsäure
1% Harnstoff (für die feinere Abscheidung des Cu).

Badetemperatur 20–25 ^{o}C – bewegt. Die Anode aus Kupfer soll 1 bis 1,5 fache der Oberfläche des herzustellenden Werkstückes aufweisen.

Es zeigte sich, das geschmiedete und gegossene Kupferstücke einen feineren Niederschlag geben. Nach ca 2–3 Stunden, wenn kein Spannungsabfall mehr erfolgt, wurde die Stromstärke auf 2–2,5 Ampere pro dm^2 Oberfläche, für den Dauerbetrieb eingestellt. Sich zeigende Fehlstellen wurden nochmals mit Cu-Staub leitend gemacht. Wo zu großer Materialzuwachs erfolgte (Ränder), wurden diese Stellen mit einem Galvanolack abgedeckt.

c) die fertigen Galvanoplastiken wurden zugerichtet und je nach Bedarf partiell galvanisch vergoldet oder versilbert.

D) Behebung der Bruchstellen und Risse, sowie Einfügender Ergänzungen.

a) die Bruchstellen an den gegossenen Messing--Leuchterarmen wurden verschweißt, da die Appliken weiterhin benützt werden und auch an exponierter Stelle an der Wand hängen (mechanisch Belastung).

b) alle anderen Bruchstellen, Risse, Ergänzugseinfügungen wurden geklebt.

Verwendet wurde der Epoxidharzkleber Araldit AY

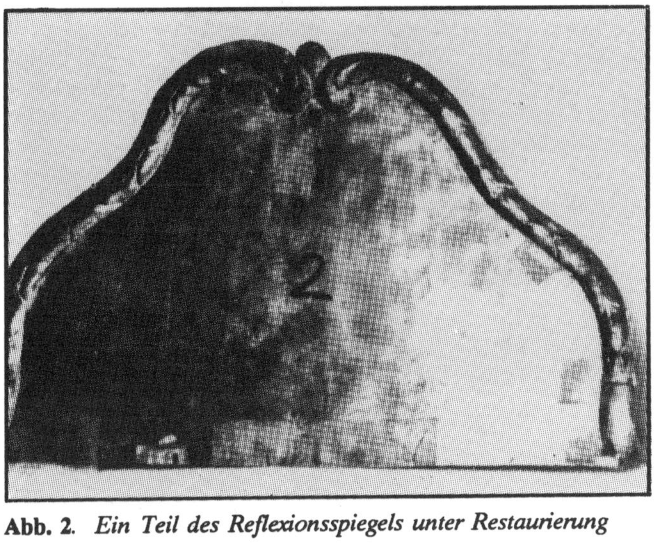

Abb. 2. *Ein Teil des Reflexionsspiegels unter Restaurierung*

103 mit dem Härter HY 956 – 100:18 Gewichtsteile der Firma Ciba Geigy.

Ausschlaggebend für die Anwendung der Klebung als Verbindungsmedium war, daß weder mechanische noch thermische Veränderungen im Materialgefüge stattfinden und auch die chemischen Änderungen an der Oberfläche minimal sind.

Das Kleben als Fügeverfahren – zur Verwendung können chem. abbindende Klebstoffe verwendet werden die durch Poymerisation, Polykondensation und Polyaddition sich in hochmolekulare Stoffe wandeln, mit einer hohen mechanischen Festigkeit. Grundstoffe können Phenolharze, Epoxidharze, ungesättigte Polyesterharze, Vinyl-Mischpolymerisate sein. Die Produkte werden meißt in Zweikomponenten geliefert, sind kalt- und warmaushärtend.

Aus diesem Überangebot selektiert sich schnell eine kleine Gruppe von geeigneten Klebern heraus. Ich bin von folgenden Anforderungen ausgegangen:

gute Fließeigenschaften
möglichst durchsichtige Farbe
geringe Viskosität
geringer Säuregehalt
niedrige Toxizität.

Wie schon oben angeführt, habe ich diesen Epoxidharzkleber Araldit HY 103, verwendet.

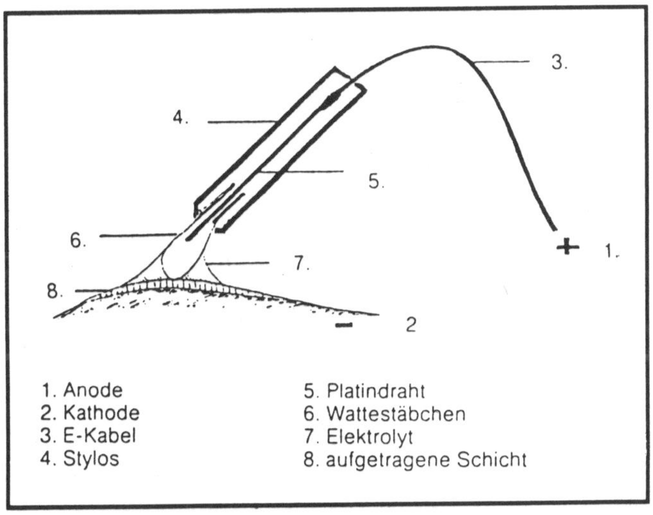

Abb. 3. *Der restaurierte Wandapplik*

Er erfüllt alle diese Anforderungen, besitzt eine harte Klebefuge und eignet sich auch zum Gießen.

Zur Auswahl stand noch das gleiche Harz gemischt mit dem Kleber HY 991, der angeblich säurefrei ist, wobei aber die Farbe hellgelb ist, und die Viskosität höher, die Klebefugen bleiben leicht flexibel.

Als Stütz- bezw. Unterfütterungsmaterial habe ich ein Polyestervließ verwendet. (Siehe: Testing Materials – in The Conservator 1979/ Mitschriften bei einem Seminar am British Museum, über dieses Thema, gehalten von S. Blackshaw und V. Daniels, 1984).

Vor der Klebung wurden die zu klebenden Flächen mit Glasfaserstiften und Trichloräthan (1,1,1, stabilisiert) entfettet. Stoßkanten zwischen den Ergänzungen und Original, wurden zusätzlich mit dünner Kupferfolie, unterlegt.

E) Lesbarmachen der Oberfläche / Oberflächenschutz.

a) Die markantesten Fehlstellen an den Appliken (erhabenen Stellen, deformierten und ausgerichteten) wurden mit dem Tampongalvanisierungsverfahren geschlossen.

Anode: ein an das Kupferkabel angelöteter Platindraht (Kathode: Stahlblech) PVC Wanne.

Elektrolyte: Starkversilberungs- und Vergoldungsbad (mittelgoldfarbe) auf zyanidischer Basis.

Die Arbeiten wurden bei ca. 4 Volt durchgeführt und unter stetigem raschen Bewegen des Anodenstiftes. Die Tamponvergoldungen wurden bei ca 10 Volt durchgeführt. Die Neuversilberten Stellen wurden in der Tiefe stellenweise mit Schwefelleberlösung (1g KS_4 auf 1 l H_2O) patiniert. Die so behandelten Stellen mußen auch sehr rasch geschützt werden, um ein Anlaufen der dünnen Versilberung zu verhindern.

b) Als Oberflächenschutz wurde Paraloid B72 in 10% Lösung aufgetragen (gepinselt), Vorder-und Rückseit. (7% Festkörpergehalt Paraloid B72 in Ethanol und

Xylol 1:1 gelöst). Weitere Lösungsmöglichkeit wäre: je ein Volumsteil Xylol, Aceton, Äthylenglycolacetatmonoäther (im Handel auch als Cellosolveacetat bekannt).

c) die Eisenhaken wurden in 15 – 20% Zitronensäure entrostet mit einer Tanninlösung stabilisiert und ebenfalls mit Paraloid B72 geschützt.

Paraloid ist die Werksbezeichnung der Firma Röhm & Haas, für Verguß und Laminierharze auf Polyacrylat oder Styrol, Butadien, Copolymerisatbasis.

Die Arbeiten wurden großteils mit Mag. Martin Klobassa, akad. Bildhauer, durchgeführt.

Seit mehr als drei Jahren befinden sich die Appliken wieder im Festsaal des Stiftes St. Florian und sie weisen keine sichtbaren Veränderungen auf.

Adresse des Autors:

Bubendorf 85
A-3354 Wolfsbach

BÁTHY Géza

Restoration of the Turkish-age treasure find from Ozora

Abstract

16th century jewerly buried in the cellar of the Ozora castle, and recently disclosed, has been restored. The find was buried wrappen in different fabrics that have been conserved. Also, important textile remains rests corroded on metal objects sufraces were found. It had to be decided whether to keep them conserved with the metal or, after documentation (photos, silicon imprints) and removal of textile remains rests, to treat silver finds to lend them the usual aspect of jewelry exhibits.

Restoration and conservation were carried out by duly selected chemical, mechanical and ultrasonic procedures. It is important to keep and carefully document observations made in disclosing undisturbed find complexes as historical information.

Monumental-archaeological excavation of the 15th century, late Gothic castle of Ozora started a few years ago. This castle is at a distance of 60 km to the SE of Veszprém. In investigating the building, excavations have been made in its surroundings and cellar, down to the level of the construction age. The treasure was found in the 16th century fill of the cellar, near another find of coins. The castle, having been occupied by the Turks for 150 years, probably a Turkish soldier might bury his plunder of treasures.

The find comprises 53 jewelry objects, 109 different pearls and thousands of silver coins. The time of burying may be determined from these latter. The find of coins consists of high-grade silver Ferdinand (ruled 1526 to 1564) deniers, and Turkish akche minted from low-grade silver.

The jewelry gives a cross section of Hungarian goldsmithery of that age. The objects comprise: a gilded silver Gothic cup, a silver spoon with twisted handle decorated with a herold figurine, a ball-headed filigree knot-pin and a big silver shank button, eleven rings, some hooks and eyes. The find also comprises 34 small round shank buttons. Curiously, when buried, these were not strung on cords but on bands partly cut from fine textile and twisted, and one a wire. The find was buried wrapped in several layers of different fabrics.

The find was undone in the laboratory, with the cooperation of Mrs Judith Bakay-Perjés who dealt with the disinfection and first conservation of the fabrics, using the usual mix of alcohol, glycerine, thymol and water. She also dealt with the further conservation of fabrics. The relative soundness of textiles detachable from the objects may be attributed to the known fungicidal effect of silver and copper ions released in electrochemical processes and transformed to salts.

The find of coins was restored in the Kecskemét Museum, in close cooperation with the researcher concerned with numismatic processing.

The jewelry was restored in the Hungarian National Museum. The aim was set in common with archaeologist colleagues. It is here that this work is related to the scope of this Conference: Are textile remains to be left on metal objects or not? Or else: Should our finds present themselves in the future as possibly authentic complex archaeological objects or as jewelry objects?

The problem of in-situ preserving textile remains corroded on metal objects surfaces may be ruled in general by the principle: The more important it is to in-situ preserve the textile remains, the more precious the inherent information, and the less important the metal object it adheres to. As a concrete example: For Roman or older textile remains found on unimportant iron scraps, the

Fig. 1 *Rings before restoration*

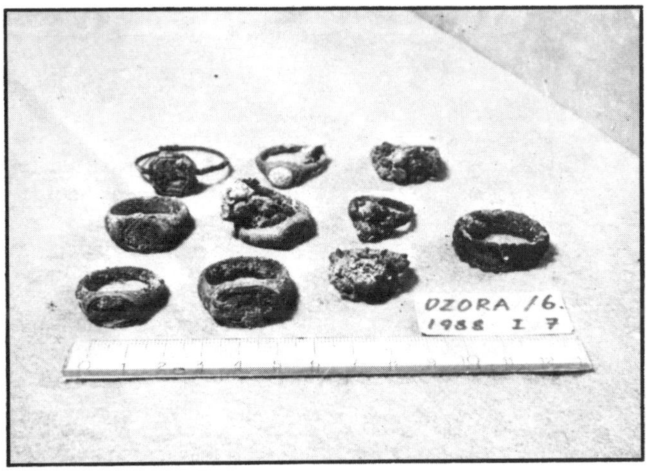

Fig. 2 *Rings after restoration*

Fig. 3 *Plain shank buttons with textile remains and big filigree button before restoration*

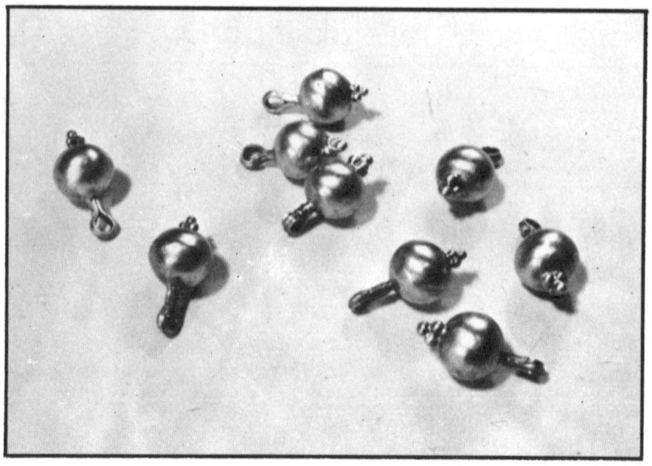

Fig. 4 *Small shank buttons restored*

pereservation of the textile should take preference in the course of the joint conservation.

In the case of the Ozora find, this problem was decided after the beginning of the very careful cleaning. Since the find offered very many textile remains of different counts easy to conserve, we did not insist on keeping the textile remains rests adhering to spoon and cup surfaces. They are recorded, instead, by photo-documentation, and by silicon imprints. It was decided to treat items of jewelry to lend them the aspect usual in museums, in spite of the unlikehood of being able entirely to reproduce the metallic brightness peculiar to jewelry kept in baronial and ecclesiastic collections.

Treasure objects were coated with a wide range of coloured corrosion products. Typical were atacamite-azurite-malachite corrosion products from the copper content of the silver. Some surfaces exhibited rather smooth black silver sulphide layers. Chemical cleaning involved pre-soaking in a lukewarm solution of sodium hexameta-phosphate, treatment with a citric acid solution of thiourea and with ammoniacal potassium carbonate - at times combined with ultrasonic cleaning. Among the mechanical methods were: microshaving, polishing with a watery solution of chalkstone powder, cleaning with metal wire and using glass-fibre brushes (Figs. 1-4).

Fig. 5 *Silver cup before cleaning*

Fig. 6 *Silver cup restored*

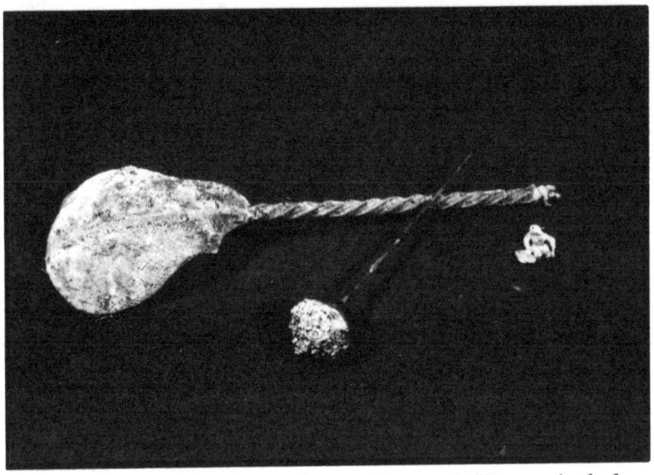

Fig. 7 *Silver spoon with textile remains and filigree pin before restoration*

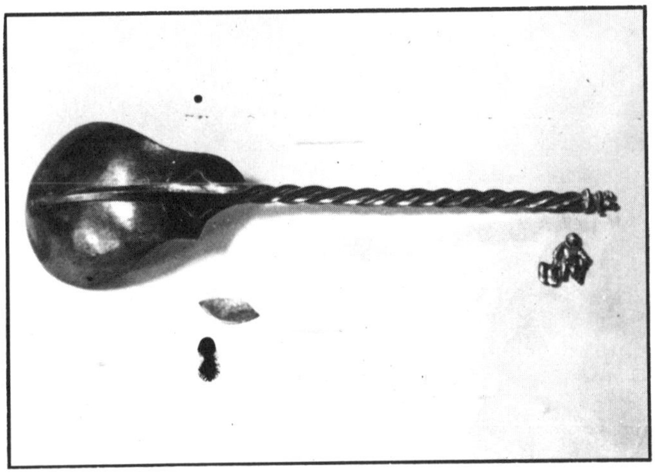

Fig. 8 *Silver spoon cleaned, before recovering broken parts*

The upper part of the cup wall has thinned already in the making and in the embossing. Here, electrolytic corrosion was more marked since coins and rings wrapped in textiles had been in contact with the cup's inside. This is why, in removing the corrosion products, the silver plate reduced to filmy thickness, corroded to crystalline, was pierced or cracked at several spots. These defects were repaired by sticking and complementing with Araldite modified with aerosil and silver dust. In the course of treatment, sulphide black galenite or niello-type modifications cemented in silver were found even in deeper layers of the metal. There was a phase where further decomposition of the surface would cause lesions and damage that would overwhelm the advantages expected from the removal of these compounds and cleaning to metal. (Figs. 5 and 6)

Fig. 9 *Big filigree button and pin restored*

Cleaning and conserving the gilded silver spoon went on as above. After final polishing with a soft glass-fibre brush, both fragments were stuck on by means of a two-component adhesive "UHU Windfest 300" modified with a minimum of aerosil and fine silver dust. (Figs. 7-9)

The herold figurine had originally held a lance in his left hand. When found, both parts of the lance were missing. It was decided to replace the missing part between hand and base by a silver wire complement to the due length and thickness, both for aesthetic and static reasons, connecting the herold figurine to the base at three, rather than two points without the lance.

Conservation was made with a 5:1 solution of Mowilith-Acrylane in mixed solvent, and with Paraloid B72 in a mixed solvent. (A 5:3:2 mixture of acetone-toluene-xylene, was expected to provide a possibly uniform film coat to prevent from spreading the speed of evaporation of different solvents.)

The treasure find fits the coin finds of Turkish age. Interesting results are expected from testing imprints of the seal rings. Disclosure and restoration of the complex of finds buried for 400 years have yielded a lot of interesting information thanks to conscientious restoring activities.

Author's addres

Hungarian National Museum
Múzeum krt. 14-16.
H-1088 Budapest

BHATNAGAR, Inder Kumar

Conservation of an excavated Chola bronze

Abstract

The conservation procedure of a bronze icon discovered in a chance excavation in the Dharwad region of Karnataka State of South India is described. The huge dimension – 75 cm high, 55 cm maximum width and solid cast nature of the image of a deity – presented a challenging job. The metal corrosion products were identified by X-ray diffraction, the spots of active corrosion were successfully stabilized.

Introduction

The Chola dynasty is considered a golden age for creative activity in the southern region of India.

The Rajarajeswara temple in granite stone with its towering gopuram and innumerable bronze icons of gods and goddesses bears testimony to the craftsmanship and skill of artisans in the Chola kingdom which existed during the 9-13th century AD. Some exquisite metal sculptures can be seen in the bronze gallery of the Goverment Museum, Madras, and in the Art Gallery at Thanjavur.

The art of metal casting was at the zenith of perfection in the reign of the Chola kings. Most of the images are solid cast but occasionally one comes across images produced by the hollow cast technique.

A four faced image of a deity was received at the Regional Conservation Laboratory, Mysore, for stabilization of corrosion products and preservation. It was a chance discovery during the excavation of a pit by a local farmer of the Dharwad region in Karnataka State of South India. On the basis of stylistic considerations and the profuse decorations which appeared after the removal of adhering clay and corrosion products, it can be assigned to the early Chola period, i.e. around the 10th century AD. It is a very heavy sculpture and when excavated it looked like a stone image. The sculpture is a solid cast in which an iron armature has been used. At one stage it was suspected to be a gold image buried in the soil for safety against theft.

Preliminary examination

Preliminary examination of the image when received revealed that it was covered with a thick layer of clay. Superficial cleaning with brushes and with pincers, surgical knives and smaller tools indicated the presence of corrosion products which were greenish in appearance. A sample of the cleaned portion indicated the presence of elements such as copper, tin, lead, iron, etc. The water extract of the corrosion products indicated the presence of chloride ions.

Analysis of corrosion products

The bluish green corrosion products were separated and subjected to X-ray diffraction analysis.

The minerals identified are as follows:

(a) Atacamite $2\ CuCl_2 \ .\ 5Cu(OH)_2 \ .\ H_2O$

(b) Malachite $Cu(OH)_2 \ .\ CuCO_3$

The spot from where the samples were taken again developed blue-green corrosion products in the presence of water vapour suggesting the presence of active corrosion popular known as 'bronze disease'.

Conservation treatment

Because of the hugeness of the sculpture – 75 cm in height and maximum width 55 cm – and the solid cast nature, it required special treatment. The immersion of the object into a huge water tank containing the solution of chemicals was found inconvenient and unsuitable. It was, therefore, decided that the object could be treated by locally wetting and then mechanically removing the adhering clay and underlying corrosion products.

As a first step, a lukewarm solution of sodium hexameta-phosphate (Calgon) was applied with the help of cotton swabs dipped in the solution. This helped to soften the clay which was later removed using blunt knives, scalpels and pincers, etc.

The thigh portion of the image had been scratched in the past to ascertain the material of the object, i.e. whether it was gold or some other metal. The removal of clay revealed the underlying corrosion product which – on wetting – became active. A 5% solution of sodium-potassium-tartarate was prepared in distilled water and the affected spots were locally treated with cotton swabs dipped in the solution and applied on the spots. When the solution had dried, some of the liquid reagent was poured on the surface of cotton attached to the spot. The reagent solution was occasionally warmed to a temperature of 45–50 ^{o}C and applied to improve the rate of reaction. Two or three applications were sufficient to distrub the corrosion products which were then removed mechanically using similar tipes of tools as used those for remo-

Fig. 1 *The statue before treatment*

Fig. 2 *The statue after conservation treatment*

ving clay from the surface. A portion of the image which was under active corrosion was also treated with about 5% sodium sesquicarbonate solution which was periodically warmed before applying to the affected spots. It was found that the action of sesquicarbonate was slower and was taking much time for cleaning. The whole image was cleaned using sodium-potassium-tartarate solution both from the front and the back. A thorough washing of the metal sculpture was carried out in running water from a tap. Final rinsing was done with distilled water and also with warm distilled water to extract any soil which might have gone inside the crevices. (Figs. 1 and 2)

At this stage the statue was kept under observation for the possible recurrence of corrosion on the image. It was found that on the face near the nose and on the broken limbs and below the shoulder, patches of greenish corrosion products appeared again. These spots were once again cleaned using tartarate solution and carefully scraping the corrosion products with blunt tools so as to avoid possible scraching of the image. The image was once again thoroughly washed to remove the chemicals which were used for treatment. The sculpture was allowed to dry in order to stabilize it. A saturated solution of benzotriazole (BTA) in alcohol was applied on the entire surface. Once again the sculpture was kept under observation for a possible out-break of bronze disease in the areas which had not stabilized. The broken limbs of the arm and the spot below the arm had patchy growth of corrosion products due to appearance of atacamite. The spot was once again cleaned, dried, and the solution of BTA was applied with the help of cotton swabs. The spots were kept wetted for 3 days. Finally it was found that there was no recurrence of the corrosion products in any of the areas of the bronze image. On complete drying, a thin layer of white petroleum jelly was applied on the entire surface and the excess was removed by rubbing with soft tissue paper.

Conclusion

This four faced image of a deity can now be seen in its true perspective and all the concealed details of necklace, crown and other carved portions are now distinctly visible. The image is kept indoors and there is no climate control in the room where it is kept. During monsoons when there is heavy downpour and the relative humidity is close to 100% there is no recurrence of corrosion on the image. Two monsoon seasons have passed and there are no sings of damage of any kind. The image seems to have become fully stabilized with the treatment given.

The conservation treatment given to this image is of routine nature but the dimensions and solid cast nature of the bronze was a challenging job which has been accomplished to the art historians and the owners of the sculpture.

Selected bibliography

Plenderleith, H.J. and **Werner,** A.E.A.: Conservation of Antiquities and Works of Art, Oxford University Press, London, 1976

Sastri Nilkanta, K.A. and **Srinivasachari,** G.: Advanced History of India, Allied Publishers Pvt Ltd., New Delhi 1986, p. 321

Swezey–Kenneth, M.: Tips and Data for Home and Workshop, Popular Science Publishing Co, New York, 1979

Rozenfeld, R.L.: Corrosion Inhibitors, McGraw Hill Inc, U.S.A., 1981. p. 98

Stambolov, T.: Corrosion Inhibitors, ICOM Committee for Conservation, 5th Triennial Meeting, Zagreb (1978)

Coburn, Seymour, K.: Corrosion Source Book, American Society for Metals, Ohio, U.S.A., 1984

Madsen, H.B.: A Preliminary Note on the Use of B.T.A. for Stabilizing Bronze Objects, Studies in Conservation. 12 (1976) pp 163-6

Madsen, H.B.: Further remarks on the Use of B.T.A. for Stabilizing Bronze Objects, Studies in Conservation, 16, (1971) pp. 120-2

Cotton, J.B., and **Scholes,** I.R.: Br. Corr. Jr., Vol. 2, 1967

Sease, Catherine: Benzotriazole: A review for Conservators, Studies in Conservation, 23, (1978) p. 76-85

Bhatnagar, I.K., **Agrawal,** O.P., and **Kumar, Virendra:** Investigations on Copper Corrosion Inhibitors. 10th International Congress on Metallic Corrosion, Vol III, 1987, p. 2895-2900

Author's address:

Regional Conservation Laboratory
The Palace Complex
57 0001 Mysore, India

GREEN, Lorna

A re-evaluation of lead conservation techniques at the British Museum

Abstract

In the early 1960's, many different conservation treatments were used in the British Museum in an attempt to stabilise lead objects. These included consolidation, ion-exchange resins and electrolytic reduction. After treatment, a coating, often wax, was applied. The effect of these treatments has been evaluated.

The body metal and corrosion products of treated lead objects have been analysed. Experimental work has been carried out using electrolytic reduction on corroded lead samples to determine the effect of this treatment and the degree of protection afforded by coatings against an aggressive environment under controlled conditions.

I. Introduction

Lead is a passive metal and corrosion resistant due to the formation of a protective layer of compact insoluble compounds, often containing PbO_2, plattnerite. However, in the presence of organic acids, thits protective film does not form, and the lead actively corrodes producing a porous coating of compounds, often basic lead carbonate, $Pb_2(CO_3)_2(OH)_2$. Other corrosion products, for example, lead formate, $Pb(HCO_2)_3$ may occur [1, 2, 3]. Lead corrosion is often voluminous and can destroy important surface detail of a lead artefact. If allowed to continue, the lead object will eventually corrode completely and disintegrate.

The condition of lead can be influenced by many factors. The function of the artefact when in use may have a bearing in its present state. For example, if it had been used in intimate contact with wood, the lead may have corroded due to the presence of organic acids. Residues of the acids may remain within the lead and any associated corrosion layers. The condition of storage or burial of the lead artefact at the end of its useful life should also be taken into account. Certain yellow brown patinas, often containing litharge, PbO, occurring on buried lead, appear to be stable and protective after excavation [4]. Alloy materials such as tin also affect the stability of lead [3].

The conservation treatment used to stabilize a lead object will have an effect on its long term preservation. Effective storage after conservation is essential if the lead is not to re-corrode.

Over the past twenty years several conservation treatments have been used on lead artefacts in the British Museum. The first stage of conservation involves removing or arresting the corrosion, the second stage involves inhibiting future corrosion. The aim of this investigation was to correlate the present condition of lead artefacts with the conservation treatments which were applied to them. To do this, lead artefacts were examined, and their conservation history traced through the conservation record system. Fourteen lead seals, belonging to the British Library, and twelve lead coins from the Department of Coins and Medals in the British Museum were examined. Electrolytic reduction, the technique presently in use at the British Museum, was evaluated on corroded lead samples. After reduction the lead samples were coated with waxes and synthetic lacquers. The coated samples were placed in a corrosive environment to assess the degree of protection provided by each coating.

2. The investigation of lead objects from the British Library and the British Museum

2.1. Examination

2.1.1. Lead Seals

In the 1960's a large number of lead seals were treated by conservators in the British Museum. Details of past treatments used on each seal were recorded.

Fourteen lead seals, representing a range of treatments were selected for examination. They had all been stored in cardboard boxes in metal lined drawers until 1986-1987 when they were trasferred to closed polystyrene containers and returned to the metal lined drawers. Two lead foil controls were also put into the drawers. One piece was put into the drawer open to the atmosphere, and the second piece was placed in a closed polystyrene cointainer in the drawer. After two years the lead in the container remains in a good condition, whereas the lead exposed in the drawer has thick white corrosion over its entire surface. This indicates that there are organic acid vapours present in the storage area. Oddy tests [5] have been carried out on the materials used in the storage area: wood, paint cardboard tags, and cardboard boxes originally used for storage of the seals. All these materials corroded lead in the tests. This illustrates that the closed polystyrene containers afford effective protection against the corrosive environment.

The body metal of the seals was analysed semi-quantitatively using X-ray fluorescence (XRF) (see Table 1). The corrosion on all of the seals was identified using Xray diffraction (XRD) as hydrocerussite (basic lead carbonate).

Table 1 *X-ray fluorescence analysis of lead seals*

Seal No.	Condition (Oct 1988)	Elements detected (%)			
		Cu	Pb	Sn	Trace
XXXVIII (167 (Half A)	Stable	<1	99	<1	Ba, Fe
XXXVIII (167 (Half B)	Corroded	<1	99	<1	
XXXVIII 36	Stable	<1	99	B.D.	Ba,Fe
XXXVIII 27	Good	<1	97	2.5	
XXXVIII 24	Good	<1	94	5.5	
XXXVIII 199	Stable	4	96	B.D.	
XXXVIII 29	Good	1	95	4	
XXXVIII 147	Stable	<1	99	B.D.	Fe
XILIV 7	Stable	2	97	<1	
XXXVIII 198	Slight corrosion	<1	99	B.D.	
XXXVIII 149	Good	<1	99	B.D.	
XXXIX 55	Slight surface bloom	<1	99	<1	
XLIV 66	Stable	3	96	<1	
XXXV 73	Actively corroding	3	96	<1	Ba

Key: B.D. – Concentration below detection limits of XRF equipment

2.1.2. Lead coins

Four lead coins from Burma and eight from India were examined. Within the group there were three categories; those with a natural stable patina not requiring conservation; those with no obvious patina which were stable and did not need conservation; those which had been conserved in the past and begun to recorrode.

The coins were stored in mahogany trays and cabinets coated with Ercalene, a nitrocellulose lacquer. The body metal of the coins was analysed semi-quantitatively using XRF (see Table 2).

Table 2 *X-ray fluorescence analysis of lead coins*

Lead object	XRF analysis	Condition (Oct 1988)
Coins Burmese 1924 12-10 14 1894 6-5 30 1924 6-5 30 1894 6-5 49	Trace Cu, 2% - 3% Sn	Stable
Indian 1970 5-12 55 1970 5-12 15 BMC 50 BMC 49	Trace Cu <1% Sn	Natural patina, slight local corrosion Natural patina, stable Natural patina, stable Natural patina, stable
Indian BMC 105 1905 10-7 88 1905 10-7 63 BMC 108	Trace Cu, <1% Sn	Local outbreaks of corrosion

XRD was used to identify the natural patina and corrosion products. The natural, stable patina from the Indian coin, 1970 5-12 55, contains cerussite, $PbCO_3$, and small amounts of hydrocerussite. The corrosion on the Indian coin, 1905 10-7 63, was identified as hydrocerrusite with a trace of plumbonacrite, $6PbCO_3 \cdot 3Pb(OH)_2 \cdot PbO$.

2.1.3 Effect of alloy and trace metals

From Table 1 it can be seen that the majority of the seals in good condition contain a small percentage of tin. Similarly the stable coins listed in Table 2 also contain a small percenage of tin. Tin has been reported to have a stabilising effect on lead. Copper is also reported as a stabilising component [3], but insufficient evidence has been gathered in the present study to confirm this.

2.2 Conservation Treatment

A number of conservation techniques for lead have been documented [6, 7, 8, 9, 10]. Many lead seals were coated with shellac before being conserved in the 1960's. One unsuccessful example of a lead seal embedded in an epoxy type resin was found. However, the main methods of conservation used were ion-exchange, ethylenediaminetetra-acetic acid (EDTA), and electrolytic reduction.

2.2.1 Ion-exchange

Ion-exchange involves immersion of the lead artefact in a solution which effectively dissolves the corrosion leaving a clean metal surface. Using this method any surface detail preserved in the corrosion layers may be lost. Amberlite IRA 400 appears to have been the commonly used ion-exchange resin.

2.2.2 Ethylenediaminetetra-acetic acid (EDTA)

A salt of EDTA, usually the sodium salt, has a similar effect on the corroded lead as an ion-exchange resin. It has been suggested that EDTA is more suitable for use on complex objects with detail preserved in the corrosion layer [11]. EDTA can attack metallic lead and this method has to be carefully controlled. The EDTA must be washed out thoroughly after treatment.

2.2.3 Electrolytic reduction - consolidative reduction

Electrolytic reduction is now usually used for the conservation of lead in the British Museum. Oxidized lead is reduced to metallic lead at the cathode by the following reaction:

$$Pb^{2+} + 2e^{-} \longrightarrow Pb$$

The electrolyte used for this process can be either an acidic or alkaline solution.

The rate of this reaction increases with current density, thus a high current density of 2–5Adm^{-2} can be used for stripping the corrosion from the metal core. The process may take only a few hours. In this situation the reduced lead appears as a sludge at the bottom of the reaction vessel. A high current density increases the risk of obtaining a porous lead structure due to the evolution of hydrogen at the surface of the lead object, the cathode. When a lower current density is used, ie. 100$mAdm^{-2}$, the reaction occurs at a much slower rate, no hydrogen bubbles are visible and the corrosion which is reduced to lead, remains on the artefact. This is known as consolidative reduction and several days are often required for completion of the reaction.

The following electrolytes have been used for reduction of lead in the British Museum:

(i) Sulphuric acid, H_2SO_4 (10%)
(ii) Sodium hydroxide, NaOH (5%)
(iii) Barium hydroxide, $Ba(OH)_2$ (saturated solution)
(iv) Sodium carbonate, Na_2CO_3 (5%) *
(v) Alkaline Rochelle salt (sodium hydroxide, NaOH (5%) with sodium potassium tartrate, $NaK(C_4O_6H_5)_2$ (15%)*

* These electrolytes are presently used by the Department of Conservation of the British Museum.

The object undergoing treatment is made the cathode and a variety of metals such as stainless steel, iron and lead, are used for the anode. After reduction in solutions (ii)–(iv) the lead is rinsed in several washes of sulphuric acid to neutralise the alkaline electrolyte and form a thin protective layer of lead sulphate. To remove residues of sulphuric acid prolonged washing in distilled water is carried out [12].

The coins were all treated using electrolytic reduction in alkaline Rochelle salt, with an aluminium foil anode. After a few minutes in formic acid, the coins were cleaned in water in an ultrasonic bath.

2.4 Protective coatings

In most cases a coating was applied to the lead artefact after conservation. The coating can have two functions, as a consolidant for friable corrosion and lead after consolidative reduction and as a barrier to protect the lead substrate from corrosive gases.

Shellac appears to have been widely used prior to 1960 to consolidate corrosion and to coat the lead seals. Cosmolloid 80H, a microcrystalline wax, was used on the lead seals after consolidative reduction until the 1970's. The use of this wax with a melting point of 80 °C has been gradually discontinued and paraffin wax, with a lower melting point of 45 °C, used in its place.

More recently a range of non-wax materials has been used to coat lead after treatment. These include Frigilene and Ercalene, both nitrocellulose lacquers, and Paraloid B72 an ethyl methacrylate methyl acrylate copolymer applied as a 10% solution in acetone.

2.5 Results

The relationship between the conservation treatment on

Table 3 *Lead seals – Conservation and Condition*

Seal No.	Condition	Conservation treatments
XXXVIII 167A	Stable	1961: wrapped in bandage, CR, alkali, Neut, Oven 1970: CR, H_2SO_4 (10%), 6 weeks, IMS, GB, C80H.
XXXVIII 167B	Corroding	1961: Cons. P.wax 1970: CR, H_2SO_4 (10%), 6 days.
XXXVIII 36	Stable	1961: LR, graphite anode, NaOH, Neut, GB. P.Wax
XXXVIII 27	Good	1961: CR, NaOH, Neut, Oven, GB, P.Wax
XXXVIII 24	Good	1961: Amberlite IER. Dried GB. P.Wax
XXXVIII 199	Stable	1961: Cons. P.Wax (to preserve design held by shellac). 1970 :CR, NaOH (5%), (leaving shellac and wax in place). 1988: Corrosion removed, CR, H_2SO_4 (10%).
XXXVIII 29	Good	1961: CR, alkali, Neut, Dried, GB, P.Wax
XXXVIII 147	Stable	1961: CR, NaOH, Fe anode, Pb cathode 8 days. Neut, Oven. P.Wax
XILIV 7	Stable	1961: CR, H_2SO_4 (10%), Washed. Oven. P.Wax (Heavily waxed).
XXXVIII 198	Slight Corrosion	1961: CR, H_2SO_4 (10%), Pb anode and cathode. Washed. Dried (IMS). GB. C80H.
XXVIII 149	Stable	1961: Cons. P.Wax 1970: (Layer had peeled off.) CR, H_2SO_4 (10%). Washed. Dried (Acetone) GB. C80H.
XXXIX 55	Slight corrosion	1961: CR, NaOH. Neut. Oven. GB. P.Wax
XLIV 66	Stable	1961: CR, NaOH. Neut. Dried, GB. Central portion cons. C80H. Whole P.Wax. 1963: Reported to be unstable - Retreated? No record of re-treatment found.
XXXV 73	Actively corroding	1961: Shellac removed with acetone. IER GB. P.Wax 1969: Soaked in petroleum ether. CR, $Ba(OH)_2$ (sat^{d}). Steel anode. Dried (MeOH). C8OH

Key

CR	– Consolidative reduction	Neut	– Neutralized
LR	– Local reduction	GB	– Glass bristle brushed
IER	– Ion Exchange Resin	Oven	– Dried in an oven
Cons	– Consolidated	C80H	– Cosmolloid 80H, mikrocrystalline wax (mp. 80 °C).
IMS	– Washed in Industrial methylated spirits.	P.Wax	– Paraffin Wax (mp. 49 °C)

Table 4 *Lead Coins – Conservation and Condition*

Lead object	Condition (Oct '88)	Conservation Treatment(s)
COINS BURMESE 1924 12-10 14 1894 6-5 30 1924 6-5 30 1894 6-5 49	Stable	None
INDIAN 1970 5-12 55 1970 5-12 15 BMC 50 BMC 49	 Natural patina, slight local corrosion Natural patina, stable Natural patina, stable Natural patina, stable	None
INDIAN BMC 105 1905 10-7 88 1905 10-7 63 BMC 108	Local outbreaks of corrosion	All treated with: CR alkaline Rochelle salt Frigilene lacquer

an object and its present stability are presented in Tables 3 and 4.

2.5.1 Conservation treatmens

Ion-exchange resins used for removal of the corrosion appear to stabilise the lead. However, this can be destructive as surface detail in the corrosion may be lost.

Evidence from lead seals treated by consolidative reduction using sodium hydroxide (5%) suggests the treatment results in an unstable artefact. However, two seals treated by this method were in a good condition, but one was heavily waxed, and the other had a tin content of 4%, both of which appear to impart stability to lead.

As yet, the assessment of the long term effects of electrolytic reduction of lead using sulphuric acid (10% in water) are inconclusive. Consolidative reduction with sodium carbonate or alkaline Rochelle salt has produced stable lead in most, but not all, cases examined.

2.5.2 Protective coatings

Shellac used as a consolidant and coating does not penetrate into the corrosion layer and often causes the corrosion to laminate and become detached. Application of shellac has, however, proved useful in some cases since it has preserved surface detail in the corrosion layer. The lead seal which had been embedded in a block of epoxy resin was actively corroding and the resin had cracked and discoloured.

Results from the objects examined indicate that Cosmolloid 80H, the microcrystalline wax, is not protective. It may have a destabilising effect on lead, but this would be unexpected as microcrysralline wax is a stable material. It is reported that "no free acids are present and non can be formed by oxidation or hydrolysis" [13]. Contact between the wax and the lead surface may be insufficient to provide a protective coating against corrosive vapours in adverse storage conditions. Therefore, this result may be due to another factor, such as storage of the lead after conservation.

Paraffin wax appears to give good protection to a lead surface. The synthetic lacquers, Frigilene, Ercalene and Paraloid B72 also appear to be protective as no corrosion has been observed on lead to date [4].

It must be noted that where a coated lead seal has corroded it may not be as a result of a corrosive environment. For example, residues of the conservation treatments may have been left in the lead and caused corrosion.

3. Experimental

Depending on the nature of the corrosive environment, lead will corrode to form a number of compounds. In the British Museum, the lead corrosion product commonly found is hydrocerussite. However, at the Science Museum, London, lead formate has been identified as the major corrosion product [15]. Only two instances of lead formate has been found in the British Museum. One, on an Egyptian pectoral of a king, EA16529, is thought to have formed from contact with a formaldehyde producing material within the showcase in which the object had been housed for many years. The other was found on a pewter dish. (The occurrence of lead formate on these objects will be investigated further.)

The following experimental work was carried out on lead which was corroded under two different environments - with acetic acid and formic acid - to produce different corrosion products. Electrolytic reduction was undertaken on the corroded lead pieces to assess any variability in performance. Coating were than applied to the conserved lead which was transferred to a corrosive environment to evaluate the success of electrolytic reduction and protective coatings. The effects of different corrosion layers on the conservation treatment were thus assessed.

3.1 Sample preparation

Lead sheet was abraded mechanically and cut into strips (50 x 20 x 2mm). A hole was drilled close to one end to allow the pieces to be handled by Nylon line threaded through the hole. All the pieces were degreased in acetone, weighed, labelled, and suspended over a bath of acetic acid (50% in distilled water) and the whole system was enclosed in polythene. A second set of samples was prepared in a similar manner and suspended over formic acid (50% in distilled water). Both sets of samples were inspected daily for signs of corrosion. The lead over acetic acid started to corrode after three days and became covered with a solid, white crust within a week. Slight corrosion occurred on the samples over the formic acid within a week, but corrosion did not develop into a crust even when left for several weeks. Samples of corrosion were taken and analysed by XRD. Hydrocerussite had formed over acetic acid and lead formate over the formic acid.

3.2 Electrolytic reduction

The circuit was arranged as in Fig. 1 with a stainless steel anode and sodium carbonate, Na_2CO_3 (10% in distilled water), as the electrolyte. The current was switched on and the corroded samples, in groups of four connected in series were immersed in the electrolyte. The voltage was increased until a current was registered on the potentiometer and then decreased until hydrogen evolution was at a minimum. The current density varied between 2 and 4 Adm^{-2} depending on the exact size of the lead sample. Reduction was complete in 2-4 hours in most cases. The washing procedures described by Plenderleith and Werner [12] were followed. Samples of wash liquor were taken for pH measurement (see Table 5). The lead was dried through industrial methylated spirits and a glass bristle brush was used to remove powdery lead formed on the surface of the lead during reduction. The samples were reweighed to assess weight loss. Lead pieces originally corroded over acetic acid suffered an average weight loss of 9.3%. The lead pieces originally corroded over formic acid lost on average 0.5% of their weight during electrolytic reduction. The relative weight losses reflects the degree to which the lead corroded over the acids. The pH of the wash liquors fell until it reached the pH of the sulphuric acid then rose to the pH of the water within two more rinses in distilled water. This suggests that the residues of the treatment are washed out of the lead.

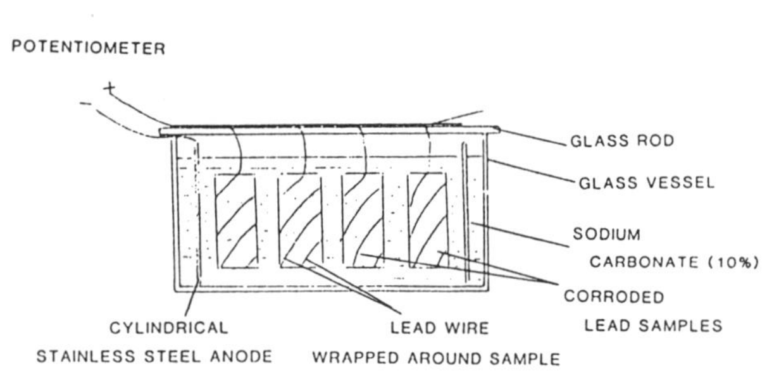

Fig. 1 *Cross sectional diagram of electrolytic reduction apparatus*

Table 5 *pH of wash liquors, after electrolytic reduction*

In H_2SO_4 (pH 5.3)		pH
Washes:	1	8.5
	2	7.5
	3	7.4
	4	6.9
	5	6.1
	6	5.4
In distilled water (pH 7.1)		
Washes:	1	6.8
	2	7.1
	3	7.3
	4	7.3
	5	7.3
	6	7.2

3.3 Coatings

3.3.1 Application of coatings

The following coatings were applied to the lead pieces after reduction:

Cosmolloid 80H (mp 80 °C), mikrocrystalline wax
Paraffin wax (mp 49 °C)
Frigilene - nitrocellulose lacquer
Poly(butyl methacrylate) (10% in acetone) *
Paraloid B72 (10% in acetone)
Polyurethane lacquer #

*Poly(butyl methacrylate) is used as a lacquer in the electronics industry as a coating for lead components [16].

#Polyurehane has been found to reduce emissions of formaldehyde from fibre boards used in the furniture industry [17]. This suggests it provides an impermeable barrier to such gases.

Each lacquer was applied, under vacuum, to four pieces of reduced lead – two originally corroded over acetic acid and two corroded over formic acid. The thickness of the coating was not controlled. When molten waxes were applied to the samples under vacuum, a thicker coating was produced. Four lead pieces were not treated by electrolytic reduction but cleaned using an air-abrasive and not coated. These pieces were used as controls.

3.3.2 Evaluation of coatings in an aggressive atmosphere

The protective effect of the coatings was evaluated in the presence of acetic acid vapours, at high and low relative humidity. Each lead piece was suspended in a boiling tube containing two drops of acetic acid (50% in distilled water) on a cotton wool plug in a small glass tube (see Fig 2). To create a high humidity a second glass tube containing 1 ml of distilled water was added. No water was added to the low humidity tubes. Humidity indicator cards were added to several of the tubes to measure the humidity inside them. Solid carbon dioxide was placed in all the tubes before they were stoppered and sealed with masking tape. To accelerate the rate of corrosion, the tubes were placed in an oven at 45 °C. (This temperature was chosen as paraffin wax melts at 49 °C.) The following controls were prepared in a similar manner:

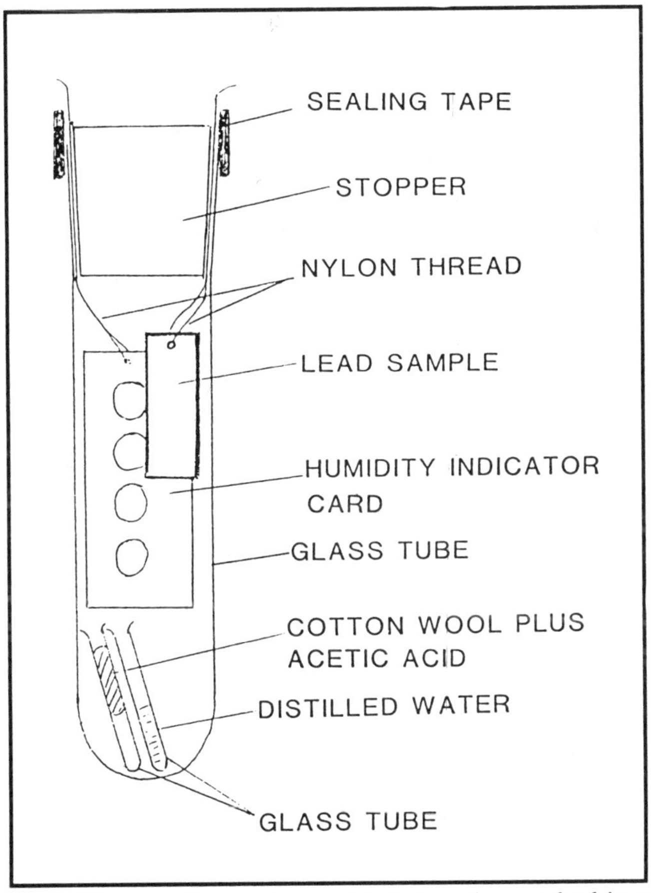

Fig. 2 *Diagram of apparatus used to test coatings on lead in a corrosive atmosphere*

Lead after reduction, no coating, acetica cid, carbon dioxide;

Lead not previously corroded, no acetic acid, water, carbon dioxide;

Lead not previously corroded, acetic acid water, carbon dioxide.

3.4 Results

The condition of the lead pieces after exposure to an acetic acid atmosphere for 72 hours is given in Tables 6, 7 and 8.

Table 6 *Condition of untreated lead control samples*

Lead Sample	Low RH	High RH
Untreated, Uncoated	Entire surface covered with corrosion	Entire surface covered with thick patches of corrosion
Untreated, Uncoated no acetic acid	A few, localized outbreaks	Entire surface covered in thin layer of corrosion

3.4.1 High humidity conditions

A film of corrosion had formed on all samples, both uncoated and coated, in high humidity conditions, within 24 hours.

3.4.2 Low humidity conditions

The results of coated samples originally corroded over acetic acid before electrolytic reduction are shown in Table 7. These samples exhibited the fastest corrosion in an corrosive atmosphere due to their pitted surface caused by extensive corrosion prior to treatment. Least corrosion was observed on the samples coated with Cosmolloid 80H and polyurethane lacquer. Corrosion occurred at faults in the Paraloid B72 and polyurethane coatings.

Table 8 shows the results using the lead corroded over formic acid prior to electrolytic reduction. These samples were not pitted after reduction and corroded at a slower rate than the other samples. The most protective coatings were Frigilene, polyurethane and Cosmolloid 80H.

3.4.3 Chemical analysis of corrosion products

Corrosion from the test pieces was identified by XRD as hydrocerussite, containing minor quantities of other components. The pink- brown corrosion formed on the

Table 7 *Condition of lead corroded over acetic acid prior to electrolytic reduction*

Coating	Low RH (35%-40%)	High RH (70%-80%)
Cosmolloid wax	Small spots of corrosion	Thin film of corrosion over entire surface
Paraffin wax	Thin film of corrosion some thicker patches	Entire surface rapidly covered with thick corrosion
Frigilene	Localized thick patches – slow to start	entire surface rapidly covered with thick corrosion
Poly(butyl mehacrylate)	Localized paches	Thick yellow corrosion
Polyurethane	Thin film, thick patches – slow to start	Thin film over one side – slow to start
Paraloid B72	Thin film with thicker patches – slow to start	Surface rapidly covered in thick orange patches
Air-abraded (not electrolytically reduced)	Entire surface patchy	Surface rapidly covered with thick corrosion
Uncoated after reduction	Spots of corrosion and thin patches	Severe corrosion - orange/pink in places

Table 8 *Condition of lead corroded over formic acid before electrolytic reduction*

Coating	Low RH	High RH
Cosmolloid was	Slight corrosion after 48 h	Entire surface covered in thin layer of corrosion
Paraffin wax	No change – darkened at edges	Tick layer of corrosion some orange patches
Frigilene	Thin layer of white corrosion Some brown spots	Thin layer, not severe
Poly(butyl methacrylate)	*Thick localized outbreaks	*Thick localized outbreaks
Polyurethane	No corrosion	Slight corossion
Paraloid B72	*Small thick localized outbreak	Entire surface covered with corrosion
Air-abraded (not electrolytically reduced)	Localized thick corrosion	Severe corrosion over entire surface
Uncoated	Extensive patches – some yellow	Entire surface – some pink and brown patches

*Corrosion developed where droplets of water condensed on the surface of the lead.

uncoated lead in high humidity was identified as hydrocerussite with litharge. Litharge was also detected in smaller concentrations in some of the yellow corrosion. Plumbonacrite, and lead carbonate hydroxide hydrate, $3PbCO_3 . 2Pb(OH)_2 . H_2O$ were also identified in several of the corrosion crusts.

4. Conclusions

The examination of previously conserved lead artefacts has confirmed that the presence of a low concentration of tin in the lead has a stabilising effect. The effects of the conservation treatment and the coating applied were less clear. The examination suggests that electrolytic reduction in sodium carbonate or alkaline Rochelle salt are the best treatments and that Cosmolloid 80H is a poor coating. The history of storage of artefacts which could affect their stability is not known precisely. However, storage conditions were probably not ideal since the corrosive effects of acidic cardboard and woods on lead were not appreciated in museums until recently.

The experimental work reported here shows that the type and extent of corrosion of the objects prior to conservation has a bearing on its subsequent stability. The exposure of the coated test pieces to an atmosphere containing acetic acid has shown that non of the protective coatings tested was truly protective. However, Cosmolloid 80H gave good protection to the samples. This conflicts with the observation on the lead seals that those coated with Cosmolloid 80H wax were in a poor condition. Cosmolloid 80H was tested for its corrosive effects on lead in 1974 [6] and no adverse effects were observed. The method of application of the wax may be an important factor. When the seals were conserved in the 1960's, baths of molten Cosmolloid 80H wax were kept in an oven at a temperature greater than 80°C (the melting point of Cosmolloid 80H wax). The seals were coated by immersion in the molten wax for 24 hours. Eventually the wax became discoloured and contaminated. This could have affected the long term stability of the wax, and its subsequent effects on the lead. When evaluating the wax as a coating in this and other studies it has not undergone the constant heating process, and may therefore not produce the same results.

In other areas of conservation, the effect of pollutant gases in the atmosphere is being countered by careful choice of materials to avoid those which give off gases which will corrode the artefacts being displayed. Removal of these gases from the environment within showcases and storage areas is now favoured instead of the use of a coatings the object. Initial experiments suggest that the use of materials which will absorb acetic acid gases in showcases and closed storage areas may prevent corrosion of lead. Future work on the problems of lead corrosion in the British Museum will concentrate on this approach.

Acknowledgements

I would like to acknowledge the assistance of Mrs S. Bradley for discussion and advice, Mr J. Cribb, Mr S. Dove and Mr D. French for bringing relevant information and objects in the collections of the British Museum and British Library to my attention, and to Mr W.A.Oddy for permission to publish.

Suppliers

Cosmolloid 80H microcrystalline wax

Picreator Enterprises
44, Park View Gardens
London NW4 2PN

Paraffin wax

BDH Chemicals
P.O. Box 15
Freshwater Road
Dagenham
Essex RM8 1RF

Frigilene nitrocellulose lacquer, Paraloid B72

Archival Aids
P.O.Box 5
Spondon
Derby DE2 7BR

Polyurethane lacquer

Jaxalac, s/mat lacquer
Sonneborn and Rieck Ltd
Jaxa Industrial Finishes
91-95 Peregrine Road
Hainault
Essex IG6 3XH

Poly(butyl methacrylate)

Aldrich Chemical Co Ltd
The Old Brickyard
New Road
Gillingham
Dorset SP8 4JL

References

1. **Turgoose S.** (1988) "The corrosion of lead and tin before and after conservation " UKIC occasional paper No. 3' Lead and Tin studies in Conservation and Technology 'Ed Miles G. Pollard S. pp15-26

2. **C. R. C. Press** (1976-1977) "The Handbook of Chemistry and Physics" 57th edition. Ed Weast R.C.

3. **Uhling** H. H. (1948) "Corrosion Handbook" Wiley and Son, London

4. **Dove S.** (1981) "Conservation of glass-inlaid bronzes and lead curses from Uley, Gloucestershire" The Conservator. No5, pp31-35

5. **Oddy W. A.** (1973) "An unsuspected danger in display" Museums Journal vol 73. Nòl, pp27-28

6. **Lane H.** (1987) "The conservation and storage of lead coins in the Department of Coins and Medals-British Museum" UKIC Jubilee conference proceedings "Recent advances in the conservation and analysis of artefacts" pp149-153

7. **Organ** R.M. (1963) "The consolidation of fragile metal objects" in Recent advances in conservation Ed Thomson G. London pp128-1347.

8. **Guida** G., **Marabelli** M., **Reindell** I. "Restoration and storage of lead artefacts" Restaurator. 4, pp227-337

9. **Lane** H. (1975) "The reduction of lead" IIC conference proceedings "Conservation of archaeology and-applied arts" pp215-217

10. **Werner** A. E. A.(1967) "Two problems in the conservation of antiquities- Corroded lead and brittle silver" in "The application of science in examination of works of art" pub Museum of Fine Arts, Boston. pp126-144

11. **Watson** J. (1983) "Conservation of lead and tin alloys using EDTA solutions" UKIC occasional paper No3 "Lead and tin studies in conservation and technology" Ed Miles G. Pollard S. pp44-46

12. **Plenderleith** H., **Werner** A. E. A. (1971) "The conservation of antiquities and works of art" 2nd edition. Oxford University Press.

13. **Werner** A. E. A. (1957) "Synthetic waxes" The Museums Journal. vol 57, pp3-5

14. **Heath** D., **Martin** G. (1987) Preprints of the IIC conference, Kyoto, "Conservation of Far Eastern Art" pp137-141

15. **Tranter** G. C. (1987) "An infra-red spectroscopic examination of corrosion samples from the Science Museum" Cookson Group Report, Perivale, Greenford.

16. Stirling Varnishes, Manchester, Priv. comm.

17. **Sparks** T. (1989) "The effect of surface coatings on formaldehyde emission" preprints SSCR & Museum Association conference "Environmental monitoring and control", pp78-87

Author's address:

British Museum
Department of Conservation
Great Russel Street
GB-London WC1B 3DG

PENNEC, Stéphane

La conservation des objects du Titanic

Introduction

C'est en 1987, grâce à des moyes techniques Français et à des capitaux Américains, que près de 2000 objets furent remontés du TITANIC.

Le bateau, coulé tragiquement en avril 1912, repose actuellement, brisé en deux parties, par 4000 mètres de fond au milieu de l'Atlantique.

Deux bateaux en surface et un sous-marin de titane, le Nautile, équipé de bras articulés, ont permis d'aller chercher délicatement sur le champs d'objets qui entourait l'épave, ces témoins d'un passé proche, d'un drame encore présent à nos esprits. Soulignons l'exploit technologique de cette opération "TITANIC 87" organisée par Taurus International.

Mais le sauvetage des objets ne s'arrète pas là. Une fois à l'air, et en l'absence de traitement, ils courent encore de graves dangers. Au contact de l'air, en effet, la dégradation, la décomposition des objets du Titanic étaient inéluctables. Les arracher à l'Océan allait rompre l'équilibre qui les protégeait. Il fallait prévenir cette destruction à plus ou moins court terme par des traitements appropriés.

Peu de temps avant le début de l'expédition, le Laboratoire Valectra, financé par Electricité de France, fût sollicité pour prendre en charge les objets qui allaient être remontés, et donnait les conseils pour les stockages temporaires sur le bateau.

C'est en septembre 1987 que les objets du Titanic arrivèrent au laboratoire de Saint-Denis. Immergés dans des bacs d'eau douce, callés par des mousses, ils avaient voyagé par bateau dans un container. Tous les objets furent sortis, inventoriés, étiquettés, puis restockés matériau par matériau, de façon plus rationnelle, permettant d'attendre en toute sécurité les traitements conservatoires qui allaient se faire sur près de deux ans.

Le laboratoire créé en 1983 afin de mettre au point et d'appliquer des techniques électrochimiques à la conservation des objets d'archéologie sous-marine, allait s'étoffer, pour l'occasion, de spécialistes de différents matériaux (papiers, textiles et bois, verres et céramiques...) en tout plus de 10 personnes qui ont collaboré et travaillé sur ces objets. Nous disposions de nombreux moyens d'analyses et d'observation (MEB, absorption atomique, chromatographie...), qui nous ont permis d'adapter à chaque cas le meilleur traitement, et de contrôler les résultats.

Les objets

Sur les 1892 objets remontés lors de l'opération Titanic 87, 41% étaient constitués de métaux, 43% de matières organiques, le reste se composant de matériaux siliceux (céramiques, verres, et pierres) mais aussi de plastique, caoutchouc et autre.

Il est difficile de les décrire dans leur ensemble, tant leur diversité est grande: du compas de navigation et du transmetteur d'ordre, aux effets personnels tel que brosse à cheveux, noeud-papillon, chaussettes, cartes de visite, montres, bijoux etc... trouvés dans des sacs de voyage, en passant par la vaisselle de bord (plats à oeufs, assiettes de faience, de porcelaine, carafes en cristal, bouteilles, argenterie), etc... Difficile également de les sérier par taille (de la lame de rasoir au hublot de 160 kg, par fonction ou par forme. Cet ensemble constitue une sorte d'échantillonnage représentatif de la diversité des objets se trouvant à bord du navire.

Les conditions d'enfouissement

Il était prévisible que les détériorations d'ordre physique (due à la pression) n'aient qu'une incidence mineure dès lors que les parois des objets ne subissent aucune pression différentielle. Tel était bien le cas pour la plupart des objets, à vrai dire fort peu déformés. Deux conséquences particulières de la pression exercée : les bouchons des bouteilles étaient systématiquement enfoncés, et les manches creux des couverts aplatis.

Les altérations chimiques, quant à elles, étaient bien réelles. Les premiers examens en laboratoire le confirmaient.

L'eau de mer constitue un milieu redoutable et complexe (1). Il comporte en solution des sels (chlorures, sulfates,...), des gaz et des substances organiques multiples. Dès qu'un objet est immergé, des processus de dégradation physique, chimique, électrochimique et biochimique s'amorcent rapidement : des phénomènes d'hydrolyse accompagnent l'action des sels et des microorganismes.

Ces processus se combinent, mais leur agressivité varie selon la profondeur et les conditions d'immersion. Par 4000 mètres de fond, dans une obscurité totale, par une température basse, en présence de peu d'oxygène les processus chimiques sont sensiblement ralentis.

Mais à de telles profondeur (2) la détérioration des objets se poursuit: les processus de dégradation restent actifs, mais de façon différente selon les matériaux.

Les traitements de conservation

La conservation des objets dans leur nouvel environnement impose, pour être menée à bien, de surmonter trois difficultés. Il faut éliminer ou neutraliser les facteurs de dégradation des matériaux constitutifs de l'objet, empêcher les altérations dues au changement de milieu, et enfin, consolider les matériaux fragiles de façon à préserver leur intégrité.

Selon les matériaux considérés, les traitements mis en place sont différents : stabilisation des métaux, dessalement des matériaux poreux, consolidation des matériaux organiques, séchage, et protection de surface.

La stabilisation des métaux

Lorsque les objets sont ramenés à l'air libre, la corrosion reprend, car l'équilibre atteint dans leur ancien environnement est rompu. Cette réaction se produit en raison de l'activité des sels instables en présence d'humidité et d'oxygène.

Le traitement de stabilisation réside précisement dans l'arrêt de ce processus de corrosion. Deux méthodes sont envisageables: par l'extraction des sels contenant les ions agressifs, ou en isolant ces derniers de l'air et de l'humidité par la création de barrières physiques ou chimiques. Pour certains métaux et alliages, bien connus en archéologie sous-marine, tels que les alliages cuivreux et ferreux, ou encore le plomb, divers procédés ont été mis au point. En revanche, les métaux récents comme l'aluminium et le nickel, leur stabilisation fait depuis peu l'objet de recherches.

Pour la déchloruration des alliages cuivreux, ce sont les méthodes chimiques qui sont les plus couramment pratiquées (3). Les objets sont immergés dans des produits susceptibles d'extraire les chlorures (eau distillée, sesquicarbonate de sodium).

Le principal incovénient de ces méthodes réside dans leur durée, qui peut atteindre plusieurs mois. Aussi nous nous sommes tournés vers des méthodes électrochimiques plus rapides. La plupart des objets provenant du Titanic ont été ainsi déchlorurés par électrolyse. Les principaux avantages de cette méthode sont le gain de temps et l'absence de modifications dans les objets et leur produit de corrosion. Ce procédé, dont les différents paramètres expérimentaux ont fait l'objet de recherches spécifiques, consiste à brancher l'objet métallique en cathode dans un montage électrolytique. On impose alors un courant suffisant pour extraire les chlorures, sans pour autant obtenir un dégagement d'hydrogène. Les chlorures ainsi dissouts se déplacent vers l'anode au sein de la solution.

Egalement sensible à la présence des chlorures, le fer nécessite aussi une déchloruration. Les méthodes utilisées dans les laboratoires sont chimiques (tels que les bains de sulfite de sodium alcalin Schweitzer, Rinuy 1981...) ou électrolytiques.

Le dessalement des matériaux poreux

Les matériaux poreux, qu'il s'agisse de céramiques, de verres ou de matériaux organiques, arrivent imbibés d'une eau chargée en sels. Ils présentent souvent des concrétions composés de sels ou d'oxydes métalliques (sulfures et oxydes de fer étant les plus courants) mais aussi des carbonates provenant d'organismes marins.

Pour assurer la conservation de ces matériaux il convient de procéder à une opération de dessalement, c'est à dire à une extraction des sels qui les menacent.

Pour réaliser ce dessalement, on procède généralement à des rinçages dans de l'eau distillée. Les sels se dissolvent mais jusqu'à un certain point seulement. On peut également utiliser des méthodes électrochimiques avec de bons résultats: l'électrophorèse est la plus adaptée. Cela consiste à placer l'objet dans un champ électrique, sous l'effet duquel les sels dissociés en ions, vont pouvoir se déplacer vers le pôle de signe opposé au leur.

Les matériaux organiques

Le corpus des matériaux organiques est sans doute l'ensemble le plus complexe par le fait que les objets se présentent souvent comme des assemblages très composites: cuir – tissu – carton – bois – métal – poils...

Les objets nous sont parvenus dans des états de conservation variables. Tant qu 'ils restent immergés, les dégradations que subissent les matériaux organiques modifient peu leur forme. Leurs structures sont affaiblies, mais aussi longtemps que l'eau se substitue aux constituants dissouts, ils paraissent en bon état. Leur manipulation s'avère des plus délicate.

Si leur forme reste inchangée, en revanche, l'aspect de ces objets se trouve modifié. Les oxydes et les sulfures métalliques provoquent un assombrissement général, surtout spectaculaire sur les matériaux clairs comme les papiers. Le nettoyage est donc une des étapes les plus importantes du traitement. Cette opération doit débarasser l'objet de ses sels en tenant compte de la sensibilité chimique et de la résistance des matériaux constitutifs.

Pour le nettoyage, l'électrophorèse a été retenu dans de nombreux cas. Ses conditions d'applications sont assez souples et bien souvent, elle est moins agressive qu'un nettoyage chimique.

C'est avec succès que cette méthode a été utilisée pour le papier des billets de banque, les objets en cuir et en bois. Il faut bien entendu, tenir compte de la résistance mécanique de l'objet considéré, et donc adapter la durée du traitement et le champ électrique.

Nous avons dans tous les cas travaillé avec un électrolyte neutre (KNO_3 à 0,25% dans de l'eau déminéralisée), un champ électrique faible (de 2 à 10 V), et des électrodes propes (titane ruthénié).

L'extraction des sels est généralement suivie par un dosage des ions en solution. Nous avons constaté que les ions hygroscopiques sont extraits les premiers, tandis que les sels métalliques ne sont extraits que plus tard.

Dans la plupart des cas, ce nettoyage s'est avéré suffisant pour restituer la lisibilité de l'objet (tout spécialement les billets de banques). Dans certains cas, il a été suivi de bains chimiques destinés à blanchir le papier (hypochlorites) ou à retirer les taches de rouille persistantes. Enfin, tous ces traitements se sont achevés par une désacidification, assurant au matériau une meilleure conservation.

Il faut souligner que le nettoyage par électrophorèse n'est pas applicable à tous les papiers. Le papier monnaie s'y prêtait particulièrement bien, car il est très stable chimiquement et très solide mécaniquement. On a noté lors d'un test effectué sur un fragment de journal la dissociation des fibres dans le champ électrique.

Nettoyage des taches d'oxydes de fer au pinceau électrolytique

Certaines céramiques, et ivoires synthétiques ont été débarassés de leurs taches d'oxydes de fer à l'aide du pinceau électrolytique.

Il s'agit d'une sorte de cellule électrolytique ponctuelle placée sur la tache. L'objet est immergé dans un électrolyte (KOH 3%). On peut également déposer une goutte d'électrolyte à la surface de l'objet, si celui-ci ne peut être immergé. La cathode est constituée de laine d'acier, et l'anode d'un fil de Nickel.

Conclusion

Cette aventure a permis à l'équipe du Groupe Valectra de trouver une occasion exceptionnelle de mesurer pleinement les possibilités des techniques électrochimiques, d'en tester l'efficacité, mais aussi d'en cerner les limites.

Ces longs mois de travail de laboratoire ont été riches d'observations, d'enseignements, de découvertes techniques, mais aussi d'émotions devant ces fragiles vestiges.

Cette moisson d'éléments compose désormais l'expérience unique d'une équipe qui veut aujourd'hui la faire connaître et la partager.

Bibliographie

1. **Legaly** Y.: Biochimie Marine, ed Masson, Paris, 1988.

2. **Florian** M. L. E.: "The underwater environment", in Pearson, Conservation of Marine archaeological objects. ed Butterworths, London, 1987.

3. **MacLeod** Ian:"Stabilization of corroded copper alloys: a study of corrosion and desalination mechanisms".in Preprints 8th Triennal Meeting ICOM, Sydney 1987.

L'adresse de l'auteur:

Groupe Valectra,
Electricité de France,
25 allée Privée
F-93200 Saint Denis

MILKOVA, Bogdanka – JOSSIFFOVA, Zvezdalina

Ein neues Verfahren zur Konservierungsbehandlung von musealen Eisengegenständen aus der Praxis des Nationalen Polytechnischen und Naturkundemuseums an der Bulgarischen Akademie der Wissenschaften in Sofia

Die Konservierung und Restaurierung von archeologischen, ethnographischen und anderen musealen Gegenständen aus Metall sind ein wesentlicher Bestandteil der Tätigkeit auf dem Gebiete der Forschung und Erhaltung von Kulturdenkmälern – Zeugen der historischen Entwicklung der Völker – für die nachfolgenden Generationen.

Das Problem der Konservierung und Restaurierung ist die Folge einer komplizierten Verflechtung zweiter Tätigkeitsbereiche, denen die Konservierungs- und Restaurierungsprozesse unterliegen: Das betrifft einerseits die Kenntnisse über den historischen und künstlerischen Gehalt der Gegenstände und zum anderen die Kenntnisse über die Struktur der Metalle, über den Wirkungsprinzip der Korrosionsprozesse und den Schutz der Oberflächen gegen Einwirkungen ihrer Umgebung.

In diesem Artikel werden die Probleme aus dem zweiten Tätigkeitsbereich behandelt, nämlich die vorhergehende Vorbereitung und Konservierung der Gegenstände aus Metall auf der Grundlage eines von den Autoren dieses Referats entwickelten Systems, das eine minimale Zerstörung des Metallkerns sichert und schnell zum erwünschten Endeffekt führt – gut konservierte museale Exponate. Ausgehend aus den Kenntnissen über die Technologie zur Erhaltung von antiken Eisengegenständen (für die Balkanländer fast bis zum Ende des XI. Jahrhunderts), wurden eine Reihe von Eisenfunden zur Bestimmung der Art und Weise der Korrosionsschäden, die eingetreten sind, untersucht. Normalerweise sind neben den Korrosionsprodukten auf der Oberfläche der Funde auch Beschläge von nach Zusammensetzung und Schlichtdicke unterschiedlichen Verunreinigungen biologischen und anderen Ursprungs (besonders kompliziert nach Zusammensetzung und Struktur bei den Meeres und ethnographischen Stücken) zu finden.

Besonders charakteristisch für die Eisenfunde ist die Tatsache, daß die Veränderungen nicht allen auf die geometrische Oberfläche des Fundes beschränkt bleiben, sondern auch auf das innere Metallvolumen übergreifen. Dies erfolgt durch Risse oder heterogene Schichten im Metall selbst, wobei letztere eine gesetzmäßige Erscheinung für die Eisenerzeugnisse sind, die auf dem Wege der direkten Reduktion bis zum Harteisen erhalten wurden.

Die Agressivität der Umgebung des Fundes, die Lagerungsdauer, das Vorhandensein von Salzen und Säuren – das sind die Faktoren, die zur partiellen oder völligen Umwandlung der Metallmasse in Korrosionsprodukten geführt haben, Korrosionsprodukte, die nur der Form nach an der Umrissen des gefundenen Gegenstands erinnern. Diese Besonderheit erfordert in vollem Maße die Anwendung der Kenntnisse auf dem Gebiete der Konservierung und Restaurierung.

Erwartungsgemäß sind die überwiegende Anzahl von Korrosionsprodukten folgende: FeO(OH) und Fe_3O_4. Sie stellen 97% der untersuchten Muster dar, unabhängig von dem Ort und der unmittelbaren Umgebung, in der sie gefunden wurden. Nur 3% der Muster sind mit einer Basiskorrosionsschicht bedeckt, die nur aus Fe_3O_4 besteht. In 17,5% der Muster wurden auch Spuren von Cl und FeOH entdeckt. Von besonderem Interesse ist die Verbindung zwischen der Struktur des Metalls und der Korrosionsprodukte. Sie dringen dicht in den Metall ein, an den Stellen mit Schlackeeinschlüssen. Charakteristisch ist, daß das Eisenoxid den Hauptanteil dieser Korrosionsprodukte in den Mikrorißbildungen bildet. Nur von außen existiert ein Gleichgewicht, ein stabiles Korrosionsprodukt.

Das war der Ausgangspunkt des von uns entwickelten Verfahrens zur Reinigung, Stabilisierung und Konservierung der Exponate. Zu diesem Zwecke wurden Röntgen – oder Defektoskopaufnahmen gemacht zur Bestimmung des Grades der Metallkernzerstörung und der Konservationsart.

Die Reinigung der Oberfläche von Rußspuren, Fetten, Korrosionsprodukten, sowie Biowirkungen (Meeresorganismen) erfolgt mittels einer Sonderlösung, zusammengesetzt aus organischen Säuren und Phosphorsäure, in Kombination mit benetzenden und globulierenden oberflächenaktiven Stoffen und Inhibitoren. In diesen verhältnismäßig "weichen" Bedingungen (pH 2,0-2,2) findet eine partielle Auflösung der Korrosionsprodukte statt, wobei sie vor allem von der Metallunterlage förmlich "abgeklebt" werden. Das Lösen aller Nichtmetallprodukte wird stark durch das Vorhandensein von oberflächenaktiven Stoffen erleichtert, wobei jedoch die Lösungsgeschwindigkeit für die unterschiedlichen Komponenten auch verschieden ist. Es ist zu vermerken, daß die Phase Fe_3O_4 in diesen Bedingungen fast unlösbar ist. Minimal ist auch die Lösungsgeschwindigkeit von FeO und der harten Komponenten auf dieser Basis. Die Inhibitionswirkung zum Metallkern ist – 98%. Sie erhöht sich beim Absenken der Temperatur. Aus diesem Grunde wird die Behandlungsdauer während des Prozesses selbst bestimmt, damit der Fund sein Aussehen als Exponat beibehält. Diese Behandlung wird beim Erreichen einer fest anliegenden Schicht von Fe_3O_4 oder von FeO im Rißinneren eingestellt. Die so gereinigten Oberflächen haben nur unbedeutende Korrosionsbeständigkeit, trotz der teilweisen Phosphatierung und Inhibierung. Aus diesem Grunde folgt die Etappe der Nachbehandlung in einer transformierenden Phosphatlösung. Dadurch sollen alle Nichtmetallprodukte als Reste der Korrosionsprodukte fest am Metallgrund anhaften und gleichzeitig in nicht hygroskopischen stabilen Kristallverbindungen umgewandelt werden. Die Behandlung erfolgt nach einer gründlichen Spülung des gereinigten Gegenstandes mit normalem oder destilliertem Wasser.

Die transformierende Lösung ist eine Variante von bekannten Zusammensetzungen und stellt eine wässrige Lösung von Orthophosphorsäure dar, die Phosphate einiger Arten von Metallionen enthält: ($Fe^{2+}Zn^{2+}Mn^{2+}Cr^{3+}$ und CO^{2+}).

Spuren von Nitrate und oberflächenaktiven flüchtigen Stoffen. Letztere wurden so gewählt, daß sie ein bestmögliches Eindringen der Lösung in den Mikrorissen der Oberfläche sichern, doch leicht durch ein spontanes Verdampfen entfernbar sind. Der mit der Lösung behandelte Gegenstand braucht nicht gewaschen zu werden. Besonders starken Einflußüben die Metall-Bivalentionen aus einigen Arten. Die Basis- und Normalphosphate, die sich an der Eisenoberfläche bilden, bestimmen die Stabilität der Korrosionsprodukte und die Passivierung der restlichen Metalloberfläche. Von den die Oberflächenschicht definierenden Verbindungen überwiegen der Basis-Eisenphosphat und der Doppel-Zink-Kobalt-Phosphat.

Der so behandelte Gegenstand wird bei normalen Temperaturen getrocknet und mit einer Oberflächenisolationsschicht versehen – ein schützender filmbildender Überzug.

Das Auftragen solch einer Deckschicht erfolgt durch Eintauchen, Abschmieren oder Bespritzen mit einer Lösung oder Schmelze des filmbildenden Stoffes.

Bekanntlich wurde in der Vergangenheit der natürliche Bienenwachs als stabiles Überzugsmaterial für Metalloberflächen verwendet. In den letzten Jahren setzte sich in der Praxis überwiegend die Nutzung von Polymeren durch, die in der Form von Lösung auf der geschützten Oberfläche aufgetragen werden. Außer der massenhaften Anwendung dieser Polymerearten mit unterschiedlichen Firmenbezeichnungen, sind die Autoren dieses Referats auf keine weiteren Untersuchungen gestoßen, die die Möglichkeit bieten, die Frist der zuverlässigen Aufbewahrung der so geschützten Oberflächen vorauszusehen. Folglich existiert die reale Gefahr einer erneuten Korrosion der gereinigten und konservierten Flächen nach längerer Zeit. Dies wird auf die immerhin vorhandene Porösität des Polymerfilms zurückzuführen sein, als auch auf seine Alterungsprozesse, die zum Filmriß oder -bruch führen.

Die Anwendung von natürlichem Wachs auch in geblichenem Zustand garantiert eine zuverlässigere Konservationswirkung, da sowohl die chemischen, als auch die strukturellen Besonderheiten des Wachses in Temperaturbereich von 0 bis 40 °C die Stabilität eines dünnen Films daraus voraussetzen. Die Richtigkeit dieser These wird auch aus Erfahrungen der Vergangenheit bestätigt.

Das Vorhandensein von Sauerstoffeinschlüssen auf der konservierenden Metalloberfläche sowie im geschmolzenen Wachs selbst kann auch eine sekundäre Korrosion verursachen, jedoch mit weniger wahrscheinlichen Wirkung. Die Frage über die Diffusion von Feuchtigkeit und Sauerstoff, was zu Korrosionsprozessen auf der gereinigten Oberfläche führen kann, bleibt offen.

Aus diesem Grunde wird dem Wachsfilm ein aktiver Inhibitor zugeführt, der im Film selbst lösbar ist. Zu diesem Zweck wurden viele Inhibitoren und Stoffe getestet. Die Vorbedingung für die Anwendbarkeit war die völlige Auflösung des Inhibitors im flüssigen und Hartwachs.

Die Schutzeigenschaften des Films wurden durch entsprechende Stahlplättchen untersucht, wobei auch die optimalen Bedingungen für das Auftragen des Films festgelegt werden konnten.

Als Ergebnis der durchgeführten Experimente wurde folgende Zusammensetzung des Filmüberzugs präzisiert: Basis – geblichener Bienenwachs, Inhibitorzusätze – von 1 bis 4% Oktadezylamin und von 0,01% bis 0.05% Karbamid. Die Temperatur der geschmolzenen Auftragsmasse darf 80 °C und die des zu deckenden Gegenstands 30 °C bis 60 °C nicht überschreiten, in Abhängigkeit der durchgeführten Konservierungs- und Restaurierungsarbeiten.

Die Überzüge aus geblichenem Wachs weisen eine Dicke auf, dic in nicht zu breiten Grenzen bei gleichen Oberflächen mit bis zu 30% bei minimaler Dicke von 15 mkm variiert. Um einen möglichst dünnen Wachsfilm zu bekommen, darf kein großer Temperaturunterschied zwischen des zu deckenden Gegenstandes und der Wachslösung zugelassen werden.

Die Prüfungen der Beständigkeit des Films aus inhibiertem Wachs wurden unter folgenden Bedingungen durchgeführt:

1. Feuchtigkeit – warm (98-100%) und veränderlicher Temperaturbereich – 20-45 °C laut Bulgarischer Staatsnorm (BDS 4972-71).
2. Atmosphäre mit 5% Chlorwasserstoff.
3. Atmosphäre mit 5% Schwefeldioxid.
4. Salznebel (laut BDS 4975-71).
5. Wasserlösung auf 20%.

Unter diesen Bedingungen wurden auch mit dieser Zusamensetzung überzogene Stahlplättchen (0,15% C, Abmessungen 100/ 50/ 0,3 mm) , im voraus von Oxiden und Fetten gereinigt, Prüfungen unterzogen.

Zur künstlichen Beeinflussung und Beschleunigung der wahrscheinlichen Alterung des Schutzfilms führte man eine Vorbehandlung der überzogenen Plättchen mittels plötzlichen Wärmerschlägen durch. Bei 50 °C vorbeheizt, kühlen sie schnell im Wasser bei 15 °C ab. Die künstliche Alterung wurde mittels Ultraviolettbestrahlung (max. bei 3 600 Nanometer) durchgeführt.

Unter denselben Bedingungen wurde dieser Typ von Plättchen mit Schutzfilmüberzug aus Paraloid B3 (B Toluol 0,1-0,5% Lösung) geprüft. Die Ergebnisse zeigen, daß die Bienenwachsdeckung sogar ohne Inhibitor um das Mehrfache beständiger ist als der Paraloidüberzug. Das Vorhandensein eines Inhibitors im Wachsfilm macht ihn außerordentlich beständig gegen allen aggressiven Einwirkungen. Das selbe tritt um so deutlicher zutage auch bei den unter Ultraviolettstrahlen gealterten Proben, die später nach Einwirkung von aggressiven Mitteln geprüft wurden. Dazu kommt noch, daß die Paraloidschichten bei Alterung zum Teil undurchsichtig werden, was dem musealen Exponat schadet, d.h. sein äußeres Aussehen beeinträchtigt.

Trotz daß die Zeit eines zuverlässigen Korrosionsschutzes mit Wachsschicht nicht genau vorbestimmt werden kann, geben die Prüfungen in einem stark aggressiven Medium und Atmosphäre genügend Anlaß zur Schlußfolgerung, daß in den weichen musealen Bedingungen diese Schicht auf sehr lange Zeit hindurch die Museumsexponate in ausgezeichnetem Zustand erhalten wird.

Die Adresse der Autoren:

National Polytechnical Museum
Samuil st. 48.
B-1000 Sofia

BAKAY-PERJÉS Judit

Technical observations on Merovingian bronze finds

There is an ever increasing number of interdisciplinary teams of museologists – scientists – restorers for processing outstanding objects or object groups of a few items [1].

For complexes of very many objects from earlier excavations, this efficient working method – viz. to make all the tests needed before beginning with conservation – cannot be realized in most museums. Actually, in the best case, tests can be made alongside with handling, so that test results are available for determing the most efficient methods of conservation and restoration.

It is considered of importance for the archaeological object restorer to be acquainted with the particular age at least to a degree, even if he cannot learn the details about the site or the circumstances of disclosure [2].

Among Frankish cemeteries [3] of the 6th and 7th centuries, those in North France (Corbie, Monceau le Neuf, Maroeuil, Picardie) have yielded thousands of objects, part of them belonging to the Museum für Vor- und Frühgeschichte in West-Berlin. Most of the finds have been acquired by purchase in Berlin, but in World War II, a major part was annihilated in a fire.

In view of this I had several motivations to be interested in handling these objects [4]. Restoration helped me to disclose some skills that seemed novel to my eyes which were accustomed to objects of other peoples of the migration period (e.g. Avars, Hungarians).

Among objects subject to conservation and restoration, buckles of two different types will be considered. Features due to manufacture and to contemporary repair in each group will be reported, of course in relation to the actual restorations skills.

Fig. 1 *Repairs after*
a) *E. Salin,*
b) c) *B. Kurt,*
d) *R. Pirling (See Ref. 6)*

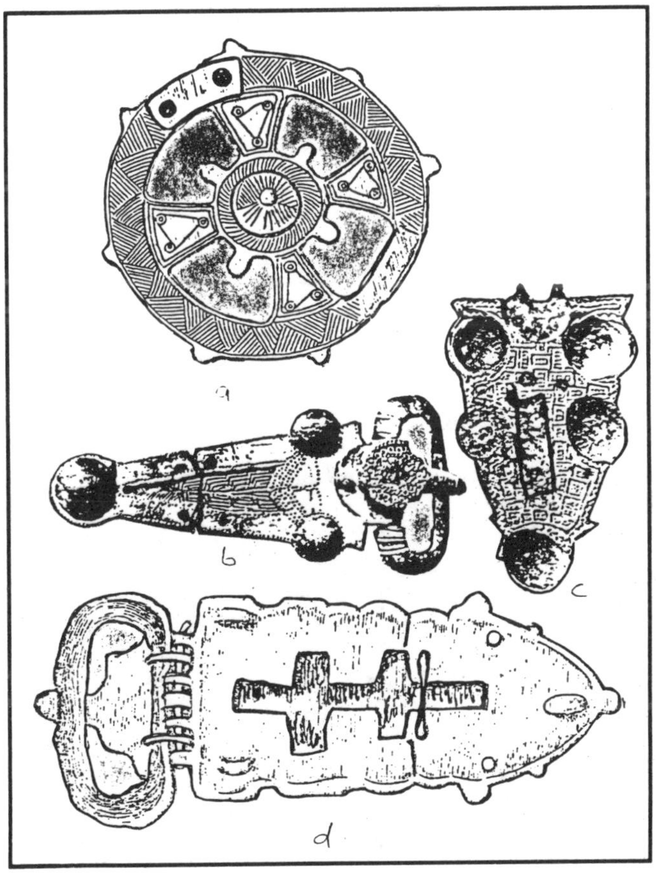

Cast bronze buckles with tinned and ornamented surfaces

Because of casting or other faults, there is a high number of belt buckles broken and repaired at that age [5], as exemplified by some exhibits [6] (Fig. 1).

For some reason, some of the buckles had broken in two. But traces of repair show the pieces not to be dis

Fig. 2 *Faces and backs of broken buckle bodies, indicating locations of fastening pins*

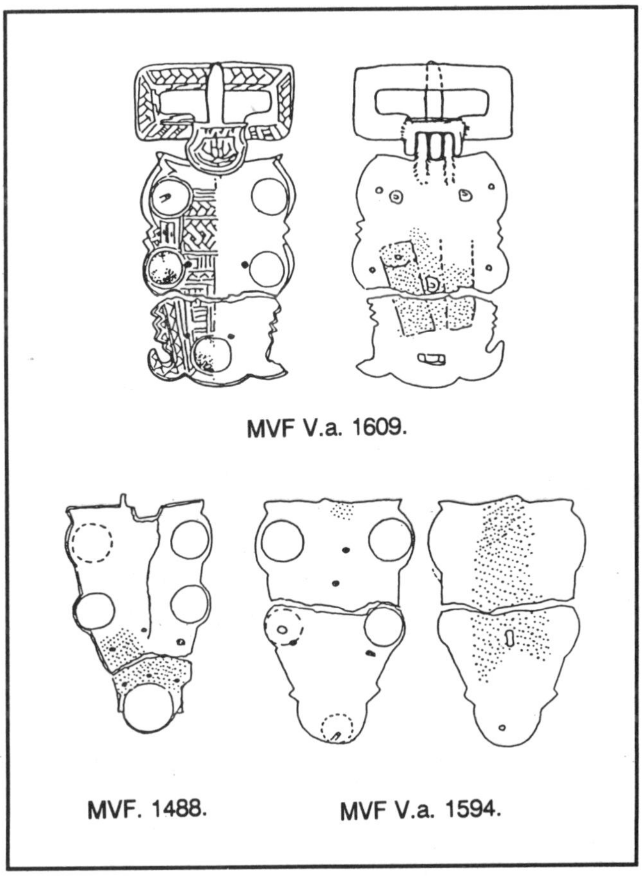

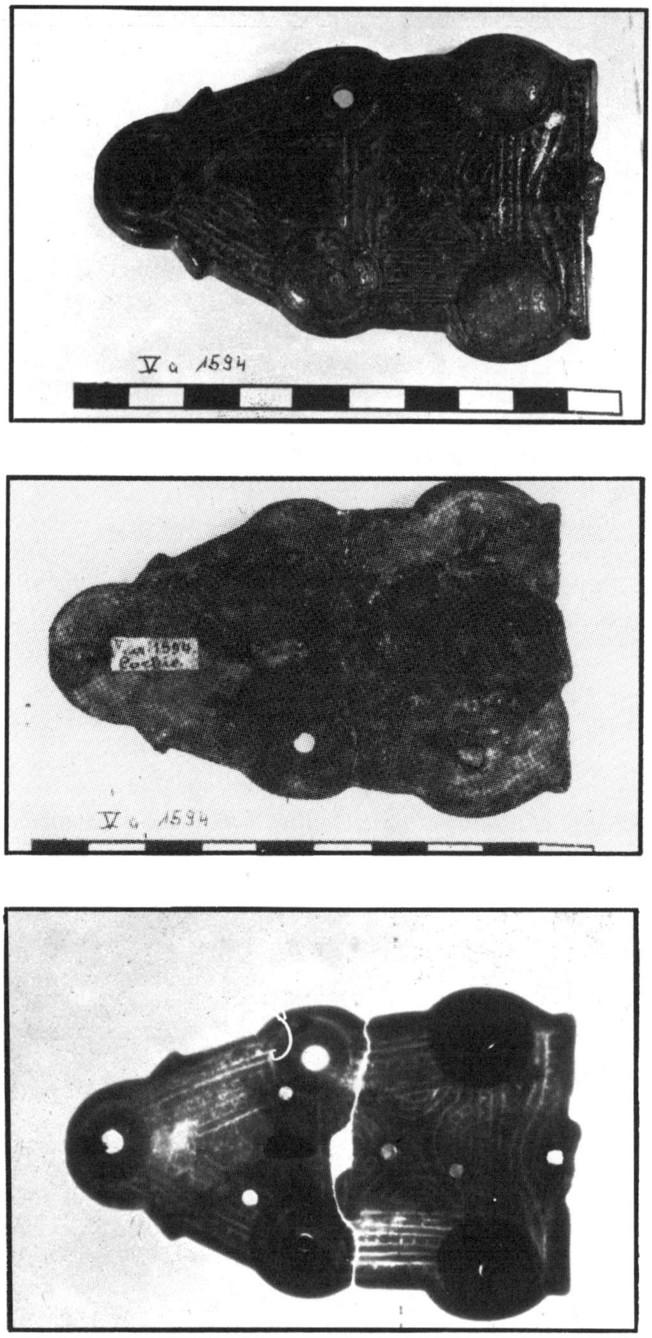

Fig. 3 *Face* **a)** *and back* **b)** *of buckle body after restoration;* **c)** *Face and back of buckle body before restoration (x-ray phyoto)*

carded but contineued to be used. Repairs -- of the presented objects - were invariacbly made on the reverse. They were clamped by one or two thin iron plates by means of bronze pins, depending on the length and size of the fracture (Fig. 2).

Cleaning of the buckles was begun by making X-radiograms to determine the positions of pins concealed by rust [7], (Fig. 3). Iron plate fully corroded on the bronze surface fully deformed and extended.

By mechanically cleaning the surface of one of the buckles, the original sizes of iron bands could be determined. This, however, was not feasible in every case, and the fastening band appears only as a mass of rust (Fig.4).

Restoration is invariably expected to restore – at least, nearly – the original condition of the object. Now,

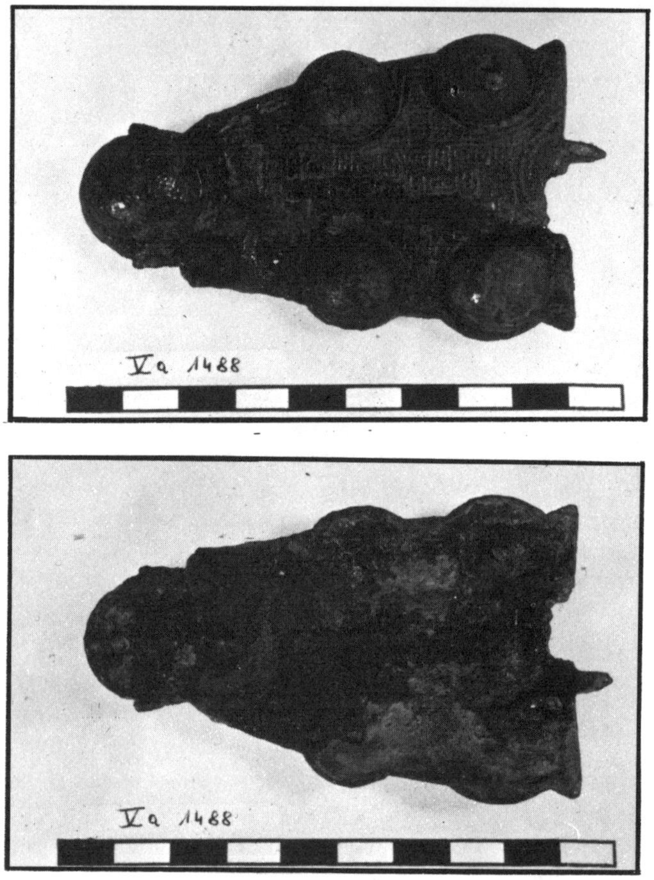

Fig. 4 *Face* **a)** *and back* **b)** *of buckle body after restoration*

what is to be considered as the original condition? Has the condition repaired by an iron band at that age, or that of the object before breaking to be referred to? Both. Thereby broken pieces have been stuck [9], of course, disclosing traces of the iron plate.

Cast, tinned, oval bronze belt buckles

These bronze buckles are special by reason of their iron cores. French researchers found earlier that an iron pin-usually pointed at both ends was put into the die, and bronze was cast around this core. After casting and stripping, primarily the upper surface of the cast was thoroughly cleaned and tinned. Thereafter the whole object was polished [10]. The middle part of the mentioned iron core or bent iron pin served at the same time for holding the dowel. Unfortunately, the available X-ray apparatus was not suitable for displaying their form within the buckle. But conclusions could be made mathematically, using physical relationships, on the size of the iron core [11]. Table 1 summarizes data on examined items.

We are aware of the approximate nature of these calculated values. Namely, it was not made of pure copper – whose specific weight has been reckoned with. But it is certain that the inner iron core is circular within the bronze body.

For most of the objects, iron being more active than copper was seen to stretch the surrounding copper due to its corrosion. It has surfaced, more or less deteriorating the beautiful, tinned surface (Figs.5 and 6).

Such objects cannot be cleaned other than by a dry, mechanical procedure [12]. Hair cracks more or less prevent perfect drying out – it is virtually impossible at present – therefore it is considered tobe the best for them tobe mechanically cleaned then conserved.

This cleaning method is also advisable in cases where the surface of the object is coated with a greyish-white corrosion layer [13] attributed not only to the corrosion of the tin coat, but to a corrosion surface of lead hydroxide or one transformed to basic lead carbonate. Being insoluble in water, it may form a protective coat on the object surface [14].

Ten among the buckles conserved up till now have been tested in the Rathgen Laboratory [15] (Analysis data see in Table 1). Buckles have been analysed by the atom absorption method [16]. All but one of the ten

Two buckles of the second group are similar to the former: lead-tin bronzes with medium lead and somewhat higher zinc and tin contents (buckles 3,5 in Fig.7). The only buckle in the third group is quite different: copper rich in zinc (buckle 8 in Fig. 7). Bismuth, cobalt, gold and cadmium contents in the objects are at the threshold values of the applied test method (Bi: 0.025, Co: 0.005, Au: 0,01, Cd: 0,001).

In general, brazen garment accessories and jewels regularly contain 5 to 10% of tin. These objects, however, are as rich in tin as 15 to 20%, otherwise exhibited only by antique mirrors and early medieval bells. Thus, a high tin content in belt buckles in a typical feature of this group of objects.

Lead contents are within the range typical of lead bronzes – except for sample 10 with 5,8% to 11,4% of lead – showing the uniformity of the castings. Zinc contents vary in wider limits, from bronze buckles poor in zinc to those nearly zinc-free. Also iron contents are readily distinguishable. In an area, they range from 0,012 to 1%. In this central area it is of uniform distribution. Also this fact hints to the use of old metals.

There is a really uniform nickel content of 0.03 to 0.07%.

Silver occurs in a remarkably high concentration sometimes exceeding 0.2%. One may wonder where this high value comes from? Maybe there is an ore rich in silver used purposefully to have buckles looking like silver. As a conclusion, nine of the ten buckles are of bronze of a somewhat uniform composition, with a very

Fig. 5 *Cracks on tinned surface of buckle frame due to corrosion of iron core*

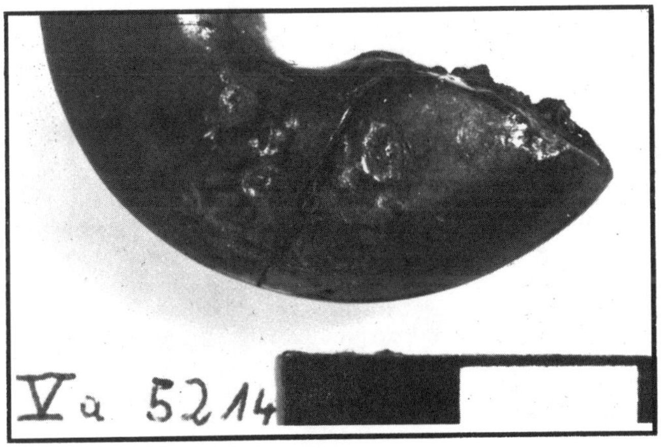

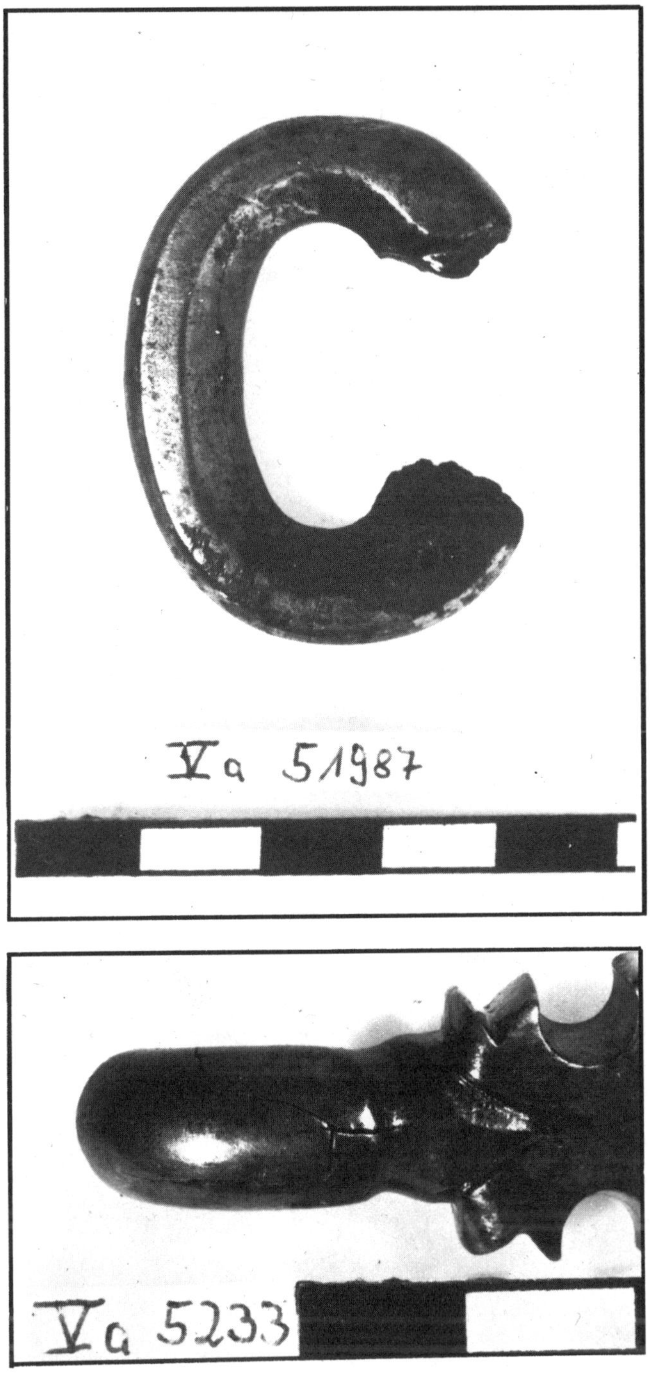

Fig 6 a) *Rust-coated area of tinned buckle frame* **b)** *Cracks due to iron corrosion in buckle pin*

The somewhat higher proportion of tin and iron contents hints to the re-use of old metal. While nearly equal concentrations of trace elements: nickel, antimony, arsenic, bismuth and cobalt, delimit the mining region whence the copper ore had come from for the objects. [17].

Notes and References

1. H.**Born**-K.**Ruthenberg**: Archäologe, Naturwissenschaftler, Restaurator = eine Arbeitsgemeinschaft? in: Archäologische Bronzen, Antike Kunst, Moderne Technik. Berlin, 1985, 10-14

Table 1. Data for buckles tested in the Rathgen Laboratory (1-10) and from computations (11-18)

Nr.	Object	Inventory	Analyses									Dimension* (mm)					Weight*
		Nr (MVF)	Cu	Sn	Pb	Zn	Fe	Ni	Ag	Sb	As	1.	2.	3.	4.	5.	(g)
1	Buckle	Va5213	73.70	17.56	7.85	0.04	0.46	0.04	0.28	0.07	0.05	44;	14,5;	10,5;	4;	34;	53,50
2	Buckle	Va5214	72.78	19.23	6.66	0.12	1.00	0.03	0.12	0.06	0.05	48;	23;	13;	7;	34;	71,00
3	Buckle	Va5205	71.63	16.50	10.06	1.40	0.08	0.03	0.12	0.07	0.10	38;	15;	8,5;	6;	27;	34,18
4	Buckle	Va5242	73.05	18.03	8.50	0.03	0.11	0.03	0.11	0.07	0.08	38;	15;	8,2;	4;	25;	26,96
5	Buckle	Va5247	73.52	16.59	5.77	2.87	0.94	0.04	0.22	0..06	0.05	39;	16;	9,7;	6	31;	40,28
6	Buckle	Va5233	77.25	16.05	6.00	0.24	0.09	0.04	0.28	0.06	0.05	36;	16;	8,5;	5;	29;	31,00
7	Buckle	Va5234	67.20	20.14	11.41	0.44	0.08	0.07	0.44	0.08	0.13	36;	16;	8;	5;	27;	25,53
8	Buckle	Va5229	72.55	0.20	1.15	25.5	0.16	0.07	0.06	0.05	0.41	38;	17;	8,5;	5,5;	29;	33,00
9	Buckle	Va5220	78.10	11.69	9.61	0.32	0.02	0.04	0.17	0.05	0.05	42;	16;	9,5;	7;	36;	43,05
10	Buckle	I.4.143.	74..92	21.14	3.22	0.29	0.11	0.05	0.16	0.02	0.09	29;	15;	9;	5;	29;	30,25
11	Buckle	Va5186a										35;	15;	8,5;	5;	28;	33,28
12	Buckle	Va5186b.										40;	18;	7,5;	5;	27;	23,00
13	Buckle	Va5187a										39;	17;	7,5;	5;	29;	29,22
14	Buckle	Va5187b										36;	16;	8;	5;	28,5;	28,90
15	Buckle	Va5187c										34;	13;	7;	4;	25;	23,42
16	Buckle	Va5198a										43;	21;	9;	6;	32;	35,65
17	Buckle	Va5198b										37;	18;	8,7;	6;	30,5;	35,53
18	Buckle	Va5198c										38;	15;	8;	5;	30;	44,48

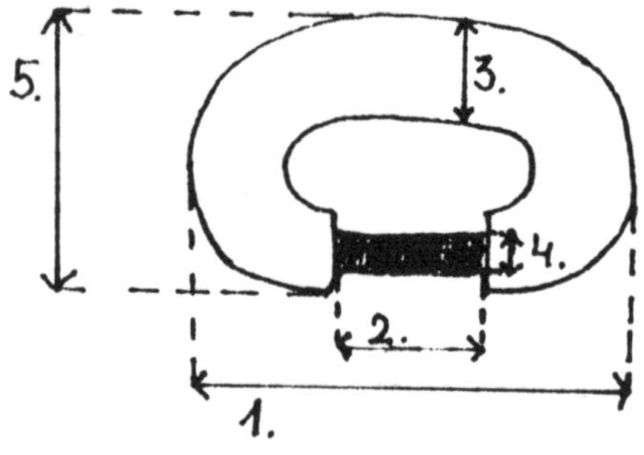

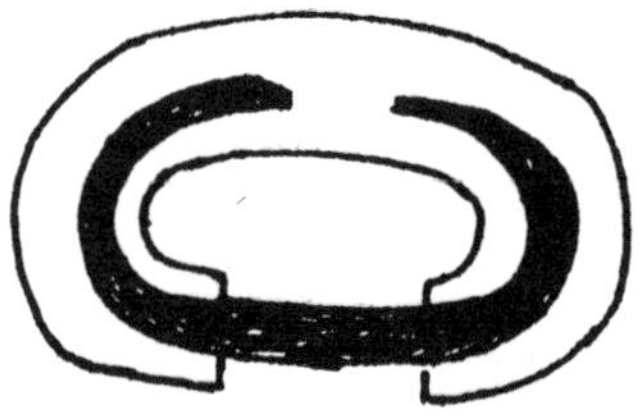

D - G = C

***Dimensions:**

1) Buckle frame lenght, mm
2) Pin holder lenght
3) Mean diameter of buckle frame, mm
4) Mean diameter of pin holder, mm
5) Overall width of buckle frame, mm

**Weight:

G: Buckle frame weight, g
D: Buckle frame weight converted to copper, g
C: Calculated weight of iron core in the buckle frame
$G_{weighed}$ = known
$D_{Cu\ weight} = V_{Cu} = V_{Cux} \cdot 8.9$ g/Cu.cm
$[V_{Cu} \cdot 8.9\ g.cm^3]$ - G = c, iron core weight

2. **B.Perjés** J.: A restaurált lelet, mint történeti forrás. (Restored Finds as Historical Sources) Múzeumi műtárgyvédelem 11, 1983, Budapest, 75

3. The northern and eastern provinces of the Roman Empire were endangered by German tribes from the 2nd century. From the east, eastern German tribes fleeing Huns or, in some cases, allied to the Huns, were in a state of flux at the period of the great migration. The north Germans were always enemies. While strongly fighting, the Romans, they nonetheless assimilated themselves in several domains. In the 5th century, power was seized by the Ostrogoths in Italy, and by the Visigoths in Iberia and South France, although they were not of an overwhelming number: on the contrary, they were decidedly in the minority.
One of the most powerful German tribes was that of the Franks living in two ethnic groups: Salic Franks in the actual Benelux region, and Ripuarian Franks in the regions of Cologne and Strasbourg. In the 5th to 6th centuries, the Franks extended their power from the Atlantic Ocean to Pannonia.

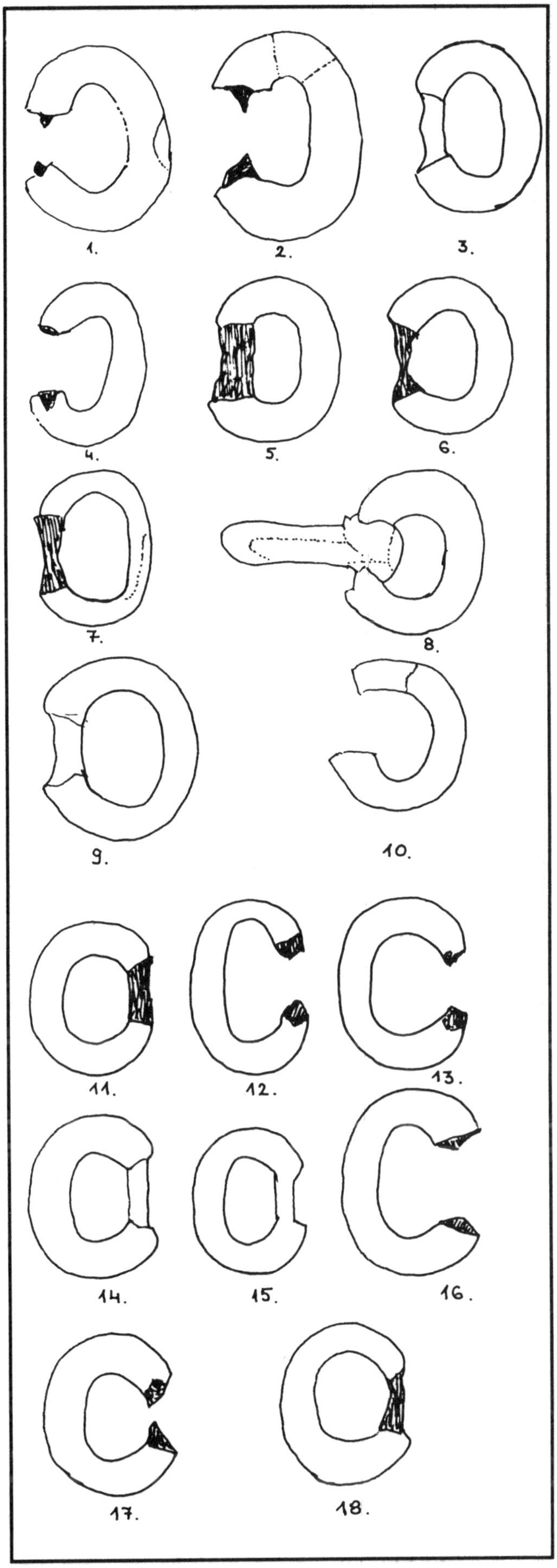

Fig. 7 *Structure sketches of tested buckles (see Table 1. rows 1 to 18)*

Merovingian kings laid bases of the later European history by creating and extending the Frankish Empire. Also the Christian culture survived savage, ferocious battles; some areas even continued to develop, mainly economically. Though, the Franks were not able to penetrate into Italy, even in the 6th century the Frankish Empire decreased; in the 7th century the Merovingian kings strengthened their power which was menaced by Arab invaders after 634.
Merovingian Frankish cemeteries clearly indicate various effects and impacts. A new Carolingian period began in West Europe in the 8th century, with foundations reaching back, however, to the Merovingian civilization.

4. From October 1988 to April 1989, I have been concerned with finds from these cemeteries as a restorer fellow in the restorers' workshop of the Museum für Vor- und Frühgeschichte. I am greatly indebted to Mr. Hermann Born and his co-workers for their help in my work.

5. Further tests are needed to decide whether breaks arose in manufacture or after a time of wearing – referred to as contemporary repairs.

6. É.**Salin**: La civilisation Mérovingienne, III. Paris, 1957, 108
B.**Kurt**: Die Frankischen Altertümer des Trierer Landes. Berlin, 1958,2. Teil, 38. Tafel
R. **Pirling**: Das Römisch-Frankische Gräberfeld von Krefeld-Gellep. 1.Teil, Berlin, 1966, 5. Tafel

7. H.**Born**: Archäologische Bronzen im Röntgenbild, in: Archäologische Bronzen, Antike Kunst, Moderne Technik, Berlin, 1985, 113

8. H.**Born**: Griechische un römische Statuetten und Großbronzen. Kunsthistorisches Museum, Wien, 1986, 176

9. Sticking and completing were made with Araldite two-component epoxy resin AY103+ HY956

10. É.**Salin**: La civilisation Mérovingienne, III.Paris, 1957, 116-150

11. ρ copper= 8.96 g/cm^3; iron= 7.86 g/cm^3
Mass = m = g/cm^3 m=ρ . V; Volume = V = cm^3/g g = V/m
Cylinder of revolution (curved surface solids) V = $\Pi \cdot r^2 \cdot m$

12. H.**Born**: Ban Chiang Bronzes: Manufacturing Techniques and Restoration, IIC, Kyoto, 1988, 131-132

13. Corrosion samples have been tested by chemists at the KMI Section of Restorers; microscopy sections by M. K. Bendefy and K.E. Török. Summarizing evaluation is in progress.

14. **Jánoska** P.: Tihany – Sajkod-i rómaikori orrszorítós zabla és lószerszámok restaurálása (Restoration of Roman-Age Barnacle Bridle and Saddlery from Tihany-Sajkod) 1978, Szakdolgozat, KMI Adattára, 5915-80.

15. J. **Riederer**: Denkschrift zur Eröffnung des Rathgen-Forschungslabors der Staatlichen Museen Preußischer Kulturbesitz, Berliner Beiträge zur Archäometrie, Berlin 1/1976, 100

16. K.**Ruthenberg**: Entwicklung der Bronzeanalyse von den Anfangen bis zur Gegenrwart in: Archäologische Bronzen, Antike Kunst, Moderne Technik, Berlin, 1985, 195

17. J. **Riederer**: Metallanalysen an merowingischen Gürtelschlißen, Rathgen-Forschungslabor, Berlin, 1989./7.1-3

Author's address:

National Centre of Museums
Könyves Kálmán krt. 40.
H-1087 Budapest

NISSEL, Dieter

Zur Verantwortung des Restaurators für die Sicherung quellenkundlicher Details – dargestellt am Beispiel des mittelalterlichen Geschützes "Faule Magd"

Einige der hier anwesenden Kolleginnen und Kollegen werden sich erinnern, daß bereits vor dem V. Internationalen Seminar für Restauratoren 1985 in Veszprém zum Stand und den Ergebnissen der Konservierung von Holzbauteilen an der Lafette dieses Geschützes berichtet wurde.

Meine heutigen Ausführungen sollen am konkreten Beispiel die Verantwortung des Restaurators unterstreichen, die er während der Konservierung für die Sicherung quellenkundlicher Details des jeweiligen musealen Objektes trägt.

Die dem Restaurator zukommende Verantwortung beim Erkennen und Sichern der unterschiedlichsten historischen Details erhöht sich dann um ein Vielfaches, wenn es sich um gegenständliche Quellen aus solchen Zeiten handelt, wo schriftliche Quellen fehlen.

Das Objekt, Frau Jedrzejewska bezeichnete es als Dokument, ist dann die einzige Quelle historischer Erkenntnis. Aus diesem Grund kommt bei der Ausbildung der Restauratoren nicht nur den Naturwissenschaften große Bedeutung zu, sondern innerhalb der Museologie auch den Hilfswissenschaften (Heraldik, Falleristik, Kostümkunde, Paläographie etc.) zur Unterstützung geschichtswissenschaftlicher Erkenntnis.

Die praktische Tätigkeit der Restauratoren belegt, daß hohes restauratorisches Niveau nur in interdisziplinärer Zusammenarbeit der Restauratoren mit den Historikern und Museologen des Museums und den Fachexperten der Sektionen von Natur- und Gesellschaftswissenschaften möglich ist.

Warum diese Betonung niveauvollen Restaurierens?

Weil die restauratorische Praxis gleichfalls belegt, daß sehr oft nur durch den Restaurator Kennzeichnungen, wie Beschlagmarken, Künstlersignaturen, Haus- und Hofmarken und heraldische Farbe sichtbar gemacht und dadurch der Sachzeuge besser erschlossen werden kann.

Bevor ich diese Behauptung belege, noch ein Wort zu dem als Beispiel gewählten Gegenstand.

Das zwischen 1979 und Herbst 1988 im Alten Zeughaus auf der Festung Königstein (Abb.1) ausgestellte mittelalterliche Geschütz "Faule Magd" verkörpert nach einem Gutachten von Dr. H. Müller, Museum für Deutsche Geschichte Berlin, einen kulturhistorischen Wert von 2 bis 3 Millionen Mark.

Nicht, weil das schmiedeeiserne Rohr (Abb. 2) aus

Abb. 1. *Stabringgeschütz "Faule Magd" im Alten Zeughaus auf der Festung Königstein bei Pirna mit teilweiser Farbrekonstruktion der ursprünglichen Bemalung auf der linken Lafettenwand. Foto: G. Thiede*

Abb. 2. *Das aus Staben und Ringen, in Kammer und Flug unterteilte 1383,3 kg schwere schmiedeeiserne Geschützrohr. Foto: G. Thiede*

Abb. 3. *Geschützlafette "Faule Magd"*
Auf die in der II. Hälfte 15. Jahrhunderts gebrauchliche Radlafette ist zur Aufnahme des schildzapfenlosen Rohres die Kastenlade mit einer die Schildzapfen ersetzenden Achse aufgelegt. Damit wird die Richtbarkeit des Rohres (mit Kastenlade) in der Radlafette gewährleistet. Foto: G. Thiede

der ersten Hälfte des 15. Jahrhunderts stammt und den damaligen Wert mehrerer hundert Kühe entsprach, sondern vor allem deshalb, weil es eins der etwa 30 in Europa erhalten gebliebenen Geschützrohre ist, deren Kaliber über 25 cm liegt und das mit großer Wahrscheinlichkeit als einziges historisch lafettiert ist (Abb. 3).

Das ursprünglich als Legstück verwendete Geschützrohr erhielt aus einem derzeit noch nicht vollständig geklärtem Grund um 1600 eine Lafettierung. Zu einer Zeit, wo der Stand der Gießereitechnik soweit war, daß 1581 in den sächsischen Zeughäusern 205 (Zwickau 52, Wittenberg 75, Leipzig 69, Pirna 9) schwere gegossene Geschützrohre existierten [1]. Zu einer Zeit also, wo die Radlafette entwickelt und die Geschützrohre zum Zwecke des Richtens mit Balanceschildzapfen ausgerüstet waren.

Zu dieser Zeit wird ein etwa 150 Jahre altes Geschützrohr mit einer Sonderkonstruktion lafettiert, an der die entwicklungsgeschichtlich in der ersten Hälfte des 15. Jahrhunderts entstandene Kasten- oder Blocklade zur Aufnahme des Rohres mit einer schildzapfenartigen ausgebildeten Achse versehen ist. Die Achse gewährleistet eine Richtbarkeit der Kastenlade (mit inliegendem Rohr) in der Radlafette. Diese kurze, nicht vollständige, aber die Besonderheiten der Lafettenkonstruktion ansprechenden Beschreibung macht deutlich, daß die Lafette der "Faulen Magd" etwa 150 Jahre Entwicklung der Artillerie und Technik verkörpert. Und darin liegt der Wert dieses musealen Objektes mit all seinen bisher unbeantworteten Fragen zur Genesis des Geschützes.

In den Wänden der Radlafette befinden sich handwerklich orientierte Schnitzerein, in den Wänden der Kastenlade Devisen (Abb. 4 und 5).

Die vor allem 1985 und 1986 durchgeführten wissenschaftlichen Untersuchungen resultieren auf dem in Frage stellen der 1907 durch Baarmann [3] erfolgten wissenschaftlichen Beschreibung des Objektes. Baarmann erwähnt darin außer den Schnitzereien und Devisen die noch heute gut sichtbaren in die linke Lafettenwand geschnitzten Majuskeln (Großbuchstaben) GLZ (Abb. 6) aber auch eine bunte Bemalung und er verweist auf eine bildliche Darstellung.

Dieses Gemälde befand sich in einem der sächsischen Schlösser, die Kopie davon im Sächsischen Armeemuseum. Nach Auskunft der Staatlichen Kunstsammlungen Dresden ist keines von beiden mehr nachweisbar [4]. Von der Polychromie war bei Übernahme der "Faulen Magd" durch das Armeemuseum der DDR 1977 nichts mehr erkennbar. Weder die geschnitzten Ornamente und Devisen, die Majuskeln GLZ noch die verlorengegangene

Abb.4/a."*Getrost mich..." Inschrift der rechten Ladenwand. Foto: G. Thiede* **b.** *"Durch Gottes Hulff..." Inschrift der linken Ladenwand. Foto: D. Nissel*

Abb. 5. *Hauptornament sowohl der Rechten als auch der linken Ladenwand. Foto: D. Nissel*

Polychromie ließen auf die Herkunft des Geschützes schließen [5].

Der Erhaltungsgrad der aus Eichenholz bestehenden Lafette erforderte eine Konservierung durch Holzverfestigung [6]

Dazu mußte aus technologischen und restauratorischen Gründen eine völlige Demontage des Geschützes durchgeführt werden, die für die Erforschung des Geschützes von Bedeutung werden sollte. Zuvor waren durch Zustandsanalysen insgesamt 42 Farbproben von der linken Lafetten- und linken Ladenwand gesichert worden.

Das Ergebnis der chemischen und chemisch-physikalischen Untersuchungen bestand in der Aussage, daß die Schnitzgründe unsprünglich eine gelbe (Auripigment), die danebenliegenden erhabenen Flächen örtlich abgegrenzt eine blaue (Smalte) Bemalung aufwiesen [7].

Dieser Sachverhalt konnte hinweisen auf:

1. die Städtefarben von Leipzig, Chemnitz (seit 1953 Karl-Marx-Stadt) und Bautzen
2. die Farben des Markgrafenthums Oberlausitz und
3. die wettinischen Hausfarben.

Auch die Farbanalysen von Proben der Lafettenbeschläge ordnen sich mit ihrem Gelbpigment in diesen Sachverhalt ein.

Innerhalb des Konservierungs- und Restaurierungsprozesses in den 80er Jahren konnten durch die Holz- und Metallrestauratoren weitere am Sachzeugen vorhandene, weder 1907 noch später ermittelte Kennzeichnungen festgestellt werden:

1. Eine zwischen den Voluten des Muschelornamentes in der linken Lafettenwand eingeschnitzte Meistermarke SB (Abb. 7).
2. Auf der Stirnfläche eines der vier, die Wände der Kastenlade zusammenhaltenden eisernen Bolzen eine eingeschlagene Krebsmarke (Abb. 8).
3. Die vierfach auf der schildzapfenartig ausgebildeten Achse eingeschlagene runde Kennzeichnung mit dem Großbuchstaben B (Abb. 9) dessen nachfolgendes Zeichen sowohl der hochgestellte Kleinbuchstabe c, aber möglicherweise auch einen durchkreuzten Kreis, das Symbol eines Zahnrades, sein könnte.
4. Eine Wiederholung der Krebsmarke beidseitig auf dem bei Baarmann als Durchsteckeisen bezeichneten, jedoch die Funktion des sonst üblichen hölzernen Ruhriegels zur Auflage des Rohres ausübenden, hier aber als Auflage der Kastenlade dienenden Eisens.
5. Die in der Lafettenwand und Stirnflächen der Achsstümpfe eingeschlagenen Großbuchstaben M, R und L.

Abb. 6. *Die in der linken Lafettenwand (Schwanzteil der Lafette) befindlichen Großbuchstaben "GLZ", gedeutet als "Großes Landeszeughaus". Foto: D. Nissel*

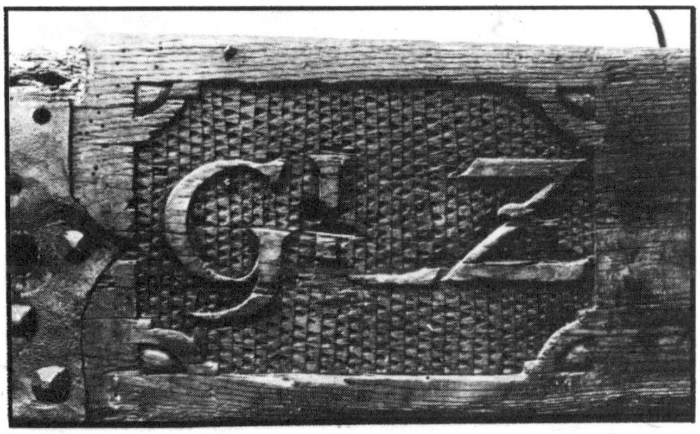

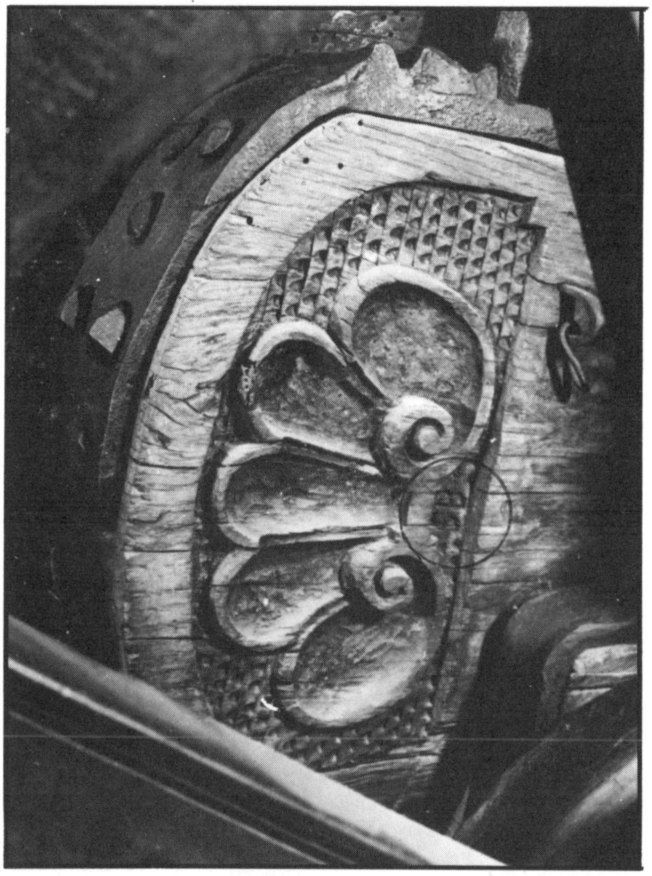

Abb. 7. *Zwischen den Voluten der Muschel in der linken Lafettenwand befindliche Bildhauerinitialien "SB". Foto: D. Nissel*

Weder die Großbuchstaben SB (vermutlich Holz- oder Steinbildhauer) noch die in Eisen geschlagene Krebsmarke konnte durch Quellen- und Literaturstudium belegbar einer Person oder einem Territorium zugeordnet werden. Die ins Holz eingeschlagenen Großbuchstaben M, R und L sind wahrscheinlich Montagekennzeichnungen. Lediglich die in der schildzapfenartig gestalteten Achse vorhandenen Beschlagmarken – die Majuskel B mit dem nachfolgenden Fragment – weisen auf steyrisches Eisen hin [8].

Inspiriert durch die Eisen- und Hammerordnung des Kurfürst August von Sachsen für die Hammermeister in Gießhübel (Berggießhübel) vom 23. April 1583 [9] und die zum Pirnischen Eisen erschienene Publikation des sich um die Erforschung dieses Eisens verdient gemachten G. H. Schmidt [10], wurden den Lafettenbeschlägen sechs Metallproben entnommen, in der Bergakademie Freiberg mit Hilfe einer Elektronenstrahlmikrosonde quantitativ untersucht und mit Eisen des Pirnischen und Schmalkaldischen Reviers verglichen. Fünf der Proben entsprechen entweder den vergleichbaren der Eisenkammer Dresden, Betrieb Königstein vor 1580 oder der Eisenkammer Pirna, Amtshämmer vor 1670 [11]. Dieses Ergebnis basiert auf dem Vorhandensein spezieller Ele-

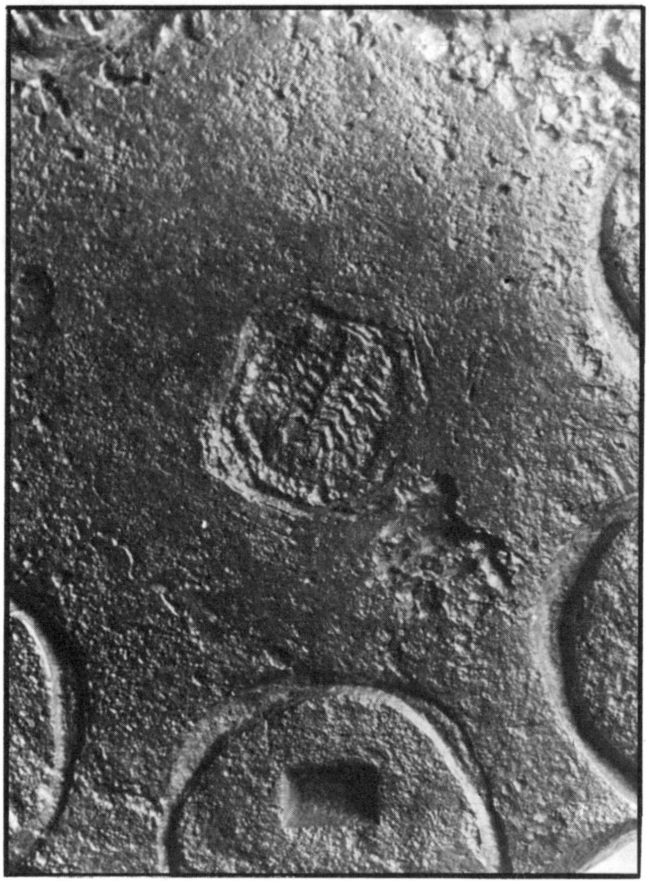

Abb. 8. *Auf einen der vier, die Ladenwände zusammenhaltenden eisernen Bolzen nach der Reinigung entdeckte Schmiedemarke eines Krebses. Foto: Sebb*

Abb. 9. *Schmiedemarke "B.." mit nicht mehr identifizierbaren nachfolgenden Zeichen auf der die Schildzapfen des Rohres ersetzenden Ladenachse. Foto: Archiv des Verfassers*

ment-Kombinationen wie Chrom-Titan, Kupfer-Zinn und Mangan-Chrom-Vanadium, die signifikant für Pirnisches Eisen sind.

Es ist einzuschätzen, daß durch die Nutzung naturwissenschaftlicher Untersuchungsmethoden und Erkenntnisse letztlich belegt werden konnte, daß die Lafette des Geschützes "Faule Magd" sächsischer Herkunft ist. Hieraus erklärt sich sicher auch die blau-gelbe Polychromie der Lafette. Die Majuskel GLZ können als Großes Landeszeughaus gedeutet werden. Die weiteren Untersuchungen zur Einordnung des Lafettenherstellers und des Eisenproduzenten werden durch dieses Ergebnis auf den sächsischen Raum eingeengt.

Diese, Ende 1987 formulierten Sachverhalte sind für die sächsische Geschichte von historischer Bedeutung, begehen wir 1989 in der DDR nicht nur den 40. Jahrestag der Staatsgründung und das Thomas Müntzer Jubiläum sondern auch das historische Ereignis der 900. Wiederkehr der Belehnung eines Wettiners mit der Markgrafschaft Meißen.

Ich glaube anhand der Konservierung/Restaurierung und der bisherigen Erforschung der "Faulen Magd" wird folgendes verdeutlicht:

1. Die Notwendigkeit der schadlosen Bewahrung unseres Kulturgutes auch im Interesse der historischen Forschung.

2. Die Verantwortung aller Historiker, Museologen und Restauratoren für die Feststellung und Sicherung der Informationen des historischen Sachzeugen, der gegenständlichen Quelle.

3. Die Bedeutung werkstoffkundlicher Untersuchungen und ihrer Aussagen für

a) die schadlose Bewahrung und

b) die, dem Sachzeugen abzugewinnenden, d. h. die Nutzung aller dem Sachzeugen innewohnenden Aussagen (quellenkundlicher Erschließung) in interdisziplinärer Zusammenarbeit, wobei den Naturwissenschaften in umfangreicherem Maße Aufmerksamkeit zu widmen ist.

Sicher gäbe es noch interessante, im Verlaufe der Forschung erkannte Details zum Geschütz anzuführen, zumal das schmiedeeiserne Geschützrohr in dieser Betrachtung aus Zeitgründen vernachlässigt wurde.

Anmerkungen

1. **Diener-Schönberg,** Alfons: Der Bestand der chursächsischen Zeughäuser zu Ende des 16. Jahrhunderts. In: Zeitschrift für Historische Waffenkunde (ZHWK), 1896, 4, 1906-1908, Nachdruck Graz 1972, S. 306 ff

2. **Müller,** Heinrich: Deutsche Bronzegeschützrohre, 1400 bis 1740, Berlin 1968

3. **Baarmann,** Otto: Die „Faule Magd" der Königlichen Arsenalsammlung zu Dresden. In: ZHWK 4, 1906-1908, Nachdruck Graz 1972, S. 229 ff

4. **Nissel,** Dieter: Die Erforschung und Restaurierung der mittelalterlichen Steinbüchse "Faule Magd" in interdisziplinä-

rer Zusammenarbeit – ein Beispiel für die Bewahrung und Pflege des materiellen historischen Erbes in der DDR. Ungedruckte Diplomarbeit, Karl-Marx-Universität Leipzig. Leipzig 1987, Anlage 2.7

5. **Lachmann, Manfred – Springer,** Mathias: Die Faule Magd. In: Militärtechnik und Gesellschaftsordnung. Das Geschützwesen im Kurfürstentum. Sachsen, Katalog. Armeemuseum der DDR. Dresden 1979, S. 10
Lachmann, Manfred: Das Riesengeschütz "Die Faule Magd". In: NMK Jg. 26 (1983). 4, S. 226 f

6. **Autorenkollektiv:** Zur Verfestigung der hölzernen Lafette der "Faulen Magd" mit Methymethacrylat. In Holztechnologie. Jg. 25 (1984), 5, S. 242 f. V. International Restorer Seminar, Budapest 1985, Bd. 2 S. 177-186
Autorenkollektiv: Die Konservierung der Holzbauteile der "Faulen Magd". In: NMK Jg. 29 (1986), 4, S.

7. **Schramm,** Hans-Peter: Untersuchungsbericht über farbchemische Untersuchungen der ursprünglichen Polychromie an der Lafette des Stabringgeschützes "Faule Magd" auf der Festung Königstein vom 28. 11. 1981, Dresden 1981

8. **Dite,** Juraj: Slowakische Jagdarmbrüste und ihre Besonderheiten. In: Waffen und Kostümkunde. Zeitschrift der Gesellschaft für Historische Waffen- und Kostümkunde. Jg. 1975, H. 1, S.

9. Eisen- und Hammerordnung. Churfürst August zu Sachsen vor die Hammermeister im Gieß-Hübel, den 23. April 1583. In: Des Codicis Augustei, oder Neuvermehrter Corporis Juria Saxoniae. Anderer Band.

10. **Schmidt,** Gunter H.: Vom Pirnischen Eisen. Aus der Geschichte der alten Hämmer und Hütten im Raum Pirna. Stadtmuseum Pirna. H. 3, Pirna 1984

11. **Derselbe:** Eisenqualität der Beschlagteile der "Faulen Magd", Schreiben an den Verfasser vom 20. 08. 1987.

Adresse des Autors:

Armeemuseum der DDR
PSF 11839
DDR- 8060 Dresden

BRUNNERT, Stephan – POSCHENRIEDER, Walter

Die Glimmentladung in der Restaurierung

Ziel des Seminars ist, die Problematik der Restaurierung von gemischten Materialien zu beleuchten, wobei der Schwerpunkt auf deren Verbindung mit Metall gelegt wird.

Läßt man hierbei die Objekte aus archäologischen Grabungen beiseite, da hier das Problem der reinen Materialerhaltung wesentlich im Vordergrund steht, so bleibt ein sehr breites Spektrum an historischen Objekten übrig. Zahllose Beispiele aus den Bereichen der Volkskunde, der Völkerkunde, der Kunst und des historischen Kunsthandwerks ließen sich hier aufführen.

Gemeinsame Aufgabe aller mit diesen Materialien befaßten Restauratoren ist es, den materiellen Bestand der Objekte zu konsolidieren, zu pflegen und zu restaurieren. Ziel dieser Arbeit ist es, die historischen Objekte als unverfälschte Quelle zu erhalten, einschließlich der Gebrauchs- und materialspezifischen Alterungsspuren.

Um dieses zu erreichen, werden mechanische und naßchemische Arbeitsweisen angewandt, wobei eine Kombination beider überwiegen dürfte. Allein auf dem Gebiet der Metallreinigung ergibt sich, um die sensible Oberfläche zu erhalten, aber dabei das Problem, daß rein mechanische Vorgehen sich als ungenügend oder zu grob erweist und sich das naßchemische als zu unterdifferenziert und zu gefährlich für die gleichfalls mitverarbeiteten hygroskopischen Materialien herausstellt. Die Trennung beider Materialgruppen für eine differenzierte Behandlung ist entweder gar nicht möglich oder bedeutet den Verlust wesentlicher historischer Substanz.

In den letzten 12 Jahren ist immer wieder auf den Einsatz von Gasen zur Restaurierung hingewiesen worden, um die Metallkonsolidierung und Reinigung zu erreichen. Ein sehr wichtiger Ausgangspunkt für diese Behandlungsweise dürften die Ausführungen 1977 von L. Barkmann sein, der die Konservierung von rostigen Eisenobjekten der "Wasa" in einer bis zu 650 Grad heißen Wasserstoffatmosphäre beschreibt.

Die zu behandelnden Eisenobjekte wurden in einer Gaskammer zusammen mit einer leicht strömenden Wasserstoffatmosphäre auf eine hohe Temperatur gebraucht, um die Reaktion (Reduktion) $Fe(OH)_2 + H_2 \longrightarrow Fe + 2\,H_2O$ beschleunigt ablaufen zu lassen. Das entstehende Knallgas wurde zusammen mit dem ausströmenden Wasserstoff auf dem Deckel der Kammer abgefakkelt. Dieses Verfahren ist jedoch mit wesentlichen Nachteilen verbunden: Zum einen wird die Metallstruktur der möglicherweise vergüteten Eisenobjekte wesentlich verändert, zum anderen werden organische Bestandteile der behandelten Objekte vernichtet. (Anmerkung: Dieses Verfahren ist ebenfalls im Zusammenhang mit der Reinigung von Silberoberflächen eingesetzt worden. Eine derartige Behandlung von Silberoberflächen bei niedrigeren Temperaturen benötigt dagegen einen sehr hohen Zeitaufwand.)

Einen neuen Ansatz in der Behandlung von Objekten mit Gasen fand 1978 das Team Daniels (Holland) Pascoe. Sie setzten zur Reinigung der hochempfindlichen Daguerrotypieoberflächen Wasserstoff/Argonplasmen ein.

Ein Plasma kann neben dem festen, dem flüssigen und dem gasförmigen Aggregatzustand als vierter Zustand der Materie bezeichnet werden. Im allgemeinen geht man zur Erzeugung eines Plasmas unter Laborbedingungen von einem permanenten Gas aus, dessen Atome durch geeignete Energiezufuhr in ein Gemisch von Elektronen und Ionen umgewandelt werden können. Die einfachste und älteste Art, ein Plasma zu erzeugen, stellt die Glimmentladung dar. Sie kann entweder durch das Anlegen einer Gleichspannung zwischen zwei Elektroden in einem Gas bei niedrigem Drücken erzeugt werden, oder auch durch Anlegen eines Hochfrequenzfeldes.

Das Team Daniels (Holland) Pascoe wählte zu Beginn seiner Versuche zwei prinzipiell gleiche Versuchsaufbauten, welche hier kurz beschrieben werden sollen. Bei den Apparaturen wurden die Glimmentladungen in Pyrex-Borsilikatglaskuppeln auf eine Aluminiumstandplatte erzeugt. In den Kammern wurde der Gasdruck durch eine Öl-Diffusionspumpe, gepaart mit einer Rotationspumpe, auf weniger als 1×10^{-2} mBar reduziert. Anschließend wurde die Kammer mit Wasserstoffargon bzw. Sauerstoffargongasgemischen bis zu einem Druck von 0,26 mBar gefüllt. Die Zündung des Plasmas erfolgte dann bei dem Versuchsaufbau durch das Anlegen einer Spannung an zwei gegen die Grundplatte isoliert aufgebaute, waagerecht in die Kuppel hineinragende Aluminiumstäbe. Die zweite Anordnung sah anstelle dieser Aluminiumstäbe zwei senkrecht angebrachte Aluminiumgitterelektroden vor.

In der Physik bezeichnet man diese Anordnung als Diodenschaltung. Um die Charakteristik eines Plasmas bei diesen Versuchsaufbauten zu verstehen, müssen grundlegende Erkenntnisse der Glimmentladung angesprochen werden, die hier nur stark vereinfacht wiedergegeben werden können.

Von den in der Reaktionskammer eingebrachten Elektroden ist eine anodisch, die andere kathodisch geschaltet. Bei geeigneten Bedingungen kommt es zur Bildung freier Elektronen. Dieser Prozeß kann im gesamten Gasvolumen (Volumenionisation) und an der Kathode (Oberflächenionisation) auftreten. Diese Elektronen bewegen sich auf die Anode zu und ionisieren bei den Stößen mit Gasteilchen diese in positive Ionen und Elektronen. Somit vergrößert sich die in Richtung zur Anode wandernde Anzahl der Elektronen mit der Entfernung von der Kathode lawinenartig. Bleibt die Anzahl der Elektronen durch die angelegte Spannung in jeder nachfolgenden Lawine gleich der Elektronenzahl der vorherigen, so dauert die Entladung selbständig an und ein elektrischer Strom fließt durch die Reaktionskammer.

In einer zylindrischen Entladungsröhre mit ebenen Plattenelektroden kann man durch die unterschiedliche Potentialverteilung verschiedene, nachfolgend genannte Bereiche erkennen.

An der Kathode sieht man eine kaum leuchtende Gasschicht, die als Kathodenfall bezeichnet wird.

Der Kathodenfall ist ein Gebiet von hoher Feldstärke. Hier gewinnen die von der auf die Kathode ausgesandten Elektronen genügend Energie, um durch Stöße Gasmoleküle zu ionisieren. Bei ausreichend hoher Spannung sind die auftreffenden Ionen in der Lage, ähnlich wie bei einem Sandstrahlgerät, Atome aus der Kathode herauszuschießen. Dieser als "Zerstäubung" bezeichnete Effekt wird später noch eingehender behandelt.

In der folgenden Zone ist die Feldstärke relativ klein. Hier erfolgt Bewegung der Ladungsträger teilweise durch Diffusion. Es sind in gleicher Anzahl positive und negative Ladungsträger vorhanden, so daß die Gesamtraumladung praktisch Null ist. Diesem Bereich, der ein quasineutrales Plasma darstellt, bezeichnet man als negatives Glimmlicht.

Vor der Anode werden die Elektronen nochmals beschleunigt, und die gesamte Raumladung wird negativ. Auch hier kommt es zu Anregungen von dort vorhandenen Atomen, so daß sich ein Anodisches Glimmlicht ausbildet.

Die hier kurz beschriebenen Bereiche sind bei einem feststehenden Elektrodenabstand, gleichbleibendem Druck und unveränderter Gleichspannung stationär. Beim Betrieb mit Wechselstrom brechen die sich ausgebildeten Bereiche periodisch zusammen, um sich in umgekehrter Reihenfolge wieder aufzubauen.

Für das englische Team erschien der Bereich des quasineutralen Plasma am geeignetsten. Man ging davon aus, daß es in diesem Bereich keine ausreichend energiereichen Partikel gibt, die eine Materialzuerstäubung hervorrufen, aber dennoch ausreichend hohe chemische Reaktionsenergie vorhanden ist. In dieser Zone wurden mehrere organische und anorganische Materialien ionisierten Gasgemischen ausgesetzt. Zum Einsatz kamen Sauerstoff-Argon- sowie Wasserstoff-Argon-Gemische. Beim Einsatz von Sauerstoff-Argon-Gemischen wurde ein von der Reaktionszeit unabhängiger Abbau organischer Substanz beobachtet sowie bei einigen Metallen, insbesondere bei Silber, eine starke Oberflächenoxydation konstatiert. Nach der Einwirkung von Wasserstoff-Argon-Plasmen auf anorganische Substanzen ließen sich bei reinen Metallen sowie Glas- und Keramikobjekten keine erkennbaren Veränderungen feststellen, während Metallverbindungen, zum Beispiel Silbersulfid oder Silberchlorid reduziert wurden. Dabei weisen die reduzierten Metalloberflächen keinerlei kristallographische Strukturen auf, sondern besitzen eine feindisperse Struktur, die aufgrund der großen Oberfläche ein hohes Reaktionsvermögen besitzt. Es besteht daher die Notwendigkeit, derart behandelte Oberflächen rasch zu konservieren. Günstig war während der ganzen Versuche bei beschriebener Versuchsanordnung der gemessene Temperaturbereich, da er in keinem Fall 50 °C überstieg. Dagegen waren die Temperaturen beim Einsatz eines Hochfrequenzerzeugers, der bei späteren Versuchen von damals zum Plasmaaufbau benutzt wurde, in mancher Hinsicht bedenklich, da die Temperaturen schnell auf 100 °C anstiegen. Die Reduktionszeiten waren dafür sehr viel geringer. Für die Behandlung von Objekten, die aus verschiedenen Materialien bestehen, seien sie organischer Natur oder mit hauchdünnen Veredelungsschichten versehen, kann die Hochfrequenztechnik allein aus diesem Grund keine Anwendung finden.

Drei Jahre später erschien eine weitere Publikation des britischen Teams, die eine neue Variante der Reinigung beschrieb. Der eingangs beschriebene Versuchsaufbau wurde so abgeändert, daß eine der Aluminiumgitterelektroden gegen das zu reinigende Objekt ausgetauscht wurde. Zur Aufrechterhaltung des Plasmas wurde ein Gleichstrom angelegt. Als Kathodenobjekte wurden dabei, soweit aus den Publikationen erkennbar, ausschließlich Daguerrotypien verwendet. Es wurde wohl beobachtet, daß durch den Wasserstoff-Argon-Beschuß eine sehr kurze Reduktionsdauer bei niedrigen Temperaturen benötigt wird, aber eine Oberflächenschädigung durch die Zerstäubung erfolgen kann. Daher wurde empfohlen, ein leichteres Edelgas, z. B. Neon, anstelle von Argon zu verwenden und mit einer möglichst geringen Brennspannung zu arbeiten.

Dieselbe Versuchsanordnung wurde 1984 von dem dänischen Team Mogens Koch/Anker Sjögren aufgegriffen, um wiederum Daguerrotypien zu restaurieren. Um die Oberflächenzerstörung durch die Zerstäubung zu vermindern, wurde der Plasmazustand durch einen Wechselstrom aufrechterhalten. Als Reaktionsgas wurde ausschließlich Wasserstoff verwendet. Die Reinigungsergebnisse von Silberobjekten waren – gemäß den Analyseergebnissen – sicher gut, wenngleich bei der benutzten Spannung von 1000 Volt zwischen Elektrode und Objekt eine gewisse Zerstäubung durch Wasserstoffionen und -molekule nicht ausgeschlossen werden kann.

Im Jahre 1985 veröffentlichte das schweizer Team J. Patscheider und S. Veprek eine aufsehenerregende Veröffentlichung über die Anwendung der Glimmentladung bei kapazitiver Ankopplung zur Restaurierung von metallenen Bodenfunden. Diese Anlage arbeitet mit hohen Leistungsdichten in der Reaktionskammer und erzeugt hohe Temperaturen. Außerdem weist sie unübersichtliche Potenzialverhältnisse auf, so daß Zerstäubungseffekte schwer unter Kontrolle zu halten sind. Dieser Versuchsaufbau ist für die Behandlung von Kompositobjekten nur sehr bedingt geeignet.

1986 stand das Team Roidl /Poschenrieder/ Brunnert vor dem Problem, drei sehr bedeutende Silberskulpturen mit Augsburger Provenienz aus der Zeit um 1625 vom Altar der "Schwarzen Madonna" in Tschenstochau zu reinigen. Die ungefaßten, sehr fein ziselierten Silberoberflächen dieser Skulpturen waren seit sehr langer Zeit nur sehr unzureichend gereinigt worden, so daß sich sehr dicke Silbersulfidschichten bilden konnten. Da eine Reinigung mit Komplexbildnern die sensible Oberfläche unwiederbringlich zerstört hätte, wurde eine sehr schonende Reduktionsmethode gesucht. Die Idee der britischen und dänischen Teams der Plasmareduktion wurden dabei aufgegriffen, die Versuchsanordnung jedoch erneut modifiziert. Die Arbeiten der vorangegangenen Teams enthalten zwar eine Reihe von sorgfältigen

Detailstudien über die Auswirkung der Behandlung, jedoch lassen sich die plasmaphysikalischen Gegebenheiten in wichtigen Punkten nicht klären. Es fehlten bei den zuvor angesprochenen Veröffentlichungen z. B. Angaben über die Erdungsverhältnisse, so daß die tatsächlichen Potentialverhältnisse zwischen Plasma und Probe nicht erkennbar waren. Es ist zu vermuten, daß diese nicht unter Kontrolle waren und ihre Bedeutung nicht erkannt wurde. Da es für die Versuchsdurchfürung unnötig und schädlich ist, verzichtete man in München auf den Edelgaszusatz bei der Reduktion. Anstatt das Plasma in der quasineutralen Säule oder die Entladung an der Kathode zur Reduktion zu benutzen, entschloß man sich, das Objekt, unabhängig von der Brennspannung, als zweite variable Kathode zu schalten. Man kann diese Versuchsanordnung als Triodenschaltung bezeichnen.

Diese Anordnung besitzt gegenüber den zwei Elektrodenanordnungen den Vorteil, daß die Verhältnisse zwischen dem Plasma und dem Objekt sehr genau bestimmt werden können. Wie bei den anderen Glimmentladungsanordnungen ließ sich die Brennspannung und der damit verbundene Gesamtstrom der Glimmentladung sehr gut messen. Das Objekt als Elektrode ist jedoch für die Aufrechterhaltung der Entladung nicht notwendig. Legt man an das Objekt eine einstellbare Spannung an und mißt, wann der Strom, der in der quasineutralen Säule auf das Objekt trifft, zu Null wird, so läßt sich daraus ungefähr das Plasmapotention bestimmen. Die Spannungsdifferenz zwischen diesem Plasmapotential und Objekt ist im allgemeinen ein Maß für die Energie der auftreffenden Teilchen. Setzt man den Strom in ein Verhältnis zur betroffenen Oberfläche, so erhält man die Stromdichte, die bei dem Münchner Versuchsaufbau zwischen 10 μA/qcm und 1 μA/qcm lag. Geht man für das Weitere von einem Wert von 100 μA/qcm aus, so entspricht dies für den Fall, daß der Strom ausschließlich von Wasserstoffionen getragen wird, einer Flußdichte von $1{,}2 \times 10^{15}$ Wasserstoffatomen/qcm/s. Bei einer Spannungsdifferenz von 200 Volt zwischen Probe und Plasma beträgt die Auftreffenergie pro Atom 100 eV. Aus der Zerstäubungskurve für Silber entnehmen wir hierfür eine Ausbeute von 10^{4}Atomen pro cm. Aus diesem Wert läßt sich errechnen, daß für die Abtragung einer Atomlage Silber ca. 2,5 Stunden benötigt werden. Eine solche Zerstäubungsausbeute würde bei Argon bereits in einer Sekunde auftreten.

Da ein Silberatom 1,8 x 10 g wiegt, entspricht die Zerstäubung mit Wasserstoff einem Gewichtsverlust von nur 2×10^{-11} g/s x qcm. Mit Argon wäre dieser Wert immerhin 2×10^{-7} g/s x qcm.

Dabei wird offensichtlich, daß die Zerstäubung bei der Münchner Versuchsanordnung keine Rolle spielt. Betrachtet man die praktische Seite dieser Reinigungsmethode, so waren die Reinigungsergebnisse bei zweidimensionalen Objekten, wie z. B. Daguerrotypien, sehr befriedigend.

Ein gewisses Problem bei der Behandlung von Gegenständen in der Glimmentladung stellen die ungleichmäßgen Teilchenflüsse dar, die vor allem bei komplizierten geometrischen Formen auftreben. Diese haben ihre Ursache in der Ausbildung des Katodenfalls. Hat der Gegenstand scharfe Kanten, so folgt die plasmaseitige Grenzschicht solchen Konturen nur mit weichen Ausrundungen. Da die Richtung der auf die Oberfläche fliegenden Teilchen im wesentlichen dort senkrecht aus dieser Grenzschicht bestimmt wird, kommt es zu Fokussierungs- und Defokussierungseffekten. In den meisten Fällen läßt sich dieser Effekt durch Masken reduzieren. Bei der Triodenschaltung tritt diese Erscheinung bei Verminderung der Spannungsdifferenz zwischen Probe und Plasma erheblich zurück. Bei höheren Drücken, wie in der Diode verwendbar, und entsprechend kleiner Katodenfallänge von mm, folgt der Glimmsaum den Konturen genauer, so daß sogar Unterschneidungen erreicht werden können und Inhomogenitäten nur noch sehr lokal auftreten. In diesem Punkt ist Erfahrung und Fingerspitzengefühl vonnöten. Die effektive Spannung liegt wesentlich höher, als bei der zuvor beschriebenen Triodenschaltung, so daß auch eine erhöhte Zerstäubung in Kauf genommen werden muß.

War die Plasmareduktion zunächst im wesentlichen zur Restaurierung von Daguerrotypien verwandt worden, so wollte das Münchener Team darüber hinaus dessen konkrete Anwendung bei kombiniert verarbeiteten Materialien untersuchen. Das Verfahren erwies sich als sehr gut geeignet, um gefaßte Silberoberflächen zu reinigen. Da jedoch eine Reaktion des Plasmas mit den lasierten Partien nicht völlig ausgeschlossen werden kann und darüber noch keine exakten Untersuchungen vorliegen, ist es empfehlenswert, die bemalten Teile mit Aluminiumfolienmasken, die ebenfalls auf Objektspannungspotential liegen, abzudecken. Diese Masken brauchen das Bild nicht zu berühren, sondern können auch in einem geringen Abstand darüber angebracht werden. Eine mechanische oder chemische Reaktion des Wasserstoffplasmas mit der Farbschicht ist dann ausgeschlossen.

Von einem gewissen Interesse könnten die Reinigungsergebnisse an Brokaten und versilberten Holzteilen sein. Bei Brokaten muß wegen der Störwirkung von Staub und Schmutz auf den Reduktionsverlauf eine Vorreinigung vorgenommen werden. Als besonders effektiv erwies sich die Vorreinigung mit Wasser und Tiovetin, einem anionischen Netzmittel. Die Reinigungswirkung von Benzin erwies sich als gering, Aceton und Alkohol greifen auch die Fasersubstanz an.

Proben, die nach der Vorreinigung noch dunkle Oberflächen zeigten, wurden danach dem Plasma ausgesetzt. Benutzt wurde im wesentlichen dazu die anfangs beschriebene Triodenschaltung. Die Reinigungsergebnisse auf den Fäden waren gut, so lange die Metallfäden ein geschlossenes Netz ohne Unterbrechung der elektrischen Leitung bildeten. Stellen, die nicht unter Spannung waren, blieben im wesentlichen unreduziert. Derartige Bereiche ließen sich jedoch durch geeignete Kontaktierung, z. B. mit Stecknadeln, ebenfalls auf das erforderliche Potential bringen. Verschiedene textile Proben, die der Glimmentladung ausgesetzt waren, wurden anschließend, soweit es zellulosehaltige Proben betraf, im Münchener Holzforschungsinstitut untersucht. Dabei wurde festgestellt, daß weder Zelluloseabbau oder Beeinträchtigung der Lignine noch eine Anätzung der Oberfläche erfolgt waren. Woll- und Seidenproben wurden in Aachen am Wollforschungsinstitut überprüft. Auch dort konnte keine unmittelbare Zerstörung des Gewebeaufbaus festgestellt werden. Ein wesentlicher

Punkt bei diesen Untersuchungen ist der starke Wasserverlust der Gewebe. Allerdings wurde in allen Fällen eine starke Wasseradsorption nach der Reduktion beobachtet, so daß sich innerhalb kurzer Zeit die vorherige Gleichgewichtsfeuchte wieder eingestellt hatte.

Recht interessant, aber für restauratorische Belange wahrscheinlich irrelevant, waren die Reinigungsergebnisse auf versilberten Holzplatten. Bei dem vorliegenden Beispiel einer Holzplatte mit wellig ausgedünnter Rückseite und strukturierter Oberseite mit verschiedenen Grundierungsschichten und Versilberungen wurde ein Teil des Fassungsaufbaus mit einer Aluminiumfolie gegenüber der Entladung abgeschirmt. Von der Blattsilberauflage wurde ein Teil mit einer Klemme kontaktiert, ein anderer Teil wurde ohne Anschluß belassen. Um das Schrumpfverhalten zu beurteilen, wurde ein feiner Bindedraht so über die Oberseite gespannt, daß er bei Berührung einen leicht singenden Ton gab. Der Versuch sollte Aufschluß geben, ob es bei größeren Holzteilen möglich ist, annähernd definierte Gasbedingungen zu erreichen oder ob die Ionisierung des austretenden Wasserdampfes dies verhindert. Des weiteren sollte beobachtet werden, ob auch die nicht kontaktierten Teile der Versilberung im Plasma reagieren. Es wurde beobachtet, daß die Pumpzeit bis zu annähernd ausreichenden Gasbedingungen sich durch den ständig ausdehnenden Wasserdampf stark verzögerte, so daß erst nach ca. 5 stündiger Pumpdauer ein Druck von 6 x 10 mBar erreicht wurde. Dieser Druck lag zwar noch weit entfernt von den Bedingungen, die ansonsten für die Reduktion von Metallteilen eingestellt worden war, trotzdem wurde eine Reinigung des Blattsilbers auch ohne die gut definierten Ausgangsbedingungen erreicht. Während der Reduktion zeigte die Glimmentladung eine wesentlich andere Farbe, die auf den austretenden Wasserdampf zurücfkzuführen ist. Die Reinigung der Oberfläche erfolgte sehr rasch. Nach ca. 5 Minuten hatte sie wieder ein silbrig glänzendes Aussehen angenommen. Der abgedeckte Teil der Versilberung war im Vorzustand verblieben, der nicht kontaktierte Teil war etwas aufgehellt. Allerdings wurde beobachtet, daß sich unter der Aluminiumfolie einige kleine Bläschen in der Versilberung aufgeworfen hatten. Augenscheinlich nicht verändert hatte sich der Bolus und die Grundierung. Letzterer wies auch keine erkennbaren Risse oder Sprünge auf. Jedoch war sehr leicht zu erkennen, daß der über der Oberfläche gespannte Draht schlaff war und der Blick auf das Stirnholz bestätigte, daß sich die Platte stark gekrümmt hatte. Diese Krümmung ist eine eindeutige Folge der Wasserdesorption. Bei nachfolgenden Versuchen wurde festgestellt, daß diese Wasserdampfabgabe bei Hölzern sehr verschieden ist. Sie war bei Fichte (4 %) am geringsten, gefolgt von Eiche (5,7 %), Linde (5,9 %) und Buche (6,7 %).

Der derzeitige Stand der Reinigung von kombinierten Materialien im Wasserstoffplasma kann folgendermaßen zusammengefaßt werden:

Für die Behandlung von reinen metallenen Kunstwerken ist das Verfahren ein sehr objektschonender Weg. Für die Reinigung von Münzen und Daguerrotypien ist es mit Sicherheit der am besten gehbare. Für die Restaurierung von Metallen mit hygroskopischen Materialien gibt es derzeit noch zuviele Probleme, als daß die Glimmentladung als bedenkenlos empfohlen werden könnte. Allerdings bietet gerade das Plasmaverfahren, welches Oberflächen hochreaktive Teilchen anbietet, ohne diese jedoch in den Oberflächen zu deponieren, viele interessante Aspekte, die ein weiteres Experimentieren und Forschen wünschenswert erscheinen lassen. Die Erfahrung hat jedoch gezeigt, daß derartige Versuchsaufbauten und -durchführungen physikalische Kenntnisse verlangen, darüber hinaus sind sie zeitraubend und damit teuer. Es wäre jedoch zu begrüßen, wenn sich wieder einmal Restauratoren und Naturwissenschaftler zusammenfänden, um auf diesem Gebiet weitere Fortschritte zu erreichen.

Literaturverzeichnis

Daniels, V. D. **Pascoe,** M. W. **Holland,** L.:
Plasma reactions in the conservation of antiquities, ICOM Committee for Conservation, 5 th Triennial Meeting, Zagreb, 1978

Daniels, V.D. **Pascoe,** M. W. **Holland,** L.:
Gas plasma reactions for the conservation of antiquities, Studies in Conservation, 24 (1979), 85–92

Daniels, V. D.:
Advances in the use of hydrogen plasma for reduction of silver tarnish
ICOM Papers 81, 19, 20–1

Daniels, V. D.:
Plasma reduction of silver tarnish on daguerrotypes, Studies in Conservation, 26 (1981), 45–49

Barger, S. M. **Krishnaswany,** S. V. **Messier,** R:
The cleaning of daguerrotypes: Comparison of cleaning methodes, Journal of the American Institut for Conservation, Vol. 22 Nr. 1 (1982)

Morgens S., **Koch, Anker, Sjögren:**
Behandlung von Daguerrotypien mit Wasserstoffplasma, Maltechnik Restauro 4, Oktober 1984

Morgens S., **Koch:**
Fotografien
Technik, Zerfall und Konservierung, S. Intern. Graph. Restauratorentag
Den Haag, 12. bis 16. Sept, 1983

Barger, S. M. **Giri,** A. P. **White,** W.B. **Edmondson,** T. M.:
Cleaning daguerrotypes, Studies in Conservation, 31 (1986), 15-28

Milch, J.:
Silberputzen auf Plasmachemischem Wege,
IPP-Pressestelle, Garching

Roidl, E. **Brunnert,** S. **Poschenrieder,** W.:
Plasmachemische Reinigung, Labor 2000 (1987)

Kapzow, N. A.:
Elektrische Vorgänge in Gasen und im Vakuum
VEB Deutscher Verlag der Wissenschaften, Berlin 1955

Veprek, S.:
Neue Methode für die Restaurierung von archäologischen Metallfunden, Neue Zürischer Zeitung, 19.03.1986

Elmer, J. Th. **Veprek,** S.:
Plasmachemische Eisenkonservierung, Arbeitsblätter für Restauratoren, Heft 1, 1986

Patscheider, J. **Veprek,** S.:
Application of low-pressure hydrogen plasma to the conservation of ancient iron artifacts, Studies in Conservation, 31 (1986), 29–37

Veprek, S. **Patscheider,** S. **Elmer,** J.:
Restoration and conservation of ancient artifacts: A new area of application of plasma chemistry, Plasma chemistry and plasma processing, Vol. 5, No. 2, 1985

Veprek, S. **Elmer,** J. Th. **Echmann,** Ch. **Jurcik Rajman,** M.:
Restauration and conservation of archeological artifacts by means of a new plasma-chemical method, Journal of the Electronical Society, Vol. 134; No.1 10, Oct. 1987

Adresse der Autoren:

Stephan Bunnert
Westfälisches Museumsamt
Schwelingstr 5 D-4400 Münster 1

Dr. W. Poschenrieder
Max-Planck-Institut,
Bolzmanstr. 2
D-8046 Jarching

ROBERTS, Frances C.

Lifting freshly excavated fragile objects

The purpose of my poster was to illustrate a technique for lifting fragile or complexly grouped objects when normal excavation proves too difficult. The use of paraffin wax is by no means new, it has been used by many archaeologists and conservators, but I have tried to bring together here some of the results of several years of practice so that those wishing to try the technique may reap the benefits.

Objects from excavations can be very fragile and degraded, whilst being of cultural importance, so that the area of damp earth in which they are buried may be taken out of its context and carried back to the museums laboratory to be excavated. Gigantic blocks of earth, weighing tons, have been known to find their way into museums in plastercasts or as timber crates. These are not the subject of this poster! A sword, a complicated collection of beads and fibula from a grave, a cranium or other bones, and such objects of reasonable size – these may be lifted in a block using paraffin wax as casing material with very little effort and at low cost. The method assumes that the earth is wet or damp and that there is a little room around the object to give working space (or that it is removed after the surroundings have been excavated).

Author's address:

Konserveringscentret
Vestergade 5-7
DK-687 Ølgod

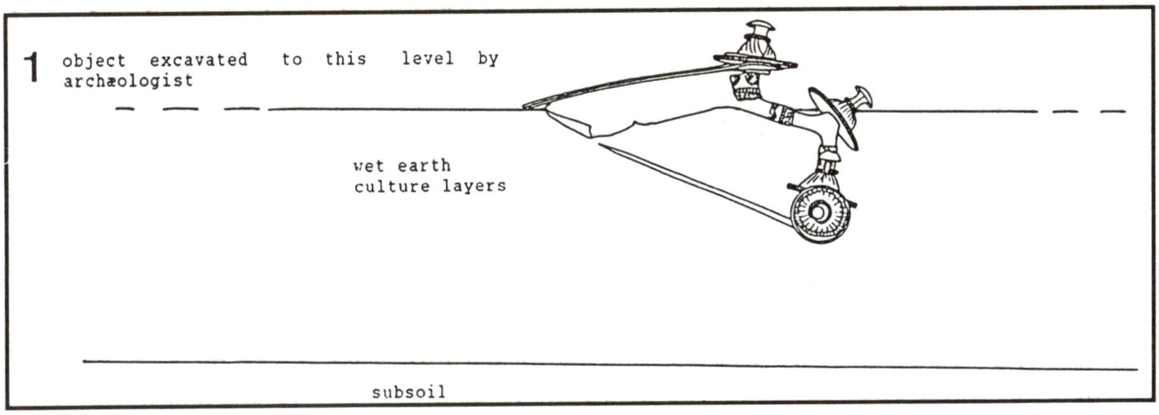

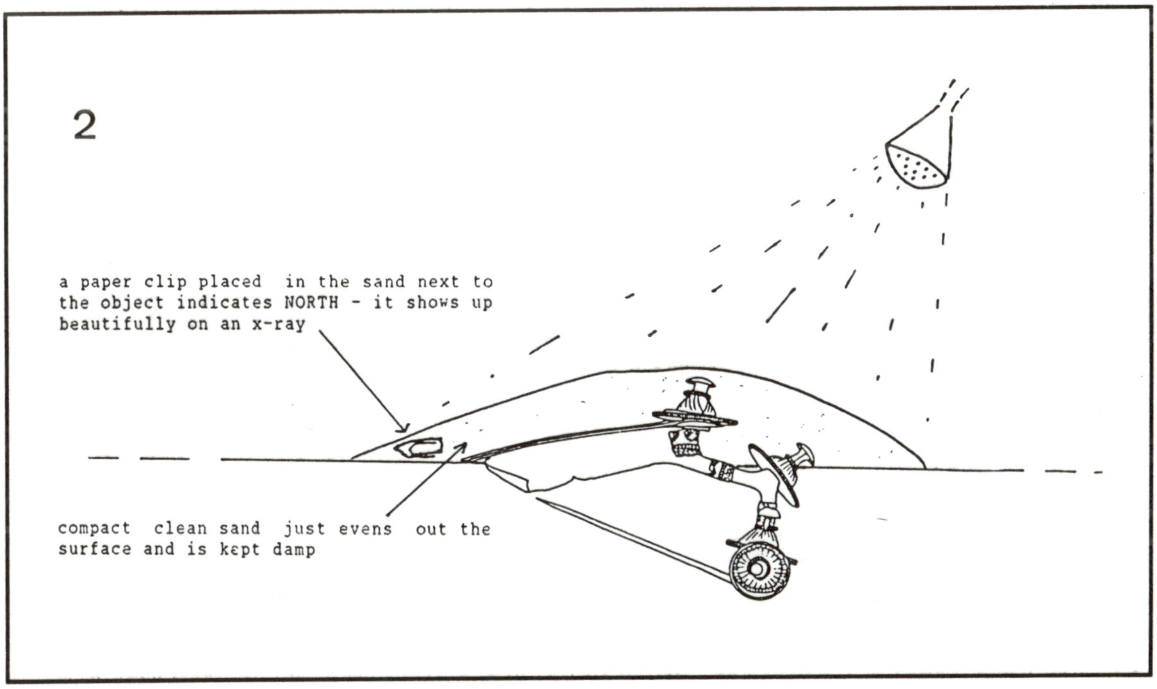

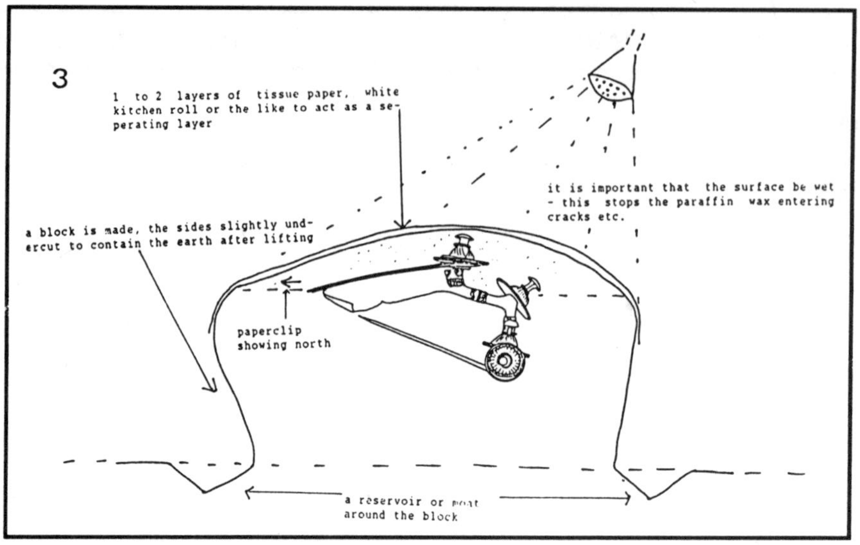
3
1 to 2 layers of tissue paper, white kitchen roll or the like to act as a seperating layer
it is important that the surface be wet - this stops the paraffin wax entering cracks etc.
a block is made, the sides slightly undercut to contain the earth after lifting
paperclip showing north
a reservoir or moat around the block

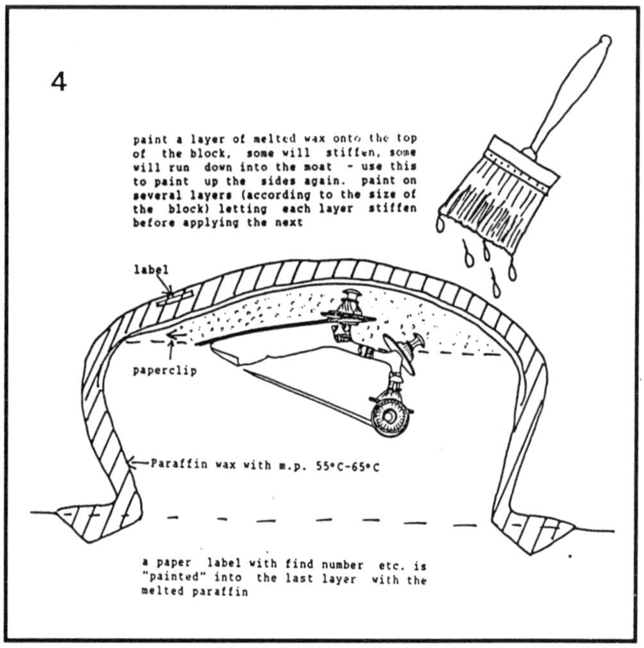
4
paint a layer of melted wax onto the top of the block, some will stiffen, some will run down into the moat - use this to paint up the sides again. paint on several layers (according to the size of the block) letting each layer stiffen before applying the next
label
paperclip
Paraffin wax with m.p. 55°C-65°C
a paper label with find number etc. is "painted" into the last layer with the melted paraffin

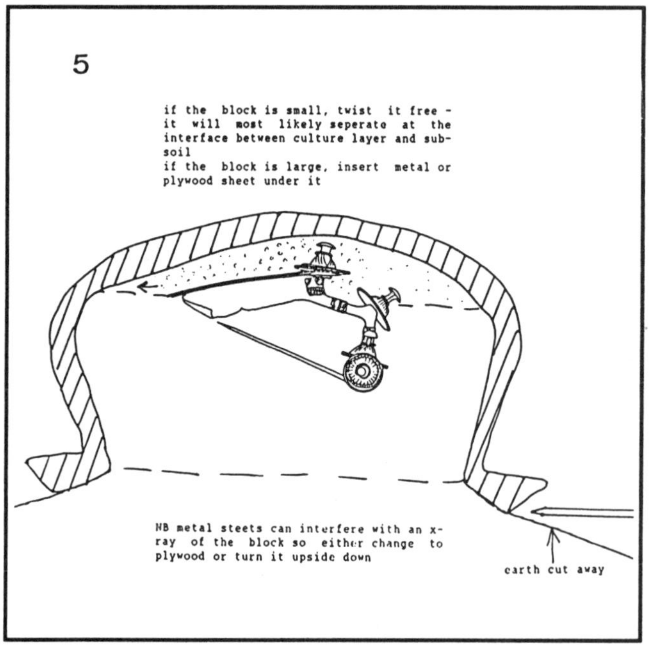
5
if the block is small, twist it free - it will most likely seperate at the interface between culture layer and subsoil
if the block is large, insert metal or plywood sheet under it
NB metal steets can interfere with an x-ray of the block so either change to plywood or turn it upside down
earth cut away

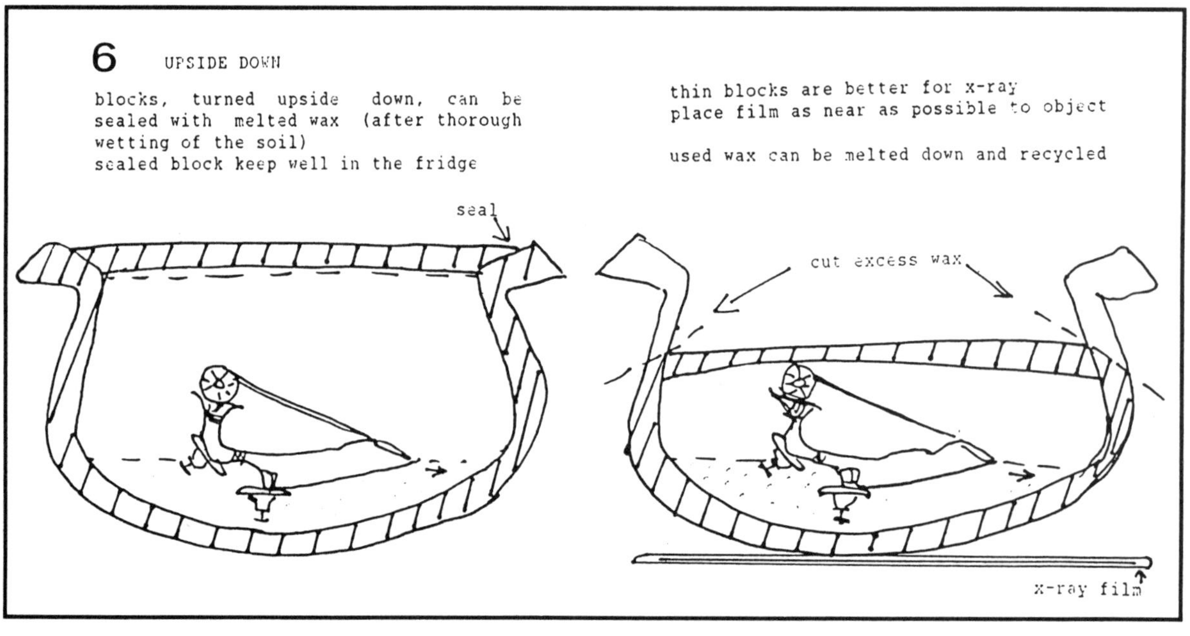
6 UPSIDE DOWN
blocks, turned upside down, can be sealed with melted wax (after thorough wetting of the soil)
sealed block keep well in the fridge
thin blocks are better for x-ray
place film as near as possible to object
used wax can be melted down and recycled
seal
cut excess wax
x-ray film

WEKER, Wladyslaw Henryk

Determination of the concentration of chloride ions extracted from corroded iron artefacts

Abstract

Determination of the concentration of Cl^- ions in solution provides important information which contributes to the advance of the desalination process. There are many analytical methods to determine chlorides of wich Flow Injection Analysis is the simplest. It can be used for various kinds of solutions.

To stabilize heavily corroded iron objects it is necessary to wash out the Cl^- ions present in corrosion layers, and to control the Cl^- content in the wash solutions.

There are many methods of removing chlorides from corroded iron objects. Mostly, they are based on the removal of chlorides from corroded layers in different extracting baths. The quantitative determination of these ions is of great importance for information about the advance of the desalination process.

The usual methods for determinig chlorides are:

- argentometric titration,
- mercurimetric titration,
- complexometric titration,
- spectrometic and spectrophotometric methods,
- gas chromatography,
- radiochemistry,
- electrochemical analysis,

All of these have disadvantages:
necessity of expensive instrumentation or tedious manual work, low reproducibility, time-consumption and laboriousness. Flow Injection Analysis (FIA) with potentiometric detection eliminates most of these inconveniences.

The principles of FIA

A system applied for potentiometric chloride determination with an indicating Cl^- selective electrode is shown in Fig. 1.

Fig. 1 *Schematic measuring system with ion selective electrode*

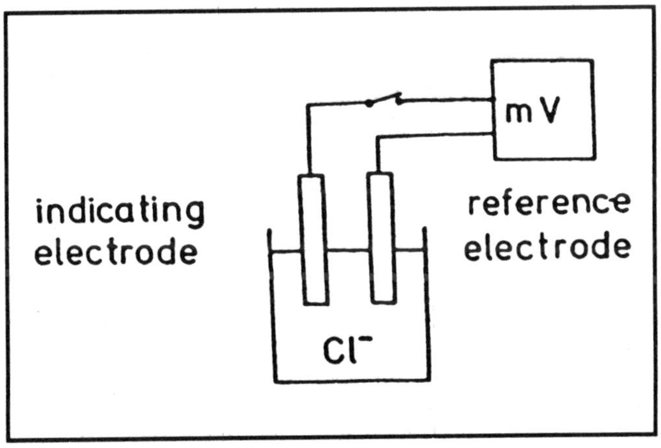

The measured potential of the Cl^- selective electrode depends linearily on the Cl^- concentration, according to the Nikolski equation:

$$E_{Cl} = E_o + 59 \ln(C_{Cl}+K_i\ C_i)\ [mV]$$

where:
E–constant,
K_i–selectivity coefficient
C_{Cl}–concentration of Cl^-
C_i–concentration of interfering ions

Some ions present in solution may interfere with electrode performance. These are:

OH^-, NH_3^-, SO_3^{--}, $S_2O_3^{--}$, Br^-, CN^-, S^{--}.

To obtain correct results we have to eliminate these interfering ions from the bath. It is easy with the flow injection system.

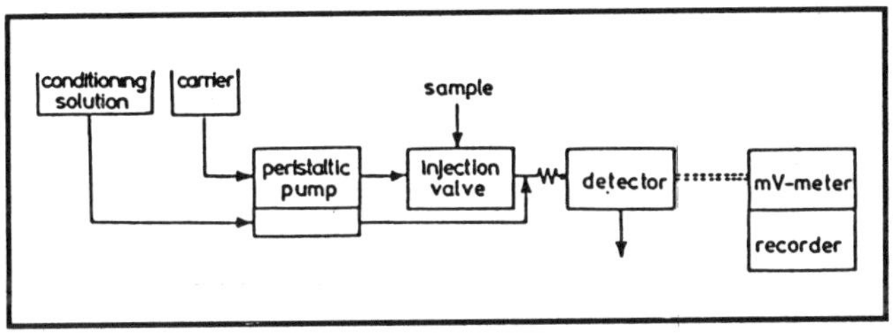

Fig. 2 *Block diagram of flow injection system*

A schematic diagram of FIA is shown in Fig 2. This arrangement allows the interfering ions to be eliminated by a chemical modification of the injected sample. A carrier of steam (H_2O) containing the sample mixes with a stream of the conditioning solution. Then they pass through the reaction coil in which they react together.

Subsequently, the stream enters a flow-through cell containing the indicating and reference electrodes. The potentials of the electrodes are measured with a mV-meter which is connected to a recorder. The manifold is made from a teflon tube of 0.9 mm diameter.

To obtain correct measurement results we must use a calibration plot based on standard samples with different concentrations of Cl^- ions. It is easy to calculate the chloride content by comparing the recorded peaks corresponding to the analysed sample with the calibration curve. The schematic way of this procedure is shown in Fig. 3.

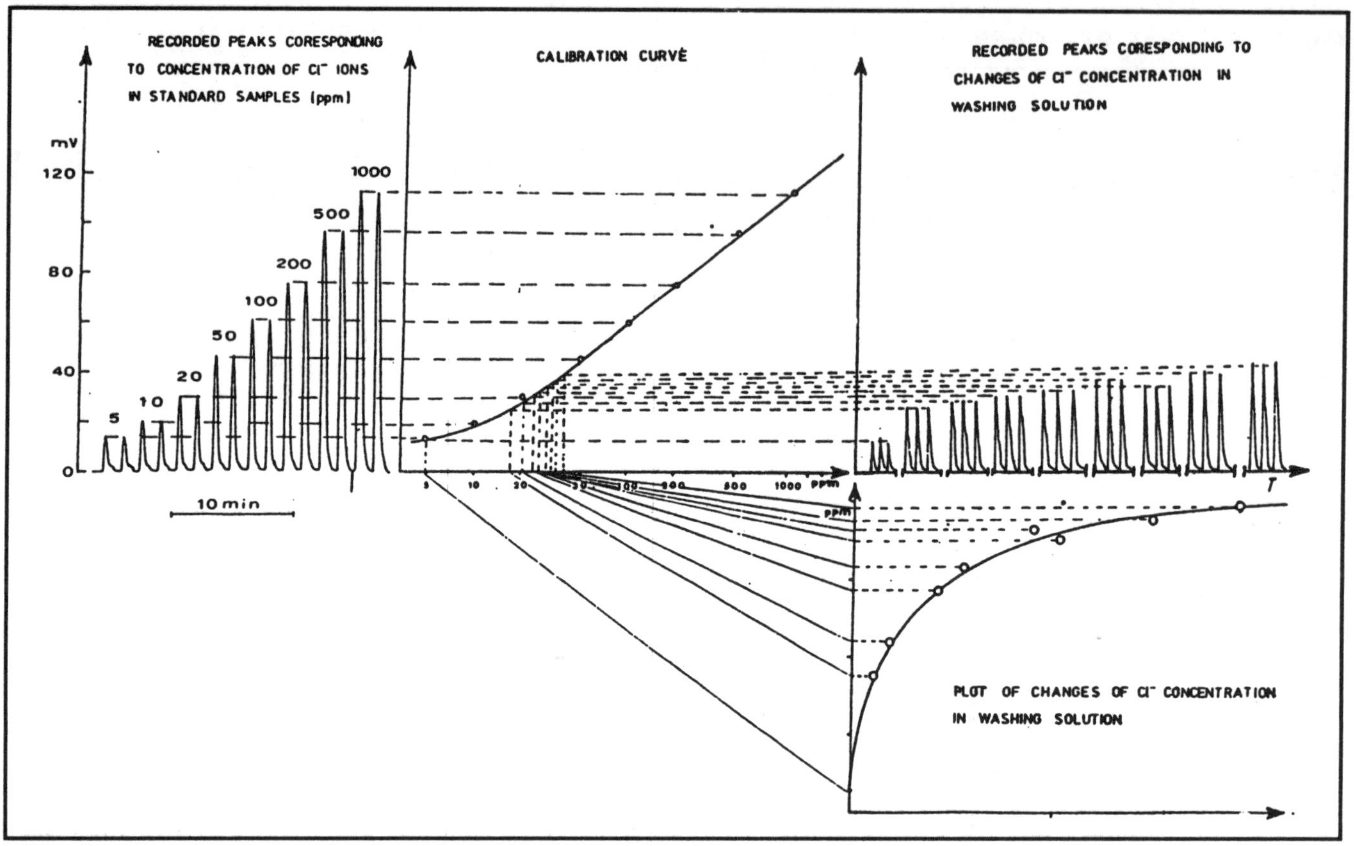

Fig. 3 *Sequence of procedures in the determination of chloride content in the washing solution (by means of FIA)*

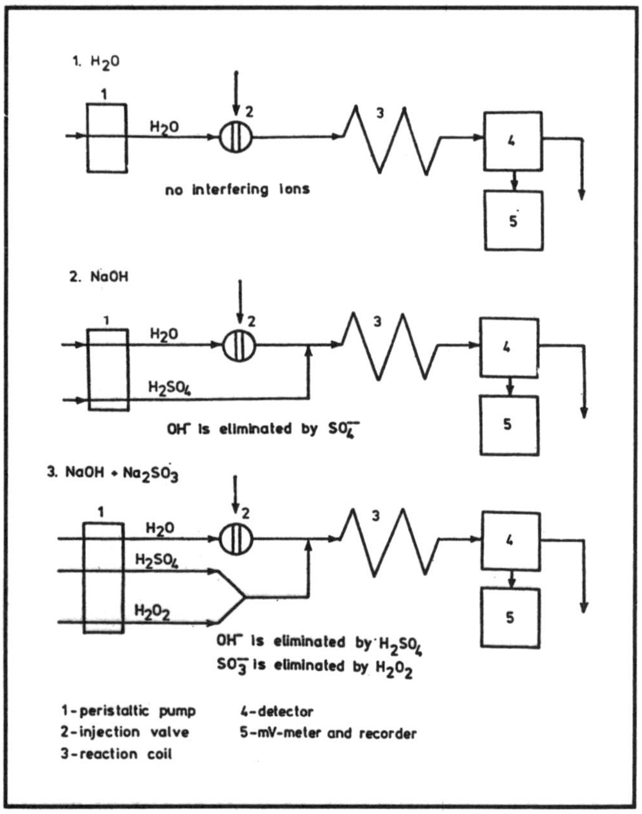

Fig. 4 *Different measuring arrangements*

The parameters of the measurements were as follows:

— the injected sample volume ~ 140 µl,
— the flow rate for carrier ~ 2 ml/min,
— the sampling rate ~ 50 samples per hour,
— the detection limit level ~ 5 ppm.

The exchange of the manifold system for different measuring arrangements is very simple. For that reason the method can be used even in conservation laboratories. Figure 4 shows some examples of different measuring arrangements.

FIA can be used in all fields of conservation where desalination is an important part of the process (for example in the conservation of stone).

Bibliography:

1 W. J. **Wiliams**, "Oznaczanie Anionow", Warszawa, PWN, 1985

2 S. **Keene**–C. **Orton**, "Stability of treated archaeological iron: an assessment", Studies in Conservation, 30 (1985), 136-140

3 J. **Ruzicka**–E. H. **Hansen**, "Flow Injection Analyses", Analytica Chimica Acta, 78 (1975), 145-157

4 W. H. **Weker**–M. **Trojanowicz**, "Application of flow-injection analysis for the determination of chloride...", Studies in Conservation, 32 (1987), 86-90

Author's address:

State Archaeological Museum in Warsaw
Ul. Dluga 52
PL-00950 Warszawa

PETUSHKOVA, Julia – LYALIKOVA, Natalie

Microbiological transformation of metals in architectural monuments

The oldest cathedrals of Russia were built mostly of white stone (calcium carbonate). Gold was the main metal used for their decoration. Iron bindings were used in Russia for the first time by Aristotle Phioravanti, who was the author of the project of the Uspensky cathedral in the Moscow Kremlin (1475-1479). In the XVII century iron nails were first used to fix the plaster layer on which levcas (chalk-and-glue priming) was applied as a base for mural painting. Metals are the basic chromophoric substance in many pigments: green ground ochre (Fe), azurite, malachite (Cu), white lead, minium, massicot (Pb), cinnabar, (Hg), etc.

All metals and their compounds are subject to corrosion, which is caused not only by chemical and electrochemical processes but by biological ones as well. Participation of microorganisms in the corrosion of metal pipes, vessels and other industrial equipment is well known. The present work deals with microbiological processes of metal transformation in architectural monuments. As far as we know, these processes have never been studied before.

Great losses of the paint layer in the Church of St. John the Divine (XVIIc) in Rostov, the Great Jaroslavl region, was caused by the process of corrosion of iron nails, which process is very active at present. Thirty nails are contained in 1 m^2 in this church. As a result of rusting and transformation of iron metal into F_2O_3 and other trivalent iron compounds the nails have considerably grown in size. In some places they project from the walls, pushing out the levcas layer together with elements of the mural painting.

Electronmicroscopic study of scrapes taken from the corroded nails has revealed the presence of bacterial cells, attached to the iron hydroxide particles. In laboratory conditions it was possible to reproduce the process of iron corrosion in the following way: the medium for iron bacteria containing iron filings was infected with microorganisms, separated from the surface of the nails. In the control variant without microorganisms oxidation of iron filings was quite insignificant in one month.

With the help of special media it was determined that corrosion is caused by mycoplasm-like organism forming hydrogen-peroxide. Detoxication of peroxide, which is a harmful metabolite, occurs during oxidation. Another groups of microorganisms which take part in iron corrosion are heterotrophic organisms, forming organic acids. From the walls of the church under investigation we separated bacteria, which decrease the pH 3 values (to 4.2) on media containing sucrose and casein. The pH of the sterile medium remained neutral (pH 7.2). The total number of heterotrophic bacteria in 1 g of the damaged levcas layer exceeded 10 000 viable cells.

We studied the cause of metal corrosion in the Pokrov Church in Phili, Moscow (XVII c.). The Tzar gates were decorated with metal plates. From the affected areas we separated bacteria which, according to their morphological characteristics, can be related to the genus Arthrobacter. It is known that members of this genus can oxidize iron under the influence of peroxides evolved by them (Dubinina, 1978). On the inner walls of this church the microbiological process of iron oxidation can be observed. The area of the church's location is characterized by a high level of ground water. Ferric iron penetrates by means of capillary infiltration and is oxidized on the wall surface at 0.5 meter from the floor, causing rusty stains on plaster shortly after the restoration. The number of bacteria cells in the oxidation zones varied from 10 to 100 thousand per 1 g of sample.

Similar stains consisting of iron oxides were observed in pre-Mongolian churches (up to XIII c.). It is known that pyrite (FeS_2) can be found in the white stone of which the Vladimir and Suzdal cahtedrals were built. Pyrite is easily oxidized by tio bacteria in acidic medium, which is impossible in limestone. We discovered that pyrite oxidation on the walls of this monument is carried out by mycoplasm-like microorganisms.

Peroxides, released by arthrobacters, mycoplasms and some other microorganisms, can oxidize not only iron and manganese but also lead. We discovered a microbiological process of lead pigment oxidation in the church of John the Divine in the Rostov Kremlin (Petushkova, Lyalikova, 1985). X-ray examination revealed the presence of brown lead dioxide (PbO_2) in the paint layer. Lead is tetravalent in this compound. All pigments used in painting contain bi-valent lead, which can be transformed into tetravalent lead only under the influence of strong oxidizers.

In spite of the fact that lead pigments are considered bioresistant, we succeeded in demonstrating their oxidation under the impact of bacteria, separated from the transformed paint layer. White precipitate of white lead, yellow precipitate of massicot and red minium in the presence of bacteria acquired a brown shade, which deepened in the course of time. In media where antiseptic was added, no pigment changes were oberved. Formation of tetravalent lead under the impact of bacteria was determined by means of qualitative chemical analysis (based on the release of chlorine and colour reaction with the Arnold reagent). X-rays confirmed the formation of plattnerite mineral PbO_2.

Apart from oxidation, other changes of lead pigment may occur. Some of the bacteria separated from the mural painting turned white lead, massicot and minium into black coloured lead suplhide. This compound is formed as a result of the reaction of hydrogen sulphide, released in the process of bacterial consumption of sulphurcontaining aminoacids, with lead compound. This process is intensively carried out by the bacteria on a nutrient medium with pepton. The same process is observed on the painting of John the Divine Church is Rostov.

Other changes in the mural painting of this church are accounted for by the transformation of copper-containing pigments. These are black stains of blue pigment. Analysis has shown the presence of black sulphide on azurite. Bacterial transformation of azurite into black copper sulphide in also observed on a pepton-containing medium. In the paintings of the Rostov Kremlin churches animal glue was used as a binding material. It could have served as a substrate for the development of microorganisms on the walls of the monument (Brysova, 1958).

Other metals, including noble ones, are subject to bacterial transformation. Heterotrophic bacteria, dissolving gold, were described in the sixties. Certain types of bacilla, grown on organic media, can transfer up to 10 mg of gold into 1 l of the solution. Microbiological study of statues of the Petergoff fountains near Leningrad, where in some places the gilding had been ruined within a few years of restoration, confirmed the participation of bacteria in the dissolution of gold. Bacteria, separated from the lower algae-covered parts of the statues in the presence of gold wire, dissolved several mg of gold during the two-week cultivation. Algae, forming organic substance in the process of photosynthesis, provided conditions for the development of heterotrophic bacteria.

The above given examples of the transformation of metal-containing pigments and metallic structures point to the considerable role of microorganisms in their deterioration.

Transformation of pigment can be stopped by means of antiseptic. We succeeded in decolorizing the stains caused by iron hydroxide formation on the plaster of the Pokrov Church in Phili using 6 % hydrogen peroxide solution. It was possible to use formaline as an antiseptic because there were no paintings on the treated walls. Corrosion of metal particles can be prevented by means of lacquer coating. Materials used for that purpose must be tested for bioresistancy, because some of them can serve as a substrate for microorganisms, thus intensifying corrosion.

The development of restoration and conservation techniques for metal-containing pigments calls for microbiological checking and further investigation of the role of microorganisms in the corrosion process.

References

Dubinina, G.A. The mechanism of oxidation of bivalent iron and manganese by iron bacteria developing at neutral acidity of the medium, Mikrobiologiya, 74 (1978), 591-599; in Russian

Petushkova, Ju.P., **Lyalikova,** N.N. "Microbiological degradation of lead-containing pigments in mural paintings." Studies in Conservation, 31, (1986), 65-59

Bryusova, V.G. "Investigation and restoration of frescoes in the Rostov Kremlin" in Papers on the Investigation and Restoration of Architectural Monuments in the Yaroslavl Region, Yaroslavl, (1958), 95-110; in Russian

Authors' address:

Petushkova, Julia
All-Union Scientific Research Institute of Restoration
Krestyanskaya pl. 10
SU-Moscow 109172

Lyalikova, Natalie
Institute of Microbiology
Academy of Sciences
60-letiya Octyabrya 7a
SU-Moscow 117811

KOZLOWSKI, Andrzej - SEKOWSKI, Stefan

Kombinierte Metall-Lackschichten als Denkmalschutz 15 jährige Erfahrungen

Seit vielen Jahren wurden in Polen Versuche unternommen um die historischen Metallerzeugnisse, die die Werke von Schmieden und Giessern aus 19. und anfang 20. Jahrhundert bilden, vor weiteren Zerstörung zu bewahren. In Vordergrund trat das Problem der Rettung von den so zahlreichen auf den alten Friedhöfen vorkommenen, gegossenen Guss- und geschmiedeten Stahlelementen. Diese Elemente, die ein Zeugnis der technischen Kultur aus damaligen Zeiten sind, unterlagen in letzten Jahren, aufgrund des plötzlichen Anwachsens der Atmosphäreverschmutzung, einem kräftigen Verderben. Nach dem Durchführen von Einleitungsinwentur und Beurteilung des Problemausmaßes haben die an diesem Forschungsprojekt beteiligten Mitarbeiter von Institut für Präzisionsmechanik in Warszawa - zu ihnen gehören die Autoren des vorliegenden Beitrags - sich vorgenommen, eine wirksamere, als die bisher angewandten Methode des Konservation von historischen Erzeugnissen aus Eisen und seinen Legierungen zu bearbeiten.

Man soll hier darauf hinweisen, dass wir uns in unserem Intitut seit ungefähr 35 Jahren mit den Schutzüberzügen beschäftigen. Die in dem Institut bearbeitete Methoden der Oberflächenvorbereitung und Erzeugung von metallischen und nichtmetallischen Schutzüberzügen wurden rasch zur Industriepraxis eingeführt.

In Jahren 1950-60 wurden in dem Institut die Methoden der Spritzmetallisation und Lacküberzüge-Erzeugung bearbeitet und geprüft. Solche Überzüge wurden zur Gewährleistung von vielen großen Stahlkonstruktionen angewandt, wie z.B. Eisenbahnbrücken, Funkstationmast 500 m Höhe u.a.

Vieljährige Beobachtungen von so beschützten Objekten, sowie Ergebnisse der beschleunigten Korrosionsprüfungen haben bewiesen, das kombinierte, mit Hilfe der Spritzmetallisation und Lackierung erhaltene Überzüge, erfolgreichen Schutz von der Korrosion der Gussenkonstruktionen für wenigstens 30 Jähre versichern können.

Es wurde Entschluß gefaßt, diese Erfahrungen zu angewanden und sie zur Denkmälerkonservation zu passen.

Um die Vorteile der neubearbeiteten Art von Konservation richtig verstehen und einschätzen zu können, ist es notwendig einige Worte sowohl dem Charakter dieser Objekte als auch den bisher angewandten Methoden ihrer Konservation zu widmen.

Historische Metallobjekte bilden zu 98% Guss- und geschmiedeten Stahlerzeugnisse. Dazu gehören verschiedenartige Zäune, Brüstungen, kleine Pforten, Gitterwerke, Denkmäler und Kreuze. Bronzetafeln bzw. Messingverzierungen sind so vereinzelt, daß wir sie augenblicklich vernachläßingen können.

Ausgesprochene Menge von Metall auf den historischen Friedhöfen bilden geschmiedeten Stahl- und gegossenen Eisengusserzeugnisse, wobei das grundsätzlich durchbrochene Erzeugnisse, mit komplizierten Formen und weit entwickelten Oberflächen versehen sind. Gerade diese Entwicklung der Oberfläche und komplizierten, unregelmässigen Formen verursachen, daß die Konservation von Metall-Friedhofdenkmäler sehr schwierig, und was damit verbunden ist, sehr kostspielig ist.

Die bisher angewandte Konservationsmethoden von Eisenguss- und geschmiedeten Stahldenkmäler beruhten auf der mechanischen Entfernung von Korrosionsprodukten und Farbresten, und ferner auf dem Auftragen von neuen Farbanstrichsatz. In der Regel besteht solcher Satz aus vorigen Grundanstrich und aus Lackschichten. Die Korrosionsbeständigkeit des auf diese Art und Weise hergestellten Schutzes hängt, bei besten Qualität der angewandten Werkstoffe, von Genauigkeit und Sorgfalt, mit welcher die Korrosionsprodukte beseitigt waren, und von Witterungsverhältnissen, bei welchen die Konservationsarbeiten durchgeführt wurden. Es ist eine bewiesene und indiskutable Sache, daß die Hinterlassung von sogar kleinsten, ungereinigten Korrosionsstelen auf dem Gusseisen bzw. dem Stahl und ihre Abdeckung mit Farbanstrich einen schnellen und gefährlichen Korrosionsprozess unter dem Überzug hervorruft. Als Folge davon werden schon nach einigen Monaten auf dem Anstrich die Blasen entstehen, es wird abblättern und sich ablösen. Sogar die genaueste Reinigung der Oberfläche von Gusseisen und Stahl kann nicht helfen, wenn der Anstrich bei niedrigen Temperaturen und hoher Luftfeuchtigkeit aufgetragen wird. Eben unter diesen Bedingungen befindet sich auf der Metalloberfläche eine, immer für die Augen unsichtbare, dünne Schicht, die sich besonders in Poren und Vertiefungen befindet, mit einem Anstrich wird auch das Korrosionsprozess unter dem Überzug hervorrufen, und - was damit eng verbunden ist - seine schnelle Zerstörung.

Wer die Schönheit der Metalldenkmäler u.a. der Friedhofsdenkmäler kennt, ist darüber im klaren, wie unglaublich schwierige Aufgabe eine genaue Reinigung aus der Korrosionsprodukte von z.B. durchgebrochenen Verzierungen in Form von geschmiedeten Blätter bildet. Schaber und Kratzbürste - das sind praktisch die einzige Werkzeuge, die für die Beseitigung von Korrosionsprodukte angewandt werden können.

Eine starke Bewaldung und Bepflanzung der Mehrzahl von historischen Friedhöfen verursacht, daß dort ein eigenartiges Mikroklima herrscht, das durch sehr hohe Luftfeuchtigkeit gekennzeichnet ist. Die hohe Luftfeuchtigkeit bleibt dort lange, sogar während der heißeren Tage. Deswegen ist der Zeitraum, wann im Polen auf den Friedhöfen die Malerarbeiten auf den Metalldenkmäler durchgeführt werden können, ungewöhnlich kurz und überschreitet fast niemals im Jahr die 3 bis 4 Monate. In der Praxis wird auch diese Bedingung nicht befolgt, und die Malerarbeiten werden sowohl bei niedrigen Frühlings - bzw. Herbsttemperaturen, als auch im Sommer - fastgleich nach dem Regen durchgeführt.

Selbst jedoch im Falle der fachgerechten und sehr genauen Durchführung von Konservationsarbeiten können heute die erhaltenen Lackschichten die Stahl- und Eisengussobjekte nicht länger als 4 bis 5 Jahre wirksam von Korrosion schützen. Die Ursachen für diesen Zustand sind neuerdings in den ausnehmend aggressiven Atmosphäre und saurem Regen zu suchen. Bei jetzigem Staubgehalt, bei hohem Gehalt von Schwefeldioxid und Stickstoffmonoxid in der Luft werden sogar die besten Farbanstrichsätze sehr rasch zerstört.

Am Beginn der Bearbeitung des neuen Konservationsverfahrens für Metall-Friedhofobjekte wurde angenommen, daß ein schnelles, billiges und im grossen Ausmass anwendbares Verfahren sein soll, das gleichzeitig die Metalldenkmäler retten und mindest während 30 Jahren schützen kann. Nach der genauen Analyse von vielen Varianten wurde schließlich folgende Vorgehensweise ausgewählt und erprobt:

— das für Konservation vorgesehene Metallobjekt wird demontiert;
— die Metallelemente werden im Grunde genommen mit Hilfe von Strahlen aus Korrosionsprodukten und Resten von Lackschichten gereinigt; die Auswahl der Art und Kerngrösse von Schleifmitteln ist vom Objektzustand, seinem Gestalt und Herstellungsart abhängig. Je feiner, dünnwandiger und durchbrochener das Erzeugnis ist, um so kleiner wird die Kerngrösse von Schleifmitteln gewählt z.B. von 2-4 mm für gewöhnliche Erzeugnisse bis 180-250 μm für sehr feine. In Ausnahmefällen wird bei sehr weit vorangeschrittenem Korrosionsprozess der sehr feinen Erzeugnisse eine elektrochemische Säuberung in Glukonatelektrolit zugelassen;
— die gereinigten Metallelemente werden gleich mit einer heissaufgespritzten Zinkschicht mit der Stärke von 150-200 μm bedeckt;
— die verzinkten Metallelemente werden mit Farbanstrichtsatz bedeckt. Je nach dem Bedarf können sie die matten, leuchtenden bzw. auch mit dem spezifischen Graphitglanz, schwarzen Oberflächen liefern;
— die so geschützten Elemente werden wieder als Ganzes an ihre ursprüngliche Stelle aufgebaut.

Mittels vorgeschlagener Methode wurde im Jahre 1975 am Institut für Präzisionsmechanik, mit dem Erlaubnis des Konservators der Denkmäler von der Haupttstadt - Warschau, einer aus den Metallobjekten, die sich auf den Friedhöfen in Warschau befinden, konserviert. Das Objekt besteht aus der gusseisernen Umzäunung und auch aus dem gusseisernem Kreuz mit dem Sockel. Der Denkmal wurde demontiert, mit Hilfe der beschriebenen Methode in Institut konserviert und danach erneut auf dem Friedhof montiert. Während der Durchführung der Arbeiten an diesem Objekt wurden noch neue Beweisgründe für die Demontage entdeckt: es wurde eine starke und weit vorangeschrittene Korrosion aller Verbindungen und der unteren Teile vom Sockel dem Kreuz festgestellt. Alle diese Stellen, sogar bei der genauesten Arbeit, kann man weder mit einem Schaber noch mit einer Kratzbürste nicht gut reinigen, dagegen aber wird die Hinterlassung der ungereinigten Fläche keinen weiteren Korrosionsangriff hemmen können und gleichzeitig unter dem Fragezeichen die Zweckmäßigkeit der durchgeführten Konservationsarbeiten stellen.

Wiederum bei der Anwendung von Demontage erlangen wir die Gewißheit, dass die Korrosionsprodukte überall beseitigt werden und die einzelne Elemente von jeder Seite einen Zinkschutzüberzug erhalten.

Ein zweiter grosser Vorteil der vorgeschlagenen Konservationsmethode bildet die Tatsache, daß nach der Demontage von Objekten sie, das ganze Jahr hindurch, völlig unabhängig von der Wetterbedingungen, konserviert werden können. Besonders aufgrund dieser Unabhängigkeit von der Wetterungsverhältnissen gibt uns diese Methode die Garantie, daß die Temperatur und Luftfeuchtigkeit keinen Einfluß auf die Qualität und Haltbarkeit der aufgetragenen Schichten ausüben werden. Ein dritter Vorteil - längere Schutzzeit.

Die vorgeschlagene Methode garantiert bei der genauen Fertigung des Überzugs, daß die Objekte vor Korrosionsangriff während 30 bis 50 Jahren völlig geschützt sind.

Trotz zweifellosen Vorteilen, manche Restauratoren in unserem Lande hatten sich auf die in unserem Institut bearbeitete Methode mißtrauisch bezogen. Es wurden hier zwei ernstliche Vorwürfe gestellt, und nämlich:

— Verletzung der Unterlage während der Entfernung von Korrosionsprodukten mit Hilfe der Strahlabreibungsmethode,
— Einführung der nichtumkehrbaren Veränderungen mit der Verzinkung des konservierten Objektes.

Zum Glück, wie es die Praxis bedeutet hatte, beide Vorwürfe waren falsch.

Zu Oberflächenreinigung benutzt man Schlacke nach der Kupfererdebearbeitung von entsprechender Kerngröße. Diese Schlacke ist so hart, daß sie erfolgreich Korrosionsprodukte entfernt, und so weich, daß sie die Stahl - oder Gusseisenoberfläche nicht beschädigt.

Die Verzinkung des konservierten Objektes führt keine nichtumkehrbare Änderungen zu, weil Zink restlos immer in alkalischer Lösung (die greift nicht Stahl und Gusseisen an) entfernt werden kann.

In der ersten Hälfte '80, aus dem Antrieb und mit Hilfe des Instituts, enstand bei dem älteren Warschauer Friedhof Konservierunganstalt von Gusseisen und Stahl, gegründet auf der Spritzmetallisation und Lacküberzüge - Erzeugung Methode.

Ähnliche Anstalte entstanden auch in vielen anderen verschiedenen polnischen Städten. In Jahren 1984-1985 wurden mit Hilfe der erwähnten Methode u.a. zwei, aus 19. Jahrhundert stammende, 16 Meter hohe, gusseiserne Spitzsäulen konserviert.

Im Januar dieses Jahres fand in Warschau inländische Konservatorenkonferenz statt, die für die Beurteilung der, durch den Institut für Präzisionsmechanik vorgeschlagener Konservierungsmethode von Gusseisen und Stahl gewidmet war. Der Zustand der Objekten, die mit Hilfe von dieser Methode vor 10-14 Jahren konserviert waren, wurde als sehr gut beurteilt. Auf Grund dieser Beurteilung hat das polnisches Ministerium für Kultur entsprechende Empfehlung erlassen, um Spritzmetallisation und Lacküberzüge-Erzeugung zur Gusseisen- und Stahl-Konservierung möglichst breit anzuwenden.

Die Adresse der Autoren:
Institut für Präzisionsmechanik
Duchnicka 3
Pl-00967 Warszawa

ANTROSHENKO, Valentina – VINOGRADOV, P. A. – GERASSIMOVA, Nina

Inhibitor-fungicide VNKH-L-20 and its use for conservation of metal objects in museums

The possibilities of applying atmospheric corrosion inhibitors for conserving metal objects have been studied in the State Hermitage Museum since the early 1960s [1]. Vapour-phase inhibitors – dicyclohexylamine nitrite and cyclohexylamine chromate – as well as contact inhibitors – sodium benzoate and oil-soluble dicyclohexylamine salt (MSDA-II, M-2) – were examined and applied for conserving articles made of iron, steel and cast iron. Later, with the above aims in view we have been studying some inhibitors of a universal type able to protect all or nearly all kinds of metal from corrosion and biodamage.

In selecting the inhibitors for museum purposes, one of the main criteria is their degree of toxicity which must be the lowest possible. Another important demand is their inertness to the non-metallic materials which often occur in combination with metal or are kept together with metal articles. An important feature is also the fungistatic activity shown by some inhibitors [2, 3]. All these demands considered, the volatile inhibitor-fungicide VNKh-L-20 can be regarded as promising in museum work. It has been developed by the Lenneftekhim company [3-6] and is industrially produced (TU 6-02-7-140-80).

The inhibitor VNKh-L-20 is a morpholine derivative, has the appearance of white powder with an odour of bitter almond; density 1.098 g/cm^3, melting point 100°C. The solubility of VNKh-L-20 at 20°C in ethyl alcohol – 16 %,in isopropil alcohol – 4 %, in acetone – 10 %, in hexane – 2,8 % (mass units). It dissolves in water only to a limited degree: 0,4 % at 40°C. The pH of 0,3 % aqueous solution is 9,5-10,5; the saturated vapour pressure of the inhibitor is $3{,}5x10^{-4}$ Pa at 20°C; $1{,}6x10^{-3}$ Pa at 30°C; $2{,}4x10^{-1}$ Pa at 60°C.

VNKh-L-20 has an anticorrosive action radius of 300-1000 mm and fungicide properties with the action radius of 500-600 mm; Its fungistatic activity is 1:1500, 1:2000 [3]. It is intended for the protection of complex objects consisting both of metallic and non-metallic materials, against atmospheric and microbiological corrosion in the course of operation, keeping, transportation under different climatic conditions (continental, marine, tropical, arctic). VNKh-L-20 protects from corrosion steel, cast iron, copper and its alloys, aluminium and its alloys, silver, lead, tin, zinc, nickel, chromium, cadmium, tungsten, solders. It does not protect magnesium and molybdenum from the vapour phase. The inhibitor can be applied in the form of powder, tablets ("tablin"), inhibited paper, aqueous suspensions. It can be introduced into porous bearers and varnish-pigmented coatings [2-8]. When the hermetic sealing is reliable the protective action of the inhibitor is efficient for up to 19 years.

In the conditions of natural tests for more than 5 years VNKh-L-20 in the vapour phase has no negative effect upon pentaphthalic, polyvinyl acetate, melamine- and urea-formaldehyde varnish-pigmented coatings. Accelerated tests in the presence of the inhibitor vapour at relative humidity 98-100 % and temperature 18-40°C (in one cycle heating for 8 hours at 40°C, then cooling to room temperature for 16 hours) during 15 days showed that as compared to the control the inhibitor produces no changes in the appearance of amber, nacre, jasper, rock crystal, mica. It protects wood, paper, cotton and silk fabric, and leather from mould growth, but in the above test conditions, under the influence of high humiditiy, the elevated temperature and the inhibitor alkaline vapour, leather darkened and became harder, wood discoloured. Polymer materials: acrylic varnish AK-113P, textolite, glasstextolite, teflon, rubber, foamed resins underwent no changes in these conditions either in their appearance or in their mechanical properties.

The data obtained permitted us to recommend the inhibitor VNKh-L-20 for protecting museum exhibits, in particular, during transit. In 1986 the inhibitor was used in packing a large group of metal works of art of West-European and Russian manufacture which were sent from the State Hermitage to a loan exhibition in Switzerland. Many objects were of silver and it was necessary to protect them from tarnishing. In some cases silver was in combination with gold. Some objects were made of steel, in one of them steel was in combination with copper. The articles were packed in carboard boxes with the utilization of micalent paper and porolon. The boxes were then placed into plywood chests laid with oilcloth from inside and painted from outside. Big objects were placed directly into chests. The inhibitor VNKh-L-20 was applied in the form of tablets 9 mm in diameter and weighting 0,36 g. The amount of inhibitor introduced depended on the size of the box or the chest reckoned on the basis of 30-50 $g/1m^3$. Tablin was put into small bags of coarse cotton-wool fabric. Bags were tied up and distributed in the package directly before it was closed. To prevent any contact of the inhibitor with the objects, the bags where put between porolon or hung up in the corners of the chest. It should be noted that the inhibitor was not introduced into boxes and chests where works of art with enamel were placed, taking into consideration the alkaline character of the inhibitor vapour. The conservation gave good results.

The inhibitor VNKh-L-20 possesses a general toxic action but belongs to substances of little danger to man (3rd class).Tablin is a form of the inhibitor which does not produce much dust, but when putting tablets into bags, one shoud work in rubber gloves, overalls, with a respirator or a strip of cotton gauze or, best of all, in a hood. The bags with the tablin, before distributing them in the boxes or chests, should be kept in a closed desiccator or in some other reliably closed vessel. On unpacking, the bags should be put away in similar vessels. Should the inhibitor get on unprotected skin it should washed away with water and soap.

Notes

1. **Gerassimova** N. G. "On the possibility of applying atmospheric corrosion inhibitors for the conservation of metal objects in the museum." - Soobshcheniya VTSNILKR N 19, Moscow, 1967, p. 112-124.

2. **Vinogradov** P. A. – **Altsybeeva** A.I. "Prospects of applying inhibitors-fungicides for protection of archaeological monuments and museum exhibits from corrosion, aging and biodamage." - In: Kratkie tezisy dokladoy Vsesoyuznogo seminara "Novye metody issledovaniya, konservatsii i restavratsii khudozhestvennykh proizvedenij", Ministerstvo kul'tury SSSR, Leningrad, 1985, p.50.

3. **Vinogradov** P. A. "Konservatsiya izdelij v mashinostroenii" (Conservation of manufactured articles in the engineering industry). Leningrad, 1986, 270 p.

4. **Vinogradov** P. A. – **Antroshenko** V. P. et. al. "Developing and study of corrosion inhibitors in tablet form." - In: Ingibitory korrozii v sudostroenii: Trudy Leningradskogo korablestroitel'nogo instituta. Leningrad, 1983, p. 79-82.

5. **Vinogradov** P. A. – **Antroshenko** V. P. – **Ermakova** T. A. "Protection of complicated articles with porous inhibiting materials." - In: Tezisy dokladov nauchno-tekhnicheskoj konferentsii "Teoriya i praktika zashchity metallov ot korrozii". Kuibyshev, 1982, p. 114-115.

6. Informatsionnyj listok N 1172-81. Leningradskij mezhotraslevoj territorial'nyj tsentr nauchno-tekhnicheskoj informatsii i propagandy. Leningrad, 1981. 3 p.

7. **Vinogradov** P. A. – **Antroshenko** V. P. – **Altsybeeva** A. I. "Protection from corrosion of articles made of magnesium alloys with inhibited varnish-pigmented coatings." - In: Obespechenie kachestva i dolgovechnosti lakokrasochnych pokrytij v usloviyakh sovremennogo proizvodstva. Leningradskij dom nauchno-tekhnicheskoj propagandy. Leningrad, 1987, p. 53-56.

8. **Vinogradov** P. A. – **Malakhov** I. V. – **Tsvik** V. V. – **Antroshenko** V. P. "Conservation of pneumo-drives automobile brake system". - In: Vtorichnye resursy dlya zashchity ot korrosii. Gorky, 1987, p. 71-72.

Authors' address:

Antroschenko, V., Vinogradov, P. A.
NPO Lenneftekhim
SU-Leningrad

Gerassimova, Nina
State Hermitage Museum
Dvortsovaya nab.34,
SU-Leningrad 191065

YAKHONT, Oleg

Quelques questions relatives à la conservation du métal dans la sculpture de musée

Résumé

Le présent exposé offre à l'intention des spécialistes quelques exemples, avec lesquels l'auteur s'est rencontré dans son travail pratique dans les musées, concernant le problème de conservation de divers métaux dans la sculpture.

Les oeuvres sculpturales réalisées à partir de divers matériaux tels que le métal, la pierre etc. exigent, comme l'expérience du travail le montre dans les musées, une attention particulière des restaurateurs. Même si une sculpture quelconque a été initialement réalisée en pierre et en plusieurs blocs, alors lors de la réalisation de cette sculpture, ou bien plus tard (lors des travaux de réparation) ses fragments ont été souvent montés au moyen de fixations métalliques. On voit que la sculpture du baroque et du classicisme commence à combiner largement le marbre blanc avec celui en couleurs, avec le granit avec du bronze foncé et patiné de divers coloris et, cela ne concerne que des parties très importantes pour la perception, avec du bronze doré (couronnes, épitaphes etc.). Quant aux éléments de raccordement, ce furent le plus souvent des pièces à partir du métal ferreux (du fer) avec, parfois un joint de plomb. On peut trouver des exemples de pareilles combinaisons dans la sculpture plus ancienne (Grèce antique, Rome, époque de Rennaissance), mais ces cas sont plus nombreux quand il s'agit des oeuvres des XVIII-XIXe siècles. Avec le temps le pièces métalliques desdites sculptures commencent à corroder activement ce qui est dû à l'humidité excessive, à des chutes de température ou à des travaux de réparation avec l'utilisation de matériaux agressifs. Par exemple, la sculpture du XVIIe siècle en marbre blanc de Poseidôn, conservée depuis long temps dans le sous-sol humide du musée-résidence d'Arkhangelskoié s'est vu d'être détériorée à tel point qu'au milieu des années 1970 ses pieds se sont cassés en se défaisant en plusieurs fragments. Après avoir fait l'examen méticuleux on a constaté que cette détérioration avait été provoquée par une corrosion intense des tiges de fer utilisées par le restaurateur du XIXe siècle qui, au début dudit siècle, avait fait l'intégration des pieds perdus. A cause de la conservation dans des conditions d'humidité très élevée de la statue en question les tiges de fer de cette dernière, n'ayant pas été isolées par un joint de plomb du mastic calcaire hygroscopique agressif, ont commencé à corroder d'une manière forte intense. En se dégénérant en oxydes les tiges ont vu agrandir leurs dimensions, ce qui a conduit à une forte contrainte. A son tour la contrainte a détérioré les liaisons cristallines dans la pierre en transformant les zones de pierre les plus proches desdites tiges en une masse meuble, tandis que la partie essentielle de la pierre s'est fendue en plusieurs fragments. Malheureusement ces erreurs de considérations technologiques, typiques pour les XVIIIe et XIXe siècles, coûtent très cher à nos jours lorsque les conservateurs et les restaurateurs n'en tiennent pas compte et continuent à garder les sculptures dans des locaux humides. De plus on aggrave l'état de ces oeuvres aussi bien que celui des murs de bâtiments anciens quand on les lave avec de la vapeur ou de l'eau chaude. Souvent après la fin de ces travaux, qui semblent être à première vue efficaces, on voit que de grands blocs de pierre d'une cathédrale ancienne, âgée de plusieurs siècles, sont rompus en pièces dans les endroits de raccordement au moyen d'étriers métalliques.

En règle générale, lorsqu'on effectue des travaux de restauration d'une sculpture affectée de cette malaise, on fait le nettoyage de la partie intérieure en la libérant des oxydes et de la pierre détruite à laquelle ils ont conféré une couleur respective. Cela faisant, on cherche à conserver au maximum la partie plastique extérieure de l'oeuvre. Si les oxydes ont touché la surface on doit les décolorer en se servant des méthodes traditionnelles [1]. Ensuite on procède à imbiber la partie détruite de la pierre, à recoller les fragments concernés, à les monter en se servant de tiges d'acier inoxidable (ou de titane) et d'un mastic à partir de la poudre de marbre avec une colle: résine epoxyde, cyano-acrylates, Paraloid B-72 et etc. [2]. Une fois tout est terminé, on doit traiter la surface de la sculpture à l'aide de la cire de polyéthylène en vue de sa protection [3].

La corrosion due au raccordement des pièces à partir d'alliages de cuivre avec des éléments à partir du fer (fixations combinées: boulon de fer, écrou d'alliage de cuivre) constitue une cause très répendue de la destruction des sculptures. Les erreurs technologiques de ce genre du siècle écoulé ne sont pas rares. Par exemple c'est le cas des pierres tumulaires (il y a dix ans que nous les avons restaurées) se trouvant dans l'église Saint-Michel du monastère Donskoi, à savoir, le monument à M.Golitsine (sculpteur N.Piménov), monument à M.Sobakina (sculpteur I.Martos) etc. Vu une corrosion intense et une dégénération presque complète du fer en oxydes, nous avons été obligés de remplacer les anciennes fixations en les substituant à de nouvelles fixations analogues mais à partir d'alliages de cuivre (comme le préfère la pratique de restaurateur). Récemment nous avons rencontré des pareilles erreurs technologiques quand nous étions en train de faire l'attribution d'une série d'oeuvres uniques du Musée national d'histoire à Moscou, il s'agit notamment des bustes du baron Karl Beaudet (sculpteur Rachette), du maréchal Ney (sculpteur Houdon), du navigateur Lissiansky (sculpteur Guichard), du philosophe Voltaire (atelier de Houdon), de l'empereur Napoléon I (sculpteur Canova) et d'autres oeuvres réalisées à partir d'alliages de cuivre. Les fixations des supports desdites oeuvres avaient été réalisées en fer, dans certaines oeuvres pour les pièces de fixation on avait utilisé les restes des carcasses de fer de moules à couler [4].

Lors de la production des travaux de restauration dans les deux cas, c'est-à-dire en travaillant avec les pierres tumulaires et avec les bustes ci-dessus, on se servait efficacement de l'émulsion de décapage VEPOS (B3IIOC) qui comprend en pourcentage en poids: essence-4, acide oléique – 0,5, eau – 44,5, morpholine – 0,5. Possédant une douce capacité de décapage, cette émulsion permet d'enlever des encrassements superficiels et des oxydes non stables sur le bronze doré. L'élimination des encrassements et des oxydes plus stables a été faite conformément aux recommandations de Plenderleith [5], c'est-à-dire d'après la méthode traditionnelle. Quant au traitement de protection, on l'a fait au moyen de la cire de polyéthylène. Les pièces dorées ont été réitérativement couvertes d'une couche de la cire colorée cette fois-ci au marron par un pigment de pastel, selon les recommandations du Conseil de restauration, en vue d'étouffer un vif éclat de l'or bien conservé qui se mariait mal avec l'ensemble de l'oeuvre ancienne. Comme les observations menées plus de dix ans le prouvent ce traitement permet de conserver parfaitement le métal contre l'oxydation même dans des locaux assez humides. Quant à la mise sur ton, on peut toujour l'enlever sans porter aucune atteinte au monument.

L'élimination des encrassements et des oxydes est une opération qui n'est pas toujours la même, elle varie en fonction de chaque monument concret. Le choix d'une méthode à appliquer dépend de la nature du monument, de l'état de son conservation, des conditions de conservation, des résultats des recherches poursuivies. Parfois ce processus ne semble pas d'être difficile quant aux méthodes utilisées, mais en revanche la responsabilité et le volume de travaux à accomplir peuvent être énormes. Cela concerne par exemple la Reine-Cloche (Tsar-kolokol) et le Tsar des Canons (Tsar-pouchka) du Kremlin de Moscou. Pour exécuter les travaux de conservation sur lesdits monuments on a mis au point une méthode simple et assurant la sécurité du travail. Le restaurateur n'a pas été seul à élaborer ladite méthode, y participaient des savants des autres centres de recherche. Bien qu'on ait été contraints d'accomplir les travaux en engageant un grand nombre de nonprofessionnels, la préparation, la conduite et le contrôle desdits travaux ont été réalisé par un restaurateur de la plus haute catégorie. Les travaux en question est une exception car leur exécution exige l'engagement de forces colossales, mais en ce qui concerne ce cas concret, les monuments en ont profité [6].

Comme l'expérience le montre les pièces de fixation dans les oeuvres sculpturales du XII-XVII^{e}s. sont souvent faites à partir d'alliages de cuivre, si elles sont de fer, alors elles sont fermées à l'aide d'enveloppes de plomb. En règle générale, si les pièces intercalaires de fer sont dépourvues d'isolation de plomb, on peut dire qu'il s'agit des travaux de réparation du XIXe siècle. A titre d'exemple on peut citer la statue en pierre blanche du XVe siècle de Saint George terrassant le Dragon du Kremlin de Moscou. Les erreurs technologiques du restaurateur du siècle écoulé, qui avait utilisé des tiges de fer comme éléments intercalaires, ont conduit en fin à des pertes et détériorations considérables dans la sculpture. Cherchant à arrêter la destruction ultérieure on a été obligé de retirer ces pièces intercalaires (le fer y a été complètement dégénéré en oxydes). Les pièces retirées ont été soumises à une étude qui avait pour but de déterminer leur composition, la technologie de fabrication et l'époque de fabrication. Les analyses ont démontré que lesdites tiges de fer avaient été faites à main dans les années 1810, les autres, dans les années 1890, leur qualité technologique laisse à désirer [7].

Les recherces effectuées soulèvent un autre problème, celui de la conservation de ces pièces fortement corrodées. Aujourd'hui, la nécessité de poursuivre de diverses recherhes d'une façon la plus intégrales (physiques et chimiques, biologiques, technologiques, d'archives etc.) nous oblige de voir sous une autre optique le problème d'ajoutures et de pièces d'insertion tardives (cela concerne égaelement les tiges de fer en question etc.) qu'on retirait et détruisait facilement auparavant; cela est devenu le problème numéro un dans la restauration. Les fragments retirés représentent une partie du monument, de son histoire et on doit les conserver telles quelles sont (sans les faire à neuf). Chaque objet de musée n'est pas seulement une oeuvre d'art plastique, mais c'est à la fois un monument d'un métier déterminé, de la technique et de la science, d'une période quelconque de l'histoire humaine, en outre chaque monument porte aussi des qualités que les siècles lui confèrent (qualités stylistiques, écologiques etc.). En d'autre termes chaque objet de musée est un ensemble, une banque de données historico-culturelles qu'on doit conserver soigneusement et d'une manière la plus complète. En éliminant complètement une partie quelconque, par exemple des tiges de fer corrodées, nous détruisons ainsi un élément témoin qui représente une certaine importance pour l'histoire du monument ainsi que pour son attribution. Peu importe que ces pièces sont bien étudiées maintenant [8], il peut arriver que dans 50 ou 100 ans les données de nos investigations seront insuffisantes ou s'avéreront contestables. De nouvelles méthodes d'analyse, plus précises et perfectionnées, viennent progressivement à notre aide, c'est pourquoi la perte des pièces en question nous privera de la possibilité d'obtenir des données intéressantes. Comme il nous semble, il faudra conserver des échantillons de contrôle types des fragments à enlever, tels qu'ils sont, sans aucun traitement ni modifications apportées. Les échantillons sont à mettre dans des capsules appropriées (de plomb ou d'un autre matériau neutre par rapport au matériu à conserver). Dans les cas où les pièces de fixation sont solides et fonctionnelles, il est désirable de les laisser sur place. A cet effet on doit préalablement les soumettre au traitement de conservation et puis les isoler d'un matériau agressif donné.

Bibliographie

1 **Plenderleith** H. J.:The conservation of antiquities and works of art, Treatement, Repair and Restoration. London, Oxford University Press, New York, Toronto, 1956

2 **Yakhont,** O.: On the restoration methods of museum stone sculpture – Second International Restorer Seminar. Budapest, 1979, p. 87-93 ; Several notes on the completion of alebaster works of art – Fourth International Restorer Seminar. Budapest, 1984, V.1, p. 242-246

3 Idem

4 **Yakhont,** O.: Quelques propositions sur la conservation des monuments dans les conditions contemporaines du milieu environnant – Symposium International "La géologie de l'ingénieur appliquée à l'étude, à la préservation et à la protection des travaux anciens, des monuments et des sites historiques." Athènes, 1988, V. 3-5/1.

5 **Plenderleith** H. J.: The conservation of antiquities and works of art.

6 **Yakhont,** O.: Exemple de restauration d'un ouvrage de l'art monumental. Conservation des oeuvres d'art et décorations en métal exposées en plein air.- Symposium – Paris, 1986, p. 291-297

7 **Yakhont,** O.: Problèmes de restitution des oeuvres d'art médiévales ayant subi plusieurs restaurations dans le passé.- Fifth International Restorer Seminar. Budapest,1985, V. 2, p. 21-27

8 **Rastorguev,** L.-**Yakhont,** O.: On the application of indirect data in restoration research. ICOM Committee for Conservation, the triennal Meeting... Sydney, 1987, V. 2, p. 571-572

L'adress de l'auteur:

VNIIR
10, Krestyanskaya pl.
SU-109172 Moscou

ORGAN, Robert M.

A survey of current treatments: a contribution towards a knowledge base for a computer's expert system

Abstract

The simplest version of an expert system is described. In order to be useful to conservators an adequate "knowledge base" is needed. Here a framework for knowdlege is suggested and some material entered under the headings: Objectives; Limitations; Procedures; Reagents; Unsuitable conditions; References (taken from recent literature, these support many of the entries).

A. An **EXPERT SYSTEM** is a computer program that enables anyone, by entering his own observations, to reach the same conclusions as would an expert who had made the same observations. Such programs are being used increasingly in industry to store the experience gained by personnel while operating manufacturing processes. They ask a series of questions. When all are answered, a logical conclusion is reported. A simple illustration follows. Program (P): What is your question? Conservator (C): Will my object be stable at high relative humidity?

P: What is the object made of? C: Corroded iron.

P: Is any other material attached to it? C: No.

P: What is its specific gravity? C: 1.8.

P: It will be stable at high relative humidity (Watkinson 1983)

This very simple example requires a type of program known as a "decision-tree" consisting of "If-Then" branches. It performs rather as do the diagrams on metals at the ends of some of the chapters in Plenderleith 1971 or on archaeological wood in Jedrzejewska 1984.

The setting up of such a program requires a "knowledge base", comprising all of the facts and feelings that the expert has, starting with his objectives of treatment and including all of the pros and cons that will worry him during his consideration of a problem. This paper is an attempt to bring together some of the needed ideas, selected from recent published case histories, and to put them into a rational framework. They are not set out as for a program. A comparable but more thorough compilation than this, not intended for use in a computer program, nor for materials other than paper, has been in course of production for some years for members of the Book and Paper Speciality Group of the American Institute for Conservation under the title "Paper Conservation Catalog".

An attempt to set up even a limited expert system would very quickly reveal our ignorance on important facts and would stimulate further useful research. The framework provided here can be put to use and extended by any conservator. He can insert his own well-tried procedures and his own name under any of the code-headings. Perhaps at a future Veszprém Seminar many such collections could be bought together for combination into a useful knowledge base.

If ever such expert systems do come to be established for conservators they are unlikely to displace the actual expert. Use of them would, however, ensure that all of the necessary questions surrounding the problem are asked and answered before important decisions are taken, whether the advice of the computer is adopted or not. Furthermore, by associating the name of its originator or extensive user with each useful procedure, as is done here, the program would make easy a reference to the actual expert if help should be needed.

B. **SOME EXAMPLES** of objects containing metals contiguous with organic material follow. In this list the materials associated with metal include: bone, ivory, leather, paper, rock crystal, textiles, wood.

B1 párta-belt, metallic rivets through leather (Koltai 1987)

B2 pins made of bronze centred in a wooden plug that attached Bronze Age dagger blades to their hafts (Gallay 1983)

B3 paper labels on tinplate (Hawley 1981)

B4 18thC steel sword having hilts decorated with leather and textile (Kyriacou c1984)

B5 furniture, wood with metal and other fittings (Howlett 1985) (Nováková 1987)

B6 wooden chairs, painted and water-gilded, with bronze plaques, plaster rosettes, silk coverings (Bickersteth 1987)

B7 naval grenade, wooden fuse in metal shell (O'Donnell 1985)

B8 wooden shrine, gilt copper nailed on (Koller 1982)

B9 metallic thread in textiles (Darrah 1987) (Barker 1980)

B10 gilded (coated tin) leather (Stambolov 1980)

B11 copper sword, 14thC BC, ivory and wood inlay, from undersea (Payton 1987) Many others can be found in a museum.

The different materials of these composite objects will have deteriorated, each according to its character. In general, treatment is undertaken in order to enhance the physical and chemical stability of the object in its current environment or to restore its appearance to that of an earlier date.

O. **THE OBJECTIVES OF TREATMENT** needed by one such object are developed after certain preliminaries, namely: penetratingly detailed examination, supported by tests, either non-invasive, such as Xeroradiography (Magliano 1988) or else made at unobtrusive representative sites; consideration of its place in history;

consideration of its function in the collection, or as a regional or national treasure. Ultimate objectives necessarily vary from object to object. One or more intermediate objectives will probably be found in the following list. See Section **R** for reagents that could be used and section **P** for procedures.

01 *Remove*:

01.1a accreted material acquired by both metal and organic materials from outside, e.g. dust, soot, grease, deposits of aerosol salts, together with any occluded pollutants that commonly accelerate deterioration of their host

01.1b stains penetrating into a surface

01.2 decayed surface. This is sometimes brittle (h) sometimes soft (s). Its support may be hard tough, and elastic (h) or inelastic, dentable, scratchable (s). The absolute values are often less important for purposes of cleaning than the relativity between surface and support.

01.2hh discolouring and brittle polymeric material (e.g. varnish) from a hard support

01.2sh greasy, pasty, viscous, gummy protectives from a hard support such as metal. Remember that metals such as tin, pewter, lead and copper are relatively soft, but they will be harder than their organic overlay.

01.2hs discolouring coatings (hard) from a soft support such as wood

01.2ss decaying dressings (soft) from a soft support such as organic material

01.3 growths of minerals (oxides, sulphides, etc), either layered or warty, from residual metal or that obscures decoration.

01.4 hard calcareous concretion from some buried or underwater finds (wood with iron)

01.5 crystals or efflorescence grown from within

01.6 internal hidden sources of corrosion e.g. sources of "bronze disease" or of rust

02 *Change* into substances that are less harmful any minerals present, such as chlorides or acetates, that accelerate corrosion

03 *Consolidate* loose surfaces or 'size' soft paper, with no or negligible change in appearance

04 *Fill* disfiguring or weakening cracks in:

04h metals **(h)**

04s organic materials **(s)**

05 *Soften* organic materials:

05.1 preliminary to closing cracks

05.2 to enable distorted shapes to be corrected

05.3 when a brittle condition appears likely to permit further cracking

06 *Reinforce*:

06h weakened metal

06s structurally weak organic material

07 *Reconstruct* missing parts or areas of material

08 *Leave evidence intact* when early users of an object have left their marks (AIC 1988) (Herskovitz 1984) (Loy 1983)

09. *Protect*, in order to stabilise condition, against:

09.1 future corrosion of metals

09.2 future deterioration of organic material

09.3 future growth of moulds or other biological agents in organic material

09.4 other disfiguring change

L. LIMITATIONS restrict our approach to our objective:

L1 *Practical:* can we do it? Do we have the necessary knowledge, material, equipment, skills?

L2 *Ethical*: should we do it? general (AIC 1988); cultural property (Pruszynski 1985); furniture (Tigreat 1984); various (Járó 1982).

Can what we shall do be undone, if it should prove to be misguided or unsatisfactory: now? ten years hence? one hundred years hence?

L3 *Legal:* may we do it? personal exposure to toxic materials is regulated in U.S.A. by the Occupational Safety & Health Act; disposal of substances such as phosphates that are harmful to the eco-system is regulated; use of materials from endangered species is forbidden, e.g. feathers, ivory (Appelbaum 1986); in Massachusetts U.S.A. there is an artists' Moral Rights Law (Walker 1985)

L4 *Economic:* can we pay for it? in staff-time? in money?

P. PROCEDURES POSSIBLE as treatment

P1 *Separate* different materials so that each can be treated individually (Franke 1980):

P1.1. mechanically, perhaps in favourable cases by reversing the method of construction but then with risk of losing weakened screws, rivets, pins, etc

P1.2 cause differential contraction:

P1.2a in metals by local or general alteration of temperature by

P1.2a1 contact with hot or cold "iron" through liquid heat conductor (Grant 1985)

P1.2a2 blast of hot air (e.g. Leistor-Labor heat gun)

P1.2a3 focussed raint heat

P1.2a4 for immovable lock of shotgun, liquid nitrogen, -196°C (Grant 1985)

P1.2b in organic materials by alteration of moisture content (O'Donnell 1985)

P1.3 weaken adhesive that joins two materials

P1.3a by injection of acetic acid into animal glue joining pieces of wood (Robertson 1988)

P1.3b six methods for separating permeable layers (Ash 1985)

If different materials are inseparable and have to be treated together, none to be injured by treatment needed for another, then:

P2 *treat in place* each material by a mechanical method that can be localized, examples arranged here roughly in order of increasing severity:

For objectives 1.1 to 1.5, is the deposit to be removed brittle or gummy?

If brittle, select from procedures P1 or P2

If gummy, select from P 2.1.7, P 2.2.1c. P2.2.3V or P3

For objective 01.1, removal of lightly-attached accreted brittle material

P2.1.1 jet of air or gentle vacuum – for feathers (Rae 1987)

P2.1.2. lambswool or feather "duster" or soft brush

P2.1.3. sticky cloth pick-up, e.g. painters' "tack rag"

P2.1.4. rolled-on powdered so-called "gum- rubber" or Draft-Klean

P2.1.5 stencil brush – for glass (Appelbaum 1982)

P2.1.6 rubwith rubber or "vinyil" eraser – for marked paper (Pearlstein 1982) (Moffatt 1981) (Michaels 1986)

P2.1.7 scraping

P2.2 removing unwanted material from above a natural interface:

P2.2.1 by cracking along the interface, using:

P2.2.1a finger friction for hard varnish over elastic paint (Boissonas 1987)

P2.2.1b blows of a light-weight hammer (Organ 1968, p155)

P2.2.1c differential contraction

P2.2.1c1 for thick beeswax, solid carbon dioxide, "dry ice", at -78°C (Alof 1987)

P2.2.1c2 for viscous coating, liquid nitrogen (Gale 1984)

P2.2.2 picking with a needle under a microscope, sometimes aided by an electric vibrator

P2.2.2a1 for brittle material (Plenderleith 1954, 1971, p204) (Organ 1968, p153)

P2.2.2a2 for soft material, with a scalpel (Hawley 1981)

P2.2.2b chiselling with an ultrasonic dental "drill" (Livens 1974)

P2.2.3 volatilization by heat

P2.2.3Va using a selected laser under selected conditions when an overlay is darker than its underlay,

P2.2.3Va1 for tarnish on silver thread (Asmus 1978a)

P2.2.3Va2 soot on hand-painted Oriental silk (ibid)

P2.2.3Va3 coffee stains on paper (ibid)

P2.2.3Va4 calcareous deposits on lead (ibid)

P2.2.3Va5 minerals on bronze (ibid)

P2.2.3Vb high-power xenon flashlamps – for sulphation on flat areas of marble (Asmus 1978b)

P2.2.3S controlled spark discharge, to peel off electrically insulating coatings from metal (Piersig 1982)

P2.2.3H heat to soften an entire susceptible organic layer so that it can be pryed off (traditional but potentially destructive)

P2.2.3Ha hot air (Leister-Labor)

P2.2.3Hb radiant heat

P2.2.3Hc steam (McCrady 1981)

P2.2.4 localised abrasion, ranging in effect from polish to attrition, by granular powders of selected size and hardness applied either (m) manually on a form-following resilient pad while dampened with lubricant (water, stearates, oil) or (p) on a power-driven rapidly rotating felt buff whose lubricant is then the waxy matrix into which the abrasive powder was compounded

P2.2.4a damp sodium bicarbonate crystals, an alkaline water-soluble abrasive – for iron (Rácz 1981)

P2.2.4b powdered feldspar (Bon Ami)

P2.2.4c metal polishes – Solvol Autosol (Kyriacou c1984)

– abrasive wadding (silica and lubricants in Duraglit, Nevr-Dull)

– the opaline silica phytoliths in cats-tail stems (Nature's Duraglit, used in China)

P2.2.4d steel wool of fine grade used with lubricant

P2.2.4e glass-bristle brush (Kyriacou, c1984)

P2.2.4f harder abrasives on a coated paper or cloth supported on a rotating rubber buff or set into a rigid resin-bonded rotating disc – for an elementary survey (Organ 1968, pp245-271)

P2.2.4g brushed gel containing abrasive and detergent (Higuchi 1974) (N.B. risk of particles or paste becoming trapped in irregularities of surface, in corrosion pits or engraving, since most abrasives cannot be dissolved but only wiped or washed away)

P2.2.5 air-abrasion, masking off surrounding areas with resilient tape or rubber sheet, using various dry abrasives ranging in hardness from walnut shell through glass beads to silicon carbide

P2.2.5a brittle varnish over sound elastic paint, micro-airbrasive (Boissonas 1987)

P2.2.5b out-door bronze, warning (Chase 1985)

P2.2.5c a sword (Kyriacou c1984), rust over silver (Aoki 1987)

P2.2.5d ethnographic, a survey of metal, basketry, bone, ivory, leather, textile, stone and wood (Gibson 1969)

P2.2.6 burnish to re-establish overall polish e.g. by agate or bloodstone or steel (chain burnisher) rubbed over a soaped surface – brass clockwork (Rácz 1981)

P 3 *apply fresh and remove spent chemical reagents:*

P3.1 locally

P3.1.1a impermeable medium, apply and remove: by brush or swab or cotton bud or by bulb pipette (squeeze to eject reagent and relax to reabsorb it, rolling does both)

P3.1.1b on a porous medium (some varieties of paper, etc): remove under vacuum through a sintered glass plate (Collings 1985)

P3.1.2 by jet of liquid, masking off lower surroundings (Barker 1953) (Organ 1968, p111)

P3.1.3 in a strip of fabric to control quantity and thickness of application (Mel'nikova 1984)

P3.1.4 as a gel, for crusts of sulphate, use bicarbonates in carboxymethyl cellulose (Barov 1988)

P3.1.5 in a poultice or pack from whose outer face the solvent portion of the reagent can evaporate – textile, for adhesive, use acetone in Sepiolite (Heuman 1987) or enzymes in methyl cellulose gel (Smith 1987)

P3.1.6 reduce or oxidise by hand-held electrode wrapped in cloth wetted by electrolyte (Organ 1968, p113) (Aldaz 1986) e.g. for redeposited copper (Goldstraw 1986); for silver (Heinrich 1984); for lead (Guida 1980)

P3.1.7 reduce by brushed-on electrolyte under aluminium foil – coins (Hiekkanen 1981) or glycerine/agar and tinfoil in a humid atmosphere – bronze antiquities (Rosenberg 1917)

P3.2 generally, but masking off surrounding areas by temporary application of a barrier film – polyethylene (Kyriacou c1984) or of electroplaters' stop-off lacquer (e.g Canning's Lacomit)

P3.2.1 by spray (Schniewind 1984)

P3.2.2 by complete immersion of the object

P3.2.3 by immersion of the object under vacuum, wood sculpture in acrylate in plastics bag (Kaftanova 1983)

P3.3 generally, but using a reagent which is specific in action, so that it will act only on the material to be treated:

P3.3a emulsifying soaps (Wolbers 1987, 1988)

P3.3b solvents for grease, Genklene (1,1,1 trichloroethane) (Kyriacou c1984)

P3.3c sulphide from silver yarn in tapestry (Columbus 1980)

R. SOME CHEMICAL **REAGENTS** OR ENGINEERING MATERIALS applied to achieve specific objectives. Arranged roughly in order of increasing risk to the unaltered remains of the object. Some of these reagents serve for the worst-possible condition of the object's material that will not always be encountered.

Remember that:

a) a thin film of tarnish on a recently neglected object presents a different problem from that of the thick crust on an ancient excavated object;

b) quality and extent of action varies with concentration and with temperature and duration of application;

c) the method of application selected from list P3 may influence the reaction – some methods enhance oxidation by surrounding air.

For an objective listed as:

01 *Removal,* reagent is **R**xxxx

For 01.1**A**, accretions:

For 01.1**S**, stains

RO1.1a water

RO1.1a1AS plain; dissolves soluble material including products of deterioration or corrosion, carries them into contact with material that it wets, deposits them as stains or crusts at the place of evaporation – some results wanted, others not! Select from this list arranged from high to low purity and low to high cost: triple distilled, distilled, ion-exchanged, reverse osmosis, domestic water: quality measured in microSiemens per cm. Chloride in water softened by treatment may be acceptable for textiles but not for metals.

RO1.1a2A steam, for paint on cast iron or wrought steel (Sekowski 1983)

RO1.1b water + detergent

RO1.1b1A from marble, aqueous detergent froth (Plenderleith 1971, p312) from wood (Morgós 1987b)

RO1.1b2AS from textiles, aqueous detergent liquids (Hofenk de Graaf 1982) (Easthaugh 1987)

Warning: some irreversible damage to silk and wool (Golikov 1987)

RO1.1c solvents, plain, e.g.:

RO1.1c1A from enamel paint, petroleum spirit on cotton (Bickersteth 1987)

RO1.1c2AS from grease on doeskin, white spirit (oil paint diluent), (Kyriacou c1984)

RO1.1c3A from ivory, ethyl alcohol or acetone (Matienzo 1986)

RO1.1c4A from parchment, aq ethanol on cotton (Dreibholz 1983)

RO1.1c5AS from textile, tetrachlorethylene is good but its residues hasten experimental ageing (Gromina 1987)

RO1.1c6A from wood, alcohol on cotton (Rácz 1981)

RO1.1c7A from finished wood, xylene on cotton (Howlett 1985)

RO1.1d solvents with detergents e.g.:

RO1.1d1AS for textiles, tetrachloroethylene with detergent is best (Golikov 1987)

RO1.1d2A from woody fibre of basketry, methylcyclohexyloleate in Stoddard solvent (Vulpex) (Krongkright 1981)

For 01.2, removing a decayed or corroded surface from its support:

RO1.2a for a painting, after appropriate study (Wolbers 1988) use emulsifying soaps (Wolbers 1988a)

RO1.2b swell in solvent and remove. Select solvent in light of a Teas plot (Toracca 1975)

RO1.2c for a painting, use shaken-together immiscible solvents (Hook 1988)

RO1.2d for paintings, use a solvent direct (Feller 1977) (Morgós 1987a)

RO1.2d1 for aged linseed oil on paper, swab with DMF (Hawley 1981)

RO1.2e reform surface with solvent and clean with a gentler solvent (Raft 1980)

RO1.2f craze with organic liquid and push off – chloroform for polyester (Organ once used this)

RO1.2g swell in steam and remove (McCrady 1981)

Compare RO1.1a2A

For 01.3, removal of growths of minerals, use

(Warning: do not treat metallic thread [Darrah 1987])

RO1.3 aqueous solutions:

RO1.3K alkalis

RO1.3Ka for tarnish on brass, ammonia in industrial methylated spirit (Kyriacou c1984)

RO1.3Kb for rust on paper, fresh warm 10% aq dithionite (Hawley 1981)

RO1.3Kc for copper sulphides, alkaline Rochelle salt (Duncan 1987)

RO1.3C chelating compounds

RO1.3Ca for corrosion products on copper, 10% NA2EDTA plus additives (Cjeka 1975)

RO1.3Cb for iron-stained (akaganeite) textiles, briefly Na2EDTA+oxalic acid + NaCl, pH2.5 (MacLeod 1984)

RO1.3AI acids, inhibited

RO1.3AIa for copper minerals, thiourea and citric acid (MacLeod 1987a)

RO1.3AB acids, buffered

RO1.3ABa for copper sulphides, oxalic acid + sodium oxalate (Duncan 1987)

RO1.3A acids, plain

RO1.3Aa for concretions, aq formic acid (Payton 1987)

RO1.3Ab for copper sulphides, citric, oxalic acids (Duncan 1987)

For 01.5, removal of crystals or efflorescence. After analysis select from RO1.3, applied by P3.1.5

For 01.6, removal of chloride. It is often necessary to use 02 first.

RO1.6a for copper (bronze disease), alkaline dithionite away from air is better than (>) citric+thiourea > 5% sod sesquicarbonate > 1% BTA > 50% acetonitrile > 1% sodium sesquicarbonate > distilled water (MacLeod 1987)

For O2, *changing* harmful substances:

RO2a chlorides from bronze:

RO2a1 aqueous or alcoholic BTA (MacLeod 1987)

RO2a2 sodium carbonate before BTA (Weisser 1987)

RO2b acetates from bronze:

RO2b1 zinc nib connected to metal of object, wetted with dilute hydrochloric acid then phosphoric acid then sodium carbonate (Rayner 1951)

For 03, *Consolidation* of

RO3S surface coatings

RO3Sa on ethnographic material, gelatine + methyl cellulose + tragacanth (Bauer 1965)

RO3Sb loose Papuan paint, B72 acrylic in toluene (Reeve 1987)

RO3Sc ethnographic matt paint, B72 acrylic in diethylbenzene (Welsh 1980)

RO3Sd flaking paint on wood, molten dammar plus beeswax (Onnen 1983)

RO3Se on paper, aqueous methyl cellulose (Hawley 1981)

RO3Sf on leather, red-rotted, re-tan with aluminium triformate in methanol (Haines 1984)

RO3Sg for dark leather, red-rotted, after re-tan apply Lankrothane polyurethane in petroleum ether then acrylic wax (Haines in Fogle 1985, p33)

RO3Sh on light leather, red-rotted, after re-tan, apply hydroxypropyl-cellulose (Klucel G) (Haines in Fogle 1985)

RO3Si illuminations on parchment, methylcellulose in methylene chloride + methanol (Plossi 1981)

RO3Sj watercolour, B72 acrylic in acetone (Burgess 1983)

RO3M massive materials

RO3Ma mud, hydroxypropylmethyl cellulose in damp ethanol (Dodds 1988)

RO3Mb decayed wood, fine acrylic dispersions in water-alcohol (Eremina 1987)

RO3Mc decayed wood, rosin plus paraffin wax in trichlorethylene (Nakhla 1986)

RO3Md decayed structural wood, experimentally designed mobile epoxy resins (Kol'tsova 1985); furniture, designed Rütapox (Simpson 1986)

RO3Me pinewood, a recirculating spray of Butvar 98 in ethanol (Schniewind 1984)

RO3Mf brushwood, PEG (Franke 1980)

RO3Mg fragile woody tissue, polyethylene glycols in methanol (Kronkright 1981)

RO3Mh for porous materials, aqueous acrylic dispersions, PRIMALs, not emulsions (Koob 1981)

RO3Mi minerals consolidated as metal

RO3Mi1 lead by electrolytic reduction in sodium sulphate (Horie 1980)

RO3Mi2 silver by electrolytic reduction in sodium hydroxide (Charalambous 1975)

RO3Mi3 silver by reduction in alkaline dithionite protected from air (MacLeod 1979)

For 04h, *Filling* craks in

RO4h metals (Untracht 1968):

RO4ha soft solders, melting points, 183-255°C (Pb-Sn-Sb), improper for silver (Karl 1987)

RO4hb fix filling piece with hard solders (Ag-Cu-Zn-Sn), melting points 300-900°C

For working musical instrument, silver on ivory (Weber 1984)

RO4hc epoxy compounds

RO4hc1 metal-filled epoxy resins

RO4s organic materials

RO4sI Inelastic fillers

RO4sIa glue-bound, particles or fibres, – for wood, gesso (Bickersteth 1987)

RO4sIb for stucco, glass microballoons with Primal AC34 (Higuchi 1974)

RO4sIc plaster of Paris (traditional)

RO4sId coloured waxes

RO4sIe "liquid wood", filled cellulose nitrate (Nacsa 1987); warning (Koob 1982)

RO4sIf mobile synthetic resins which cure solid in place and can then never be removed later

RO4sIg for wood, epoxy resin filled with hardwood sawdust (Morgós 1987b)

RO4sE Elastic fillers

RO4sEa for wood, microballoons in silicone rubber (Barclay 1987)

For 05, *Softening* organic material before reshaping or rearranging it:

RO5a for varnish, solvents (Raft 1980)

RO5b for birchbark, methyl or ethyl alcohol (Gilberg 1986)

RO5c for leather, Pliantine (Kyriacou c1984), leather dressings (Tuck 1983)

For 06, *Reinforcing:*

RO6h weakened metal

RO6ha bronze to epoxy support with acrylic (Eichhorn 1986)

RO6hb tin to acrylic with acrylic Piafex (Pasch 1985)

RO6s weak organic parts

RO6sa PVA as glue to a support (Franke 1980)

RO6sb Pliantex as glue to nylon fabric (Kyriacou c1984)

RO6sc paper, to strengthen cellulosic fibres directly, graft copolymerisation of acrylic monomers (Burstall 1986), ethylene oxide (Sears 1982)

RO6sd "size" on soft paper

RO6se for leather, apply gelatine, egg white or polyurethane (Van Soest 1984)

For 07, *Reconstruction:*

RO7a remake part in similar material by original method (Bickersteth 1987) (Nováková 1987)

RO7b for silver, electrotype replica (Northrop 1980) (Karl 1987)

RO7c for silver, replica in Viapal (Koltai 1984)

RO7d for outdoor bronze, acrylics (Beale 1983)

RO7e for leather, fibre filler in pigmented PVA (Raphael 1984) (Van Soest 1984)

For 09, *Protecting.* Only initial treatments are considered here, not the continuous vigilance essential for long-term protection.

09.1 against corrosion of metals

RO9.1a1 wax bronze, avoiding beeswax, lanolin or acidic paraffin waxes or oils (Burmester 1985)

RO9.1a2 lacquer bronze with acrylic or silicone medium, more durable than wax (Riederer 1984) (Weil 1982)

RO9.1a3 Incralac acrylic on bronze exposed outdoors becomes less soluble in ten years (Erhardt 1984)

RO9.1a4 coat iron annually with wax-lanolin mixture in White Spirit (Child 1986)

RO9.1a5 paint iron with urethane-alkyd pigmented with micaceous iron oxide (Williams 1987)

RO9.1b1 treat monumental bronze with sulphide to allow existing chloride to leach out (Block 1987), on high-tin bronze a sulphide layer is not permanent (Kalish 1975)

RO9.1b2 change white lead carbonate corrosion crust on lead objects to black lead sulphide with hydrogen sulphide (Brunnert 1985)

RO9.1b3 treat iron on a wooden chest with tannin (Stadler 1987)

RO9.1c reduce active corrosion products to more stable mineral

09.2 against deterioration of organic materials:

RO9.2a against surface grime, synthetic resins (Feller 1984)

RO9.2a1 on wood, wax (Stadler 1987)

09.2b against ageing of organic materials:

RO9.2b1 of leather on a support, add fat dissolved in solvent through cracks in the surface (Van Soest 1984, p28)

RO9 2b2 of paper, deacidify by selected method (Baynes-Cope 1984)

RO9.2b3 of inked paper, spray on size plus deacidifier made of hydroxypropylcellulose and methyl magnesium carbonate in industrial methylated spirit (Hermans 1988)

RO9.2b4 of textiles, incorporate alkaline finish – magnesium or methyl magnesium carbonate (Kerr 1982)

RO9.2b5 for textiles, to lessen deterioration of linen, deacidify with magnesium salts (Peacock 1983)

RO9.2c against rapid change of relative humidity by coating ivory with polyvinylidene chloride latex (Lafontaine 1982)

09.3 against moulds

RO9.3a apply fungicide e.g. p-nitrophenol (Herskovitz 1984)

09.4 against damage by insects

RO9.4a treat with pyrethroid insecticide of low toxicity to humans (Morita 1987)

S. CONDITIONS OF TREATMENT **SUITABLE** FOR SOME MATERIALS AND INCOMPATIBLE WITH OTHERS:

S1 *cellulosic organic materials:* damage caused by

S1.1 a too long duration or sudden or severe change in moisture content while wetting with water, drying with desiccant liquids (e.g. ethanol) or during change of ambient RH. Lessen shock by preliminary swelling in high relative humidity (Block 1987a)

S1.2 change in temperature sufficiently severe or prolonged to change moisture content.

S1.3 making acid any inlays or paper containing carbonate

S1.4 change of pH to value outside range of 5 to 9 increases rate of degradation: most rapid at extremes

S1.5 materials such as glue having pH below 3.5 which damage wood in contact with it (Pizzi 1986)

S1.6 alkaline complexed Cu, such as is present after certain alkaline reagents have been applied to copper alloys

S2 *protein organics* (leather, silk, wool): damage caused by

S2.1 pH outside range 3 to 6 (Van Soest 1984)

S2.2 organic solvents that remove essential fats from leather to below the recommended minimum of 5% (Van Soest 1984)

S2.3 heat, causes shrinkage

S2.4 absorption of metallic ions accelerates decay – for Cu and Fe, fix ions by chelation with Cupferron (Sanders 1986)

S2.5 wetting – vegetable tanned leather blackens and curls irreversibly (Appelbaum 1982)

S2.6 leather dressings in contact with textiles (Appelbaum 1982)

S2.7 some compounds used to chelate metal ions, such as thioglycollic acid, react with proteins

S3 *other organic materials:*

S3.1 linseed oil hydrolyses at pH <5,5, >8,0

S3.2 heat may cause change

S4 *metals:*

S4.1 exposure of freshly cleaned iron or steel to dust or corrosion-accelerating vapours (Vernon 1927)

S4.2 contact of free fatty acids (oleic, stearic):

S4.2a in leather with Cu, Zn, Sn, not Fe, will corrode these metals (Stather 1935)

S4.2b in ageing linseed oil or beeswax may cause corrosion of bronze after more than ten years in the open air (Burmester 1985)

S5 *composites:*

S5.1 Contact of iron with wood of pH less than 5 when RH is higher than 75% permits the salts from corroding iron to damage the wood (Smith 1982)

S5.2 At 90% RH, contact with preservers added to wood causes metals, especially ferrous metals, to corrode (Barnes 1984)

References

The object described may not itself be composite but the method described may be applicable. These are by no means the only references available. A recent specimen has been selected as proof that the method is in line with current thinking and that the materials are available. Research studies or surveys have been preferred. AATA references are included, when available, since ABSTRACTS may be accessible even if actual publications are not.

AIC 1988
AIC Code of Ethics, Annual Directory, p21
Respecting the integrity of the object...Limitations on aesthetic reintegration
Annual Directory, American Institute for Conservation, Washington DC

Aldaz, A.–**Espana,** T.–**Montiel,** V et al 1986
A simple tool for electrolytic restoration of archaeological metal objects
Studies in Conservation, **31,** 4, 175-176 London.

Alof, Marion 1987
The Brancacci Chapel
Conservation News UKIC 33 (July) 11-13 London

Aoki, Shigeo 1987
Study of iron objects found in a Tsukubasan tomb, Gunma prefecture, and their conservation
Hozon Kagaku **26,** 35-41, Tokyo, AATA 25-823

Appelbaum, Barbara 1982
Mixed media textiles
AIC Textile Conservation Group Newsletter 5, 2 (Nov) 1-3
American Institute for Conservation, Washington DC AATA 20-494

Appelbaum, Barbara 1986
A note
AIC Newsletter **11,** 2 (May) 19
American Institute for Conservation, Washington DC

Ash, Nancy–**Hamburg,** Doris; et al (Eds) 1985
Backing Removal
Section 24 in Paper Conservation Catalog, Book and Paper Group,
American Institute for Conservation, Washington DC AATA 25-1618

Asmus, John F. 1978a
Laser Technology for Surface Preparation in the Arts
Technology and Conservation, **3,** 3, 14-18, Boston, Mass

Asmus, John, F. 1978b
Properties of laser-cleaned Carrara marble surfaces
In book: Decay and Preservation of Stone, pp81-83, Geological Society of America, Louisville, Kentucky AATA 17-656

Barclay, R. L. & **Grattan,** D. W. 1987
A silicone rubber/microballon mixture for gap filling in wooden objects
ICOM Conservation Committee 8th Triennial Meeting, Sydney, Preprints I, 183-187. Marina del Rey AATA 25-575

Barker, A. D. 1980
Gold Lace and Embroidery
Northern Scottish Costume and Textiles, Publication 1, pp14
Bradford UK

Barker, H. & **Organ,** R. M. 1953
A simple water-circulator for museum use
Studies in Conservation, **1,** 2, 82-86. London

Barnes, H.M.–**Nicholas,** D. D. & **Landers,** E. W 1984
Corrosion of Metals in Contact with Wood Treated with Water Borne Preservaties
American Wood Preservers' Association 8th Annual Meeting, Proceedings, 80, 10-25 Stevensville MD AATA 23-1183

Barov, Zdravco 1988
The reconstruction of a Greek vase: the Kyknos Krater
Studies in Conservation, **33,** 4, 165-177 London

Bauer, Wilhelm P. & **Koller,** Manfred 1965
Festigung loser Bemalungen auf ethnographischen Holzobjekten (Consolidation of friable paint on ethnographic objects)
Der Präparator, **2,** 4, 235-238. Bochum BRD

Baynes-Cope, A. D. 1984
Deacidification: a statement from the technical committee
The Society of Archivists, Journal **7,** 6 (Oct), 402-403
Sheffield UK AATA 22-438

Beale, Arthur 1987
A low cost reversible alternative for compensating damaged patinas of outdoor bronze sculpture
American Institute for Conservation Annual Meeting, Vancouver BC,
Objects Specialty Group Abstract, Preprints 205. Washington DC

Bickersteth, Julian 1987
The restoration of two Egyptian Revival chairs
Institute for the Conservation of Cultural Material (Inc), Bulletin, **13,** 3/4, 77-87 Canberra

Block, Ira 1987a
Kinetic Modelling of the Degradation of Textiles
Recent Advances in the Conservation and Analysis of Artifacts,
Institute of Archaeology Jubilee Conference, Papers, 203-205. London

Block, Ira–**Jensen,** N.–**Sommer,** S. M & **Buckingham,** W.F 1987b
Chloride leaching from outdoor bronzes
ICOM Conservation Committee 8th Annual Meeting, Sydney, Preprints, **3,** 1057-1062. Marina del Rey AATA 25-827

Boissonas, Pierre B. & **Percival-Prescott,** W. 1987
Removal of varnish and glue from painted surfaces using microfriction
ICOM Conservation Committee 8th Annual Meeting, Sydney, Preprints, 1, 137-144. Marina del Rey AATA 25-445

Brunnert, Stephan 1985
Die Anwendung von Schwefelwasserstoff bei der Bleirestaurierung
(The use of hydrogen sulphide for the restoration of lead)
Arbeitsblätter, **18,** Heft 2, Gruppe 2, 161-168 Mainz FRG
AATA 23-1191

Burgess, Helen D. & **Charette,** Carmen L. 1983

The use of fixatives to protect fugitive colourants during conservation treatments
American Institute for Conservation Annual Meeting, Baltimore, Preprints, 129-139. Washington DC AATA 20-1569

Burmester, Andreas & **Koller,** Johann 1985
Die romanischen Domtüren am Dom zu Augsburg
(The Romanesque doors of the cathedral at Augsburg)
Maltechnik **4,** Restauro 9-23 Munich FRG AATA 23-1195

Burstall, M. L. – **Butler,** C. E. & **Mollett,** CC 1986
Improving the properties of paper by graft polymerisation
The Paper Conservator, **10,** 95-100 Leigh UK

Cjeka, Jirí 1975
A simple method for conservation of zinc and copper printing blocks
ICOM Conservation Committee 4th Annual Meeting, Venice, Preprints, 25/1/1-9. Paris AATA 13-536

Charalambous, D. & **Oddy,** W. A. 1975
The 'consolidative' reduction of silver
Conservation in Archaeology and the Applied Arts, IIC Stockholm Congress, Preprints 219-227. London AATA 12-1215

Chase, W. T. & **Veloz,** Nicolas F. 1985
Some considerations in surface treatment of outdoor metal sculptures
American Institute for Conservation Annual Meeting, Washington DC, Preprints, 23-35 Washington DC AATA 22-2242

Child, R. E. 1986
Waxoyl and after
Conservation News UKIC 31 (Nov) 13 London

Collings, T. J. 1985
The Vacuum Probe – A New Tool for Paper Conservation
Paper Conservation News 34 (June) 3 Leigh UK

Columbus, Joseph V. & **Miller,** Barbara A. 1980
Treating a tapestry containing tarnished silver
Book: Conservation and restoration of textiles, Como, pp168-172 AATA 19-1433

Darrah, Josephine A. 1987
Metal Threads and Filaments
Recent Advances in the Conservation and Analysis of Artifacts, Institute of Archaeology Jubilee Conference, Papers 211-221 London

Dodds, Wendy 1988
Consolidation of mud on a World War I uniform
Institute for the Conservation of Cultural Materials (Inc) Newsletter 25 (Mar) 7 Canberra

Dreibholz, Ursula 1983
A treasure of early Islamic manuscripts on parchment. Significance of the find and its conservation treatment
American Institute for Conservation Annual Meeting, Baltimore, Preprints 30-37 Washington DC AATA 20-1583

Duncan, Steven J. & **Ganiaris,** Helen 1987
Some Sulphide Corrosion Products on Copper Alloys and Lead Alloys from London Waterfront Sites
Recent Advances in the Conservation and Analysis of Artifacts, Institute of Archaeology Jubilee Conference, Papers 109-118 London

Easthaugh, Nicholas 1987
Some experiments comparing the performance of detergent formulations based on anionic and non-ionic surfactants under conditions relating to conservation use
ICOM Conservation Committee 8th Annual Meeting, Sydney, Preprints, **1,** 357-364. Marina del Rey AATA 25-653

Eichhorn, Peter 1986
Die Restaurierung des Bronzekessels aus dem keltischen Fürstengrab von Hochdorf – mit Anmerkungen über die Herstellungstechniken.
(Restoration of the bronze cauldron from the grave of a Celtic nobleman at Hochdor – with notes on techniques on manufacture)
Arbeitsblätter **19,** Hefte 1, Gruppe 2, 169-185 Mainz FRG AATA 23-2826

Eremina, Irena M. 1987
New materials and the method for conservating icons with a ruined board
International Restorer Seminar, Veszprém, Papers, 153-156 Budapest AATA 25-1704

Erhardt, D. – **Hopwood,** W. – **Padfield,** T. & **Veloz,** N. 1984
The durability of Incralac: examination of a ten year old treatment
ICOM Conservation Committee 7th Annual Meeting, Copenhagen, Preprints, **2,** 22/1-22/3 Paris AATA 21-2179

Feller, R. L. – **Stolow,** N. & **Jones,** E. H. 1977
On Picture Varnishes and their Solvents
Book: Case Western Reserve University, Cleveland OH

Feller, R. L. 1984
Thermoplastic Polymers Currently in use as Protective Coatings and Potential Directions for Further Research
Institute for the Conservation of Cultural Material (Inc), Bulletin, **10,** 2, 5-18 Canberra AATA 22-1876

Fogle, Sonja (Ed) 1985
Recent Advances in Leather Conservation
Book: Foundation of the American Institute for Conservation, Washington DC, AATA 23-2173

Franke, Helmut & **Zumpe,** Reiner 1980
Die Restaurierung des Schatzfundes von Ralswick/Kr. Rügen
(Restoration of the treasure trove of Ralswick/Rügen)
Neue Museumskunde, **23,** 4, 280-284 Leipzig GDR AATA 22-1036

Gale, Frances & **Robbins,** John C. 1984
Removal of Interior Coatings at the Statue of Liberty
Bulletin, Association for Preservation Technology, **16,** 3/4, 63-65 Ottawa

Gallay, Gretel 1983
Zu Fragen Bronzezeitlicher Nietung in Nordwestfrankreich
(Questions of rivetting in northwestern France)
Archaeologisches Korrespondenzblatt **13,** 1, 49-57 Mainz FRG AATA 22-1040

Gibson, Bethune M. 1969
The use of the airbrasive process for cleaning ethnological materials
Studies in Conservation, **14,** 4, 155-164 London AATA 10-598

Gilberg, Mark R. 1986
Plasticization and forming of misshapen birch-bark artifacts using solvent vapours
Studies in Conservation **31,** 4, 177-184 London

Goldstraw, Ruth 1986
Anodic dissolution of redeposited copper from three gilded brooches
Conservation News UKIC 31 (Nov) 13-14 London

Golikov, V. P. & **Ustinov,** S. V. 1987
Complex experimental investigation of the effect of cleaning compositions on the fibres and dyes of museum textiles
ICOM Conservation Committee 8th Annual Meeting, Sydney, Preprints, **1,** 373-377 Marina del Rey AATA 25-666

Grant, John H. A. 1985
Liquid nitrogen for a frozen gun lock

International Insitute for Conservation – Canadian Group, Newsletter, **11,** 2 (Dec) 6 Ottawa

Gromina, T. N. 1987
The influence of some restoration processes on museum textile materials
ICOM Conservation Committee 8th Annual Meeting, Sydney, Preprints **1,** 379-380 Marina del Rey AATA 25-668

Guida, G. – **Marabelli,** M. & **Reindell,** I. 1980
Restoration and storage of lead artifacts
Restaurator **4,** 3/4, 227-237 Copenhagen AATA 20-986

Haines, Betty M. 1984
The conservation of leather bookbindings
Adhesives and Consolidants, IIC Paris Congress, Preprints 50-54, London AATA 21-1588

Hawley, Janet K. – **Kawai,** Elizabeth, A. & **Sergeant,** Christopher 1981
The Removal of Rust Stains from Arctic Tin Can Labels Using Sodium Hydrosulfite
International Institute for Conservation – Canadian Group, Journal **6,** 1/2, 17-24 Ottawa AATA 20-2153

Heinrich, Peter 1984
Die Restaurierung eines silbernen Vortragkreuzes (Reliquars) (The restoration of a silver reliquary cross)
Arbeitsblätter **17,** Heft 2, Gruppe 3, 70-77 Mainz FRG AATA 22-2267

Hermans, Johan & **Mahoney,** Patricia 1988
Deacidification and sizing
Conservation News UKIC 36 (July) 8 London

Herskovitz, Robert 1984
Conservation treatment of two rawhide ore buckets
American Institute for Conservation Annual Meeting, Los Angeles, Objects Specialty Group Abstracts, Preprints 119. Washington DC

Heuman, Jackie & **Garland,** Kate 1987
A poultice technique for the removal of cellulose nitrate adhesive from textiles
The Conservator UKIC 11, 30-33 London

Hiekkanen, Kaija 1981
Rengöring av mynt genom modifierad galvanisk reduktions metod (The cleaning of coins by a modified method of galvanic reduction)
Meddelelser om Konservering **3,** 4, 153-157 Brede Denmark AATA 19-682

Higuchi, Seiji & **Nakasato,** Toshikatsu 1974
Conservation treatments of central ornaments in the former Niigata prefectural assembly hall
Science for Conservation 13 (Mar) 17-26 Tokyo AATA 12-478

Hofenk de Graaf, Judith H 1982
Some recent developments in the cleaning of ancient textiles
Science and Technology in the Service of Conservation, IIC Congress, Washington DC, Preprints 93-95 London AATA 19-1471

Hook, John 1988
The Use of Immiscible Solvent Combinations for the Cleaning of Paintings, Journal, American Institute for Conservation **27,** 2, 100-104, Washington DC

Horie, C. Velson 1980
The Consolidative Reduction of Lead
Conservation News UKIC 12 (July) 9-10 London AATA 20-2159

Howlett, F. Carey 1985
Balancing Aesthetic Concerns with Ethical Considerations: Technical Solutions used in the Conservation of an Early 19th Century Five-Drawer Chest
American Institute for Conservation Annual Meeting, Washington DC, Preprints 68-78 Washington DC AATA 22-1994

Járó, Márta (Ed) 1982
Problems of completion, ethics and scientific investigation in restoration
Book: International Restorer Seminar, Veszprém, Budapest AATA 20-106

Jedrzejewska, Hanna 1984
Conservation of Archaeological Wood: A systematic classification of cases
ICOM Committee for Conservation 7th Triennial Meeting, Copenhagen, Preprints **1,** 7.12-7.15 Paris AATA 21-1660

Kaftanová, Alena & **Simunková** Eva 1983
Pruzkum, zpevneni a restaurovani drevene plastiky P. Maria Bolestna z Volyne
(Investigation, consolidation and restoration of wooden polychrome statue of V.P. Maria Grievous from Volyne)
PICT Scientific Papers **S10,** 73-81 Prague AATA 21-419

Kalish, M. K. 1975
Investigation on protective properties of artifical patina on bronze artifacts
ICOM Conservation Committee 4th Triennial Meeting, Venice,
Preprints 25/5/1-25/5/10 Paris AATA 13-1198

Karl, Hannelore 1987
Report on the restoration of a silver incense burner of 17th century
Arbeitsblätter **20,** Heft 2, Gruppe 3, 89-92 Mainz FRG AATA 25-2234

Kerr, Nancy – **Hersh,** Solomon P. & **Tucker,** Paul A. 1982
The use of alkaline-buffering agents to retard the degradation of cotton textiles
Science and Technology in the Service of Conservation, IIC Congress, Washington DC, Preprints 100-103. London AATA 19-1482

Kol'tsova, T. Ya; et al 1985
Epoxide compositions for wood impregnation
Khimiya Drevesiny 6, 89-93,126 Riga Latvia AATA 24-681

Koller, Manfred & **Klummer,** Gottfried 1982
Der Schrein des hl. Amandus im Benediktinerstift St. Peter zu Salzburg und seine Restaurierung
(St. Amand's shrine in St. Peter's monastery in Salzburg and its restoration)
Salzburger Museumsblätter **43,** 2, 13-15 Salzburg AATA 25-289

Koltai, Orsolya 1984
Római kori hagymafejes ezüst fibula restaurálása
(Restoration of a Bulbiform Silver Fibula of the Roman Age)
Múzeumi Műtárgyvédelem **13,** 275-284 Budapest AATA 23-1228

Koltai, Orsolya 1987
Restoration of a mediaeval párta and párta-belt
ICOM Committee for Conservation 8th Triennial Meeting, Sydney, Preprints **3,** 1073-1077 Marina del Rey AATA 25-121

Koob, Stephen P. 1981
Consolidation with Acrylic Colloidal Dispersions
American Institute for Conservation Meeting, Philadelphia, Preprints 86-94 Washington DC, AATA 19-120

Koob, Stephen P. 1982
The instability of cellulose nitrate adhesives
The Conservator UKIC **6,** 31-34 London, AATA 19-1006

Kronkright, Dale Paul 1981

New Directions in Native American Basketry Conservation
American Institute for Conservation Meeting, Philadelphia, Preprints 95-108, esp 106. Washington DC, AATA 19-791

Kyriacou, Celia undated, c1984
The History and Conservation of the Swords at Drummond Castle
The Laboratories of the National Museum of Antiquities of Scotland **2,** 118-125 Edinburgh AATA 22-2278

Lafontaine, Raymond H. & **Wood,** Patricia A. 1982
The stabilization of ivory against relative humidity fluctuations
Studies in Conservation **27,** 3, 109-117. London AATA 19-1958

Livens, R. G. 1974
Analysis of Welsh Bronze Age Implements
Antiquity **47,** 190, 139 Cambridge UK AATA 11-418

Loy, Thomas H. 1983
Prehistoric Blood residues: Detection on Tool Surfaces and Identification of Species of Origin.
Science **220** (17 June) 1269-1271 Washington DC AATA 20-1450

MacLeod, Ian D. & **North,** Neil A. 1979
Conservation of corroded silver
Studies in Conservation **24,** 4, 165-170 London AATA 17-760

MacLeod, Ian Donald 1984
Conservation, Applied Science and Why Treatments must be Monitored
Institute for the Conservation of Cultural Material (Inc), Bulletin, **10,** 2 (June) 19-41 Canberra AATA 22-2286

MacLeod, Ian Donald 1987
Conservation of corroded copper alloys: a comparison of new and traditional methods for removing chloride ions
Studies in Conservation **32,** 1, 25-40 London AATA 25-879

Magliano, Patrizia & **Boesmi,** Bruno 1988
Xeroradiography for paintings on canvas and wood
Studies in Conservation **33,** 1, 41-47 London AATA 25-1717

Matienzo, L. J. & **Snow,** C. E. 1986
The chemical effect of hydrochloric acid and organic solvents on the surface of ivory
Studies in Conservation **31,** 3, 133-139 London AATA 24-1097

McCrady, Ellen 1981
Steam Cleaning Book Spines
Abbey Newsletter **5,** 6, 79 New Carollton MO AATA 19-299

Mel'nikova, E. P. – **Smirnov,** S. A. & **Lebel,** M. N. 1984
On gradual removing Corrosion Layers from Metal, Stone and Ceramics
ICOM Committee for Conservation 7th Triennial Meeting, Copenhagen, Preprints **2,** 22.28-22.29 Paris AATA 21-2228

Michaels, Jan & **Boyd,** John 1986
Damage and conservation: a book conservator borrows an SEM
The Paper Conservator **10,** 73-94 Leigh UK

Moffat, Elizabeth & **Laver,** Marilyn 1981
Erasers and related dry cleaning materials
Canadian Conservation Institute, Analytical Report ARS No. 1738, Ottawa

Morgós, András 1987a
Complementing test series for confining solubility area in the Teas' triangular solubility diagram
International Restorer Seminar, Veszprém, Papers, 55-65 Budapest AATA 25-1813

Morgós, András 1987b
Nagymértékben hiányos és csökkent szilárdságú fatárgyak restaurálása. Két rossz állapotú láda restaurálása
(Restoration of gravely deteriorated wood objects – restoration of two timber chests in poor condition)
Múzeumi Műtárgyvédelem **17,** 159-172 Budapest AATA 24-740

Morita, T. – ~**Tujii,** Y & **Matsunaga,** T. 1987
Application of a new type of pyrethroidal compound on ethnographic textiles
ICOM Committee for Conservation 8th Triennial Meeting, Sydney, Preprints, **1,** 211-214 Marina del Rey

Nacsa, Mihály 1987
Csalikulacs restaurálása
(Restoration of a *csalikulacs* [double flask])
Muzeumi Műtárgyvédelem 17, 151-158 Budapest AATA 24-744

Nakhla, S. M. 1986
A comparative study of resins for the consolidation of wooden objects
Studies in Conservation **31,** 1, 38-44 London AATA 23-2319

Northrop, David 1980
Conservation and repair of a silver-clad glass decanter-stopper
The Conservator UKIC 4, 29-30 London AATA 171590

Nováková, Miroslava 1987
Conservation d'un cabinet du 17ième siècle
(The conservation of a 17th centuy cabinet)
ICOM Committee for Conservation 8th Triennial Meeting, Sydney, Preprints **3,** 1221-1222 Marina del Rey AATA 25-601

O'Donnell, Elizabeth B. & **Julian,** Maureen M. 1985
Conservation of a disassembled British 18th century naval hand grenade
Proceedings, 16th Conference on Underwater Archaeology, Society for Historical Archaeology 80-81 Ann Arbor MI AATA 24-1215

Onnen, Ewald 1983
Die Konservierung des Hochaltars und der Kanzel der Wallfahrtskapelle St. Marien in Heinersreuth/Oberpfalz
(The conservation of the main altar and the pulpit of the pilgrimage church of St. Marien at Heinersreuth/Oberfalz)
Arbeitsblätter **16,** Heft 1, Gruppe 25, 11-15 Mainz FRG AATA 21-1671

Organ, R. M. 1968
Book: Design for Scientific Conservation of Antiquities
London and City of Washington AATA 7-1450

Pasch, Astrid 1985
Rekonstruktion einer Goldblechscheibenfibel und Untersuchungen zu den Herstellungstechniken
(Reconstruction of a gold-plated disk-shaped tin brooch and analysis of the techniques of production)
Restaurierung und Museumstechnik **6,** 5-22 Weimar GDR AATA 23-1246

Payton, Robert 1987
The Conservation of Artefacts from One of the World's Oldest Shipwrecks, The Ulu Burun, Kas Shipwreck, Turkey
Recent Advances in the Conservation and Analysis of Artifacts, Institute of Archaeology Jubilee Conference, Papers 41-50 London

Peacock, Elizabeth E. 1983
Deacidification of degraded linen
Studies in Conservation **28,** 1, 8-14 London AATA 20-572

Pearlstein, E. – **Cabelli,** D. – **King,** A. & **Indictor,** N. 1982
Effects of Eraser Treatment on Paper
Journal, American Institute for Conservation **22,** 1, 1-12 Washington DC

Piersig, W et al 1982
Method of removing surface coatings

Fertigungstechnik und Betrieb **32**, 8, 501-502 Berlin DDR AA-TA 21-1385

Pizzi, A.–**Vosloo,** R.–**Cameron,** F. A. & **Orovan,** E. 1986
Self-neutralizing acid-set PF wood adhesives
Holz als Roh– und Werkstoff **44,** 6 229-234 New York AATA 24-771

Plenderleith, H. J. 1971
Conservation of Antiquities and Works of Art
Book: London

Plossi, Mariagrazia Zappalà & **Crisostomi,** Paolo 1981
Consolidation de la couche picturale des enluminures avec polymères synthetiques purs
ICOM Committee for Conservation 6th Triennial Meeting, Ottawa, Preprints 14/7/1-14/7/15 Paris

Pruszynski, Jan Piotr 1985
Regulaja ustawowa ochrony zabytków
(Legal regulation of the protection of cultural property)
Ochrona Zabytków **148,** 1, 28-31 Warsaw AATA 23-270

Rácz, Jenő 1981
Restoration of the crucifix clock of the Esterházy treasury from the 17th century
ICOM Committee for Conservation 6th Triennial Meeting, Ottawa, Preprints 23/7/1-23/7/9 Paris AATA 18-1695

Rae, Allyson 1987
Cleaning of featherwork
Recent Advances in the Conservation and Analysis of Artifacts,
Institute of Archaeology Jubilee Conference, Papers 243-248 London

Raft, Karen 1980
An examination of the value of the re-forming technique in practices Studies in Conservation **25,** 3, 137-140 London AATA 18-478

Raphael, Toby 1984
Leather completion: restoration of missing areas
American Institute for Conservation Annual Meeting, Los Angeles, Objects Specialty Group Abstracts, Preprints 121 Washington DC

Rayner, H. C. 1951 in: Brown, Burnett, Chase, et al (Eds) 1977
Corrosion and Metal Artifacts. – A Dialogue Between Conservators and Archaeologists and Corrosion Scientists
Book: NBS Special Publication 479. National Bureau of Standards, U.S. Government Printing Office, Washington DC AATA 15-1646

Reeve, Barbara H.C. 1987
Conservation of an Orator's Stool from Papua New Guinea
Institute for the Conservation of Cultural Material (Inc), Bulletin, **13,** 3/4 (Dec) 89-106 Canberra

Riederer, Josef 1984
The restoration of bronze monuments in Germany
ICOM Committee for Conservation 7th Triennial Meeting, Copenhagen, Preprints 2, 22.30-22.32 Paris AATA 21-2244

Robertson, Godfrey 1988
Veneer salvage
Conservation News UKIC 36 (July) 27 London

Rosenberg, G. A. 1917
Antiquités en fer et en bronze
Book: Gyldendalske, Copenhagen

Sanders, Sherri 1986
Preventing Metal Catalysed Degradation of Leather
pp79-86 in book: Hallebeek, P.B. (Ed) ICOM Symposium on Ethnographic and Water-logged Leather, Central Research Laboratory for Objects of Art and Science Amsterdam AATA 25-2153

Schniewind, Arno P. & **Kronkright,** Dale P. 1984
Strength evaluation of deteriorated wood treated with consolidants, Adhesives and Consolidants, IIC Paris Congress, Preprints 146-150 London AATA 21-1681

Sears, K. D.– **Ewing,** G. D. & **Quimby,** G. R 1982
Hydroxyalkylation strengthens high porosity pulps
Tappi Journal **65,** 5, 141-144 Atlanta GA

Sekowski, Stefan 1983
Konserwacja metalwych zabytków cmentarnych
(The preservation of cemetery metal monuments)
Ochrona Zabytków **16,** 1/2, 117-120 Warsaw AATA 21-2252

Simpson, Elizabeth & **Payton,** Robert 1986
Royal wooden furniture from Gordion
Archaeology **39,** 6, 40-47 Boston Massachusetts AATA 24-810
Smith, C.A 1982
The corrosion of metals by wood
Anti-corrosion **29,** 3, 16-17 London AATA 20-1072

Smith, C.A. 1982
The corrosion of metals by wood
Anti/corrosion **29,** 3, 16/17 London AATA 20/1072

Smith, Leslie Melville 1987
Use of enzymes in textile cleaning
American Institute for Conservation Annual Meeting, Vancouver BC, Preprints 124-131. Washington DC AATA 25-2049

Stadler, Jonny 1987
Die Restaurierung einer Wappentruhe des frühen 17. Jahrhunderts
(Restoration of an early 17th century chest with coat of arms)
Arbeitsblätter **20,** Heft 2, Gruppe 12, 34-41 Mainz FRG AATA 25-1855

Stambolov, Todor 1980
The conservation and restoration of gilt leather
In book: Konservering og restaurering af laeder, skind og pergament
Proceedings, Lund Conference, 1978, 19, 309-313 AATA 20-2323

Stather, F. & **Lauffmann,** R. 1935
Metalschädigungen durch freie Fettsäuren und Lederschädigungen durch fettsaure Metallverbindungen
(Metal damage caused by free fatty acids and leather damage caused by fatty acid-metal compounds)
Collegium **787,** 541-545 Frankfurt FRG AATA 22-1111

Tigréat, Michel 1984
La restauration (Restoration)
Métiers d'Art 26/27, 104-105 Paris AATA 22-2023

Torraca, Giorgio 1975
Solubility and Solvents for Conservation Problems
Book: International Centre for Conservation Rome AATA 12-726

Tuck, D. H.
Oils and lubricants used on leather
Book: Leather Conservation Centre, Northampton UK AATA 23-2796

Untracht, Oppi 1968
Metal techniques for the craftsman
Book: New York AATA 22-1123

Van Soest, H. A. B.–**Stambolov,** T. & **Hallebeek,** P. B. 1984
Conservation of leather
Studies in Conservation **29,** 1, 21-31 London AATA 21-1203

Vernon, W. H. J. 1927
Second experimental report to the atmospheric corrosion research commitee (British Non-Ferrous Metals Research Association)
Transactions, Faraday Society **23**, 113-183 London

Walker, Richard W 1985
You can't chop an art-work up in Massachusetts
Art News **85,** 4, 21-22 New York AATA 24-228

Watkinson, D. 1983
Degree of mineralization: its significance for the stability and treatment of excavated ironwork
Studies in Conservation **28**, 2, 85-90 AATA 20-2212

Weber, Rainer & **Kawinski,** Heinrich 1984
Restaurierung eines Elfenbeinzinken
(The restoration of an ivory zinken)
Arbeitsblätter **17**, Heft 1, Gruppe 13, 35-47 Mainz FRG AATA 22-1002

Weil, Phoebe Dent, et al 1982
The corrosive deterioration of outdoor bronze sculpture
Science and Technology in the Service of Conservation, IIC Congress, Washington DC, Preprints 130-134 London AATA 19-1904

Weisser, Terry Drayman 1987
The use of sodium carbonate as a pre-treatment for difficult-to-stabilise bronzes
Recent Advances in the Conservation and Analysis of Artifacts, Institute of Archaeology Jubilee Conference, Preprints 105-108 London

Welsh, Elizabeth C 1980
A consolidation treatment for powdery matte paint
American Institute for Conservation Annual Meeting, San Francisco, Preprints 141-150 Washington DC AATA 17-1386

Williams, Jo 1987
Protective coatings for ironwork
Conservation News UKIC 33 (July) 15-16 London

Wolbers, Richard 1987 Report by Cox, Ruth Barach
A New Cleaning Tool: Water-based Emulsions Developed by Richard Wolbers
Washington Conservation Guild Newsletter **11**, 1, 2-3 Washington DC

Wolbers, Richard 1988a Report by Southall, Anna
New approach to cleaning painted surfaces
Conservation News UKIC 37 (Nov) 43-44 London

Wolbers, Richard C 1988b
Aspects of the examination and cleaning of two portraits by Richard and William Jennys
American Institute for Conservation Annual Meeting, New Orleans, Preprints 245-260 Washington DC

Author's address:

9501 Nowell Drive
USA-MD 20817 Bethesda

SÉD Gábor

Training of metal restorers in Hungary

In Hungary, metal restorers have been trained in an officialy organized form since 1963.

Initially, education consisted in one-month, so-called "Priming Courses" imparting primarily the skills needed to preserve the usual archaeological ceramic and metal objects.

Metal restorer specialists were first trained on two-year courses launched in 1976. Although by correspondence, these four-semester courses were suited to impart the special subject matter. Since 1974 there has been a section of Restoration of Public Collection Objects at the Institute of Restoration and Preservation of the High School of Fine Arts, imparting at its first four courses a wide range of subject matter on every kind of material.

The subsequent specialization course qualified special metal restorer artists at university level.

The outlined training modes are subject to continuous development. Actually, all three forms of education are being updated. Professional demands for, and correlation between levels are unambigous.

All three courses are by correspondence.

Mandatory practical work in excess of the effective training time is checked by consultants.

Priming courses have been developed into **skilled worker** training.

Training time: 1 year (two semesters of 2x200, in total, 400 lessons).

Subject matter: basic knowledge in every group of materials.

Professional requirements: Primarily, preservation of archaeological ceramic and metallic materials.

Scope of the skilled worker examination:

- Final examination in professional and theoretical matter.
- Examination work (preservation and restoration of a metal and a ceramic object, of archaeological origin – possibly, standard specimens).

Qualification: skilled worker in object restoration.

The course for **restorer specialists** takes another 400 lessons, (two semesters of 200 lessons each) for students having passed skilled worker examinations.

There are no overall subjects any more but only special theory and practical work for different kinds of materials (metal, wood, silicate, paper, textiles, leather).

Professional requirements: possibly deep-going acquaintance with crafts in the given domain, skilled handling of the collection exhibits in conformity with the learned methods.

Subjects of the special examination:

- Examination in the special theory:
- Examination work: restoration of an art object raising bot restoration and reconstruction problems.

Qualification: Specialist ... restorer.

By 1991, a five-year course announced a priori as specialized will be launched at the Section of Object Restoration of the Hungarian High School of Fine Arts.

Professional requirements: Top-level theoretical and, possibly, also professional-practical acquaintance with every kind of material.

The adopted speciality (e.g.metals) is concerned with an as wide range of manufacturing, materials, related natural sciences, arts history, etc. aspects as possible, relying on the subject matter of the preliminary courses.

It is expected to undertake independent research and design work, and the development of new procedures and methods.

Subject of the state examination:

- Thesis: any of the aspects above
- Examination work: restoration, reconstruction of an object exhibiting values of the craft, artistic workmanship.

Qualification: ... restorer artist (for metal specialists: metal restorer artist).

At any of the three education levels: specialization in metal restoration comprises:

1. For skilled workers in object restoration:

Theory: based on natural sciences imparted in secondary school oriented to mass preservation;

Practice: craftmanship exercises primarily concerned with archaeological standarad material.

2. For metal restorer specialists:

Theory : relevant special knowledge in history, techniques, arts, natural sciences (higher technician level);

Practice: craftmanship to the level of essential metal workmanship. Independent restoration, reconstruction work.

3. For the Section of (Metal) Objects Restoration at the Hungarian High School of Fine Arts:

Theory: The entity of relevant essential knowledge in history, arts, natural sciences, techniques, etc., at university level;

Practice: knowledge of crafts at a level of workmanship, artistic reproduction.

Graduate restorers must cope with any special problem involving independent research work, individual artistic reconstruction if needed.

The curricula for three educational levels are given as Appendices.

Appendix 1

Subjects and programme for one-year skilled worker training of museum object restorers (40 lessons a month, totalling 400 lessons).

Skilled worker training of object restorers is expected to impart the theory and practice of fundamental skills needed to protect and handle the given objects. During training archaeological metal and ceramic objects are emphasized.

Programme of the Course for Skilled Worker Training

	Number of Lessons		
	Theory	Practical	Total
Introduction to Museology			
– History of Museums, Special Disciplines of Museology, Museums Lows, Network of Museums	4	-	4
– Collection, Scientific Processing, Registration	4	-	4
– Popular Education by Museums	1	-	1
– Arrangement of Exhibition	1	-	1
– Photography	2	8	10
Environment of Objects of Art			
– Labour Safety Prevention of Accidents	2	-	2
– Air Conditioning, Illumination Techniques	6	2	8
– Storage	4	-	4
– Storage, Disinfection	-	24	24
– Safety in Museum	2	-	2
– Wrapping, Shipping of Objects	2	2	4
History of Culture and Techniques	12	-	12
History and Ethics of Conservation	3	-	3
Documentation	2	-	2
Applied Knowledge in Natural Sciences			
– General and Inorganic Chemistry	9	-	9
– Organic and Synthetics Chemistry	7	-	7
– Biology	3	-	3
Labour, Accident and Fire Safety	1	1	2
Materials and Conservation			
– Metals	10	62	72
– Ceramics	10	62	72
– Glass	4	12	16
– Stone	2	6	8
– Bone	2	6	8
– Wood	4	20	24
– Textile	4	12	16
– Leather	4	12	16
– Paper	4	12	16
Duplicating	2	8	10
Excavations /Field Work	-	40	40
Overall:	111	289	400

Appendix 2

Programme of the Course for Metal Restorer Specialists

First semester

Part I

Accident Safety

General Chemistry (basic concepts, basic laws chemical modifications, major groups of compounds, knowledge of atomic structure) lectures
chemical calcutations
laboratory exercises
Metallography I (structure, physical and chemical properties, production of metals, fundamentals of electrochemistry, corrosion of metals)
lectures
laboratory exercises

Organic and Synthetic Chemistry (overall survey, compounds for metal preservation, synthetics based preservatives, adhesives and complements)
lectures
laboratory exercises

Library (use of libraries, introduction to Hungarian and foreign special literature)

Visits to museums and workshops

Examination: Inorganic Chemistry
Metallography I
Organic and Synthetic Chemistry

Part II

Fundamentals of the Goldsmith's Trade (working and ornamentation of the most frequent metals in museums: goldsmith's techniques, gems)
lectures
workshop exercises

Metallography II (corrosion of copper, treatment of museum objects of, or based on copper)
lectures
workshop exercises

Examination:
Fundamentals of Goldsmithery
Practical: Metal restoration practice

Second semester

Part I

Metallography III (corrosion of iron, handling of iron objects; corrosion of tin and lead, handling of tin and lead objects)
lectures
workshop exercises

Analytics (identification of metals current in the practice of museums)
lectures
laboratory exercises

Visits to museums and workshops

Examination: Metallography III Practical: Practice of metal restoration (two marks)

Part III

Work discussion (discussion of problems of cleaning, preserving, restoring the object selected for examination work)

Visits to museums and workshops

Final examination

Appendix 3

Curriculum of the Course at University Level

Subject	Number of lectures	practicals
Year I		
History of Culture	12	
History of Philosophy	12	
Universal History of Arts	12	
History of Techniques, Technology,Knowledge of Materials (Ceramics, Mosaics, Glass)	12	32
Fundamentals of Chemistry (General and Inorganic Chemistry)	28	40
Knowledge of Materials (Silicates)	6	
Phisycs – Material Testing	8	
Biology (Microbiology, Phytology)	20	
History of Restoration	6	
Preservation (Ceramics, Mosaics, Glass)	8	64
Drawing		40
Photography		16
Documentation	4	
Restoration of Special Materials (Pearls, etc.)	6	
	134	192
Year II		
History of Culture	12	
History of Philosophy	12	
Applied and Folk Arts	12	
Ethnography	8	
History of Technique, Technology, Knowledge of Materials (Metals)	12	32
Fundamentals of Chemistry (Organic and Polymer Chemistry)	28	40
Knowledge of Materials (Metals)	8	
Materials Testing	4	
Biology (Zoology, Anthropology)	16	
Preservation (Metals)	16	64
Drawing		40
Preservation and Restoration of Special Materials	6	
	134	176
Year III		
Museology	12	
Arts Aesthetics	12	
Archeology	12	

Appendix 3 (continued)

Curriculum of the Course at University Level		
Year III		
Subject	Number of lectures	Number of practicals
Pedology	8	
History of Techniques and Technology, Knowledge of Objects (timber, paper, textiles, leather)	24	80
Knowledge of Materials (timber, paper, textiles, leather)	24	
Preservation (timber, paper, textiles, leather)	32	160
Materials Testing (timber, paper, textiles, leather)	16	16
	140	256
Excavation exercises: one week		
Year IV		
History of Culture (referring to metals)	16	
Politology	12	
History Techniques, Technology, Knowledge of Objects (Special)	16	32
Special Preservation-Restoration (Metals)	32	160
Museum Ambience	8	
Labour Safety and Sanitary Regulations in Museums	8	
Preservation, Research Planning, Cost Forecasting, Design of Restoration Workshops	8	
Descriptive Geometry	12	
Modelling		40
	112	232
Exercises in an Outdoor Ethnographic Museum: one week		
Year V		
History of Culture (special)	8	
Specialized Preservation – Restoration (Metals)	8	120
Consultations on the Diploma Work	48	
	64	120

State Examination: History of Techniques, Knowledge of Materials, Preservation-Restoration
Note: In the near future, the outlines of the curricula given above will be issued in detail.

Author's adress:

National Centre of Museums
Könyves K. krt 40.
H-1087 Budapest

ASHLEY-SMITH, Jonathan

A new conservation course at the Victoria and Albert Museum

By October of this year (1989) a new post-graduate conservation course will have begun. This is a joint venture involving the Royal College of Art (RCA) and the Victoria and Albert Museum (V&A), both situated in South Kensington, London [1]. There will also be active assistance by staff from Imperial College of Science and Technology (IC). This is a three-year course leading to a degree of Master of Arts awarded by the RCA.

Most of the students' practical experience will be gained in the studios of the Conservation Department at the V&A. The aim of the course is to develop excellence in one chosen specialist discipline within conservation. The academic element of the course will, however, be quite broad allowing the student to develop interest in the management of historic objects or some other cross-disciplinary activity. One of the specialist options on this course will be "metalwork and jewellery", and one of the four students chosen to inaugurate this course was selected for this discipline. Surprisingly, there is no other course in Britain which offers specialist training in metal conservation.

This paper will explain the historical development of the course and, as much as can be foretold, how it will progress.

This will involve:

(i) a survey of the types of training that have been attempted in the past.

(ii) different approaches to determining course content.

(iii) the limitations of our methods of selecting candidates for the course.

This will be a *personal* view of training and not a rigorous objective assement.

Types of Training

(i) apprenticeship
(ii) apprenticeship plus teaching
(iii) self-directed learning
(iv) apprenticeship plus learning

The new RCA/V&A Masters course is a refinement of a Diploma course that had been run by the V&S alone since about 1970. Over the 18-year period of its existence the old Diploma course underwent many changes.

Metalwork was one available option and during the eighteen years *four* metalwork specialists were produced.

Apprenticeship

In its initial form the Diploma course was conceived as a type of apprenticeship. It consisted of nothing more than four years, each of 48 week spent in one specialist section of the Conservation Department.

Apprenticeship, at its best, means structured and progressive learning by example in a working environment. But it pre-supposes an able, practical expert whose desire is that *eventually* the student will be as good, if not better, than himself. In many real cases it can mean that a barely-competent "master" is reluctant to pass on information, for fear that the apprentice will rapidly outstrip him and expose the limitations of his expertise and skill. (The frequent use of the male pronoun indicates specific examples familiar to the author.)

In the right circumstances the apprenticeship model has many benefits since it emphasises practical skills. The students learn by practical experience, as well as by observation of their master and of their peers. If the student learns and practises *all* the skills needed in the working environment it can be very effective. It has the advantage that the student is unlikely to gain knowledge that is *not* relevant to the actual task that needs to be performed. In a progressive workshop the syllabus is always up-to-date.

In reality, certain skills and knowledge are not passed on in this way. Students are excluded from management and administration, much of which goes on behind closed doors. The apprentice never understands the totality of the subject. Apprentices will not learn the fundamental principles that underlie what they practice, so they never learn, and rarely question, *why* they are doing something. Many of the fundamental principles derive from the science of the materials used. The relevant materials science is not easily gained through apprenticeship.

Apprenticeship + Teaching

Some of these deficiencies were recognised early on in the Diploma course and by 1973 a series of lectures was arranged. Initially, outside lecturers were used but eventually it fell to the Conservation Department's own scientists to "dump" huge "chunks" of scientific knowledge onto unreceptive students, most of whom came from a practical arts background. Eventually a ratio of 60% practical workshop experience to 40% lectures and private study was established. The formal evaluation of students at the end of the course showed that very little of the lecture material had been absorbed and informal observation of students in their later careers tends to confirm this. There was no evidence that the students had learned how to learn.

Self-Directed Learning

This next section is necessarily personal since it describes how *I* learned about metal conservation. I was recruited into a post where I was to do 50% practical metalwork conservation, and 50% technical examination and chemical analysis of metal objects. My supervisor was Norman Bromelle, who did the most sensible thing any supervisor could do. He left me entirely alone for four years.

For the practical work I was put into an apprenticeship position but with virtual equality of rank with my tutors. I made time to experiment with metalworking techniques (my wife received a succession of pieces of jewellery and doll's house equipment at birthdays and anniversaries). I was given the freedom of the National Art Library, the libraries of the Science Museum and Imperial College, and of course the British Museum Library. I was allowed to select elements from the metallurgy and materials science course at Cambridge University and discovered a tutor there who had an active interest in archaeo-metallurgy. I was given time to experiment with techniques such as microscopy and x-radiography. I attended conferences in Britain and was given money to travel to the United States. All this, and not a single lecture and no *apparent* guidance or control.

Needless to say I judge the learning experience to have been a great success. As a model for developing metalwork conservators it has one possible disadvantage: no boundaries were set for what was *relevant* knowledge. I must admit that what I found *extremely* interesting was not always relevant. If self-directed learning is to be used, the student must be given some form of "map" that demonstrates not only the extent of the territory to be discovered, but which *suggests* the boundaries that it would be uneconomic to overstep.

Recently, when I was boasting about the success of my self-education, it was pointed out to me that I had come in equipped with the right "baggage". That is a science degree and five years of post-graduate research. The important elements of this are probably that I was "mature" (I was 27), and that I had experience of working to my own goals. The fact that I had selected a science course for my first degree may indicate a particular learning style acquired at an early age. All these factors must be considered when selecting students who will be left to do much of their own learning.

Apprenticeship + Self-Directed Learning

Based on experience of these different methods, the plan for the new course is to combine the benefits of each while trying to avoid any of the disadvantages. The students' practical skills will be developed in a working environment with the active and open encouragement of an able 'master' (in the case of metalwork, mistress). The students' academic skills will be developed by placing the student in an environment full of accessible information. The course will have a minimum of structure, a 'map' in the form of an overall syllabus. In the first year half the time available for academic study will be taken up with formal lectures, but it is assumed that the degree of formality will decrease in the second year as the students needs are individually assessed and as they become more independant. Guidance on how to use the available information will be given by tutors rather than teachers. The degree of self-determination is controlled by the student's selft-motivation and by the fact that they will be in a working environment with a working purpose that is not totally educational. There will be the overall arm's-length guidance of a liberal course leader.

To be more specific: the student will chose one specialist discipline from the list, or possibly one that crosses over two disciplines – upholstery would be an example. For the first two years of the three year course the student will spend three days each week, for the 36 weeks of the academic year, in the relevant specialist studio at the V&A. For 10 week of the academic vacation the student will spend 5 days a week in the studio. In the third year 5 days a week will be spent in the studio. In three years this gives a total of 546 practical studio days (for a healthy student, not prone to illness or despair). What the average conservator's practical day actually consists is dealt with in a following section. These practical days will be spent mostly at the V&A but also in placements at other institutions or freelance worskhops in Britain or abroad. This will depend on the needs of the student and on the cooperation of the reader (and the availability of finance).

One 'master' will never have more than three students. In many conservation courses the students greatly outnumber the teaching staff. On this course the reverse will apply. The ratio of staff (trained specialist conservators) to students will be in the region of 1:1 to 3:1 depending on the specialisation (we have more textile conservators than metalwork conservators).

The wealth of accessible information which the motivated student could tap exists in the minds of the 35 conservators and 4 scientists in the Department. It exists not only in the form of the libraries that were available to me 15 years ago, but in the form of a growing library within the Conservation Department and in the access to computerised databases such as CIN. The resources of the RCA add another dimension, with available knowledge of traditional and modern techniques of manufacture and with equipment for photography, holography and computerised manipulation of images. Involvement with IC brings further access to scientific knowledge and expensive analytical equipment.

During the course, 144 days are set aside for tutorials and private study, mainly in the first two years (lunch breaks, evenings and weekends are also available for private study).

Each student will have three tutors or guides; one for conservation, one for art-history and one for materials science. The conservation tutor will be the head of the specialist studio so that contact with the student would be almost daily. The art-history tutor would be a member of the curatorial staff of the V&A or a specialist from the RCA Design History Course. Contact would probably be limited to three tutorials a year. The science tutor would, for most disciplines, be a member of staff at IC. In the case of textiles, the tutor will be David Howell author of another paper in this volume. The overall balance of each student's development is the responsibility of the Course Leader.

The Course Leader was appointed in September 1988, full year before the first student would arrive. He is Alan Cummings, a chemistry graduate from Imperial College who later trained, and more recently taught, at the Courtauld Institute Conservation Course. He has several years' experience as a freelance paintings conservator.

The maximum number of studens at any one time will be 12. These 12 will be at various stages of the three-year course and will be in a number of different specialist disciplines. The Course Leader has the job of making these 12 feel like a cohesive group rather than isolated individuals. In many cases it will be possible to have a first year and a third year student working in the same discipline, so that there can be development by peer example and encouragement.

Course Content

At the time of writing (August 1989) there is no great certainty about the actual content of the metalwork course. The 'map' the students will be given has not been drawn. For reasons described later it was not known until June that metalwork would be one of the disciplines that would be covered in October.

There are three possible approaches to syllabus preparation. The first is to write lists. Below is a list which is derived from the United Kingdom Institute for Conservation's "Basic Syllabus" (published last year). [2]

Basic Syllabus

The discipline of conservation
Materials science for conservation
Preventive conservation
Documentation and Information
Health and safety
Management

Each heading in the list has a series of sub-headings which are each further subdivided. The difficulty is to know when to stop.

A more holistic approach is to look at the tasks a trained conservator is actually asked to perform. Then ask whether the tutorials, the reading and the work experience will equip the student to tackle these tasks competently. Since much of the training takes place in a working environment, if the tasks that a conservator is obliged to undertake change with time, then the syllabus changes automatically/silently. If the working environment is progressive, up-to-date, well-managed and well-equipped (as the V&A surely is!) then there is no need to hold faculty meetings to decide whose lecture series should be extended, modified or dropped completely. If word-processors become a part of the working environment then the student will learn word-processing along with the staff. It doesn't call for a new course unit on the use of computers.

The activities of the conservator *are* changing. The Metalwork Section is now very different to what it was when I worked in it twelve years ago. The most obvious change is the instrusion of management tasks. As the professional status of the conservator has increased, so have the responsibilities of management and organisation.

Management Tasks

Written presentation
Oral presentation
Assessing priorities
Estimating
Staff relations
Training

The conservator is now called upon to make many examinations of objects which do not result in treatment but merely assess and report on the condition of the object. This information may be needed to assess suitability for loan or to advise on purchase or storage. The conservator is required to be decisive, to report clearly and to be aware of the financial implication of any decision.

The conservator is called upon to present information in a cogent and persuasive way, for the benefit of one object or for the welfare of a whole collection. The importance of the skills of presentation cannot be overstressed. They are rarely taught at graduate level. They can be developed by the tutorial system and by "encouraging" the students to give presentations to small but critical groups of their fellow students and work colleagues.

Presentation is an essential part of the curriculum.

The nature of practical work is also changing. There is less severe intervention, much less restoration involving metalworking skills and greater interest in stabilising pigmented and patinated surface layers.

There are also new 'academic' tasks relating to technical examination, authenticity and attribution.

If the student is treated as an adult in the working environment most of these new skills can be gained through the apprenticeship element. They can be reinforced by tutorial discussion.

The test of how much detail is necessary in a syllabus can probably be tested by determining whether, at the end of the three years, the student could answer these four questions.

Metal Objects

What are they?
Why do they fall apart?
How can they be put back together?
How can we stop it happening again?

On looking at the lists and sublists of subject headings in the exhaustive first draft syllabus it might be concluded that, for instance, the student should know about metallic crystal structure, because this dictates properties such as brittleness or susceptibility to types of corrosion (why they fall apart). Crystal structure also affects hot and cold working properties (how they can be put back together again). Using this test it might be realised that a deep knowledge of the different types of crystalline dislocation does not help answer any of these questions, and so should be excluded from the syllabus.

Selection

How will the superhuman beings who will benefit from the three-year ordeal be selected? The first principle in making a good selection is to make sure you have a large number of candidates to select from. This year (1989) we had some 40 applicants for 4 places. Next year we hope for three or four times that number.

The primary aim of the selection is to end up with people who will make good students. In the film "The Karate Kid" the wise Mr. Miyagi says "No such thing as bad student. Only bad teacher."[3] This is true, but since we cannot be confident that we are good at teaching, we should invest time in selecting students who are good at learning. A good student will flourish despite indifferent teaching. In the inaugural year as with previous selections for the Diploma course, we did not say we have a vacancy in metalwork and then take the best of the applicants who happened to choose that option that year. We aimed to select the applicants who would make the best students and then we hoped to clear enough studio space in the disciplines they happened to have chosen. This is one reason that is wasn't known until June that metalwork was one of the topics for the first intake.

What qualifications are we looking for? The temptation is always to recruit in one own image, or at least in one's erroneous image of oneself. Thus *I* would prefer to recruit only people with science degrees, artistic ability and at least one earring! Unfortunately although we get a few earrings, we get very few applicants with science degrees.

The ideal for the future of the course is that it should be truly *post*-graduate. The students should ideally come from graduate conservation programmes, desperate to develop their expertise, and to explore new avenues of technique or research that they have discovered an interest in. But he majority of courses in conservation finish with a diploma which is too low an initial qualification. Or they start at a Masters level, in which case it is difficult for the student to get a grant for a second Masters course. Moreover, one only gets the products of someone else's selection and their criteria may not the same.

There is obviously a minimum level of ability. It follows from the discussion of a conservator's tasks that as a prerequisite they must be able to read, write well and speak fluently. But we are looking for *qualities* rather than qualifications.

Other Qualities

Compatability

The succesful candidate is going to be part of a small group of students who will be required to assist one another. Each student will have to be an active component of a hard-working practical studio team. Compatibility is essential. They must be able to get on with people under often trying conditions. On the day of the selection interview we try to involve as many people as possible. Someone meets the candidate and takes them to the studio, another shows them around the studio and answer their questions, another sits with them as they nervously wait for the interview. Then they finally face the stress of an interview panel of four people. After the interview, the opinions of all these people are taken into account. There is a surprising degree of unanimity. However, if a candidate is excessively rude, manic, arrogant, disdainful, agressive or uncommunicative they will not be accepted however good their qualifications may appear.

Maturity

The average age of successful applicants to the Diploma course was about 25. Maturity is an attribute of character rather than chronology, but I doubt we would ever take anyone under 20. There is no upper age limit, but we cannot get a Government bursary if the student is over 40.

Specialist Knowledge

Probably the easiest trap to fall into (and I have done it again and again) is to assume that it is an overwhelming advantage for someone to appear to have some part of the huge amount of knowledge you expect them to accumulate. At the end of three years the student's inability to learn any more may be a greater handicap than an initial lack of specific knowledge.

At our interviews candidates are asked to look at one or two problem objects and then say whatever they can about provenance, condition, materials, techniques, possible treatments or suggestions for storage and display. Candidates with no previous specialist experience often show much more open, imaginative and intelligent approaches to problems than those with previous specialist training.

Specialist Motivation

Another trap is specialist motivation. A candidate who at 25 loves only metals and knows that he wants to work exclusively with metals for the rest of his life is clearly a lunatic. This person will probably not develop the flexibility and broad imaginative outlook which is needed in the manager or the innovative researcher. I developed my love of metals at 27 when I saw the advertisement for a metals scientist/conservator at the V&A. If the advertisement had been worded differently I could have been a textiles conservator.

If there are difficulties in recruiting science graduates because none apply, then it is wise to search for someone with an interest in materials. People use materials in their environment all day and everyday, but only a few are interested in the nature and properties of those materials. Conservators have to have such an interest. At interviews I ask what the candidate knows about materials they have recently used. In the recent interviews for the Masters course, for instance, I knew that the metalwork candidates had all recently made some object out of brass. While each was able to give a definition of an alloy, only a couple were able to say which two elements were alloyed to make brass. If graduate students start with that little interest in composition then it is going to be much more difficult to make them understand how changes in composition will affect hot and cold working properties or rates of corrosion. A continuing desire to find out the names and natures of the materials of everyday life seems to me a reasonable test for the potential materials conservator. It also detects a desire to learn.

Practical Skills

At interview each candidate is asked to present examples of practical work (it is taken on trust that these objects are the work of the person attending the interview). The media presented may not be representative of the medium of the chosen discipline. One of the recently chosen textile students impressed us with skillful drawings and beautiful ceramics. The work must, however, be tidily and completely executed.

Even so it may be that exmination by portfolio discriminates against certain classes of potential students. Scientists expecially may be very dextrous yet have never accummulated a demonstrable portfolio. As a research chemist I worked with materials that would detonate with shock or burst into flames in contact with air. I knew that I had good fine motor control yet I had only a very amateurish and meagre selection of displayable examples of my artistic skill.

It is for this reason that the Conservation Department at the V&A has set up experiments to validate a series of abstract aptitude tests which can be used to judge practical potential rather than assess the present standard of learned skills. These tests were first mentioned at the ICOM Conference in 1988 and will be the subject of an article in "The Conservator" next year [4].

There are several reasons for going into this type of detail about a new conservation course. To announce great ideals publicly strengthens the resolve to realise them. One unselfish reason for announcing the course to an international gathering is to encourage foreign applicants. A selfish reason for warning about the course is that in two years' time we will be looking for laboratories to visit. And in three or four years' time we will be looking for temporary or permanent placements for our studets. Please be ready for them.

References

1. Application forms can be obtained from:
The Registrar
Royal College of Art, Kensington Gore
London SW7 2EU

Individual enquiries to:
Alan Cummings, Senior Tutor (Conservation)
Faculty of Humanities, Royal College of Art
41-43 Jay Mews, London SW7 2EU
01-584 5020 ext. 389

2. J. M. **Cronyn** and E. M. **Pye**, The Training of Conservators: Unity in Diversity. Preprints for the 30th. Anniversary Conference. UKIC, London 1988.

3. The Karate Kid. Columbia Pictures. 1984.

4. S. **Buys** and J. V. **Murrell.** The Conservation. (To be published).

Author's addres

Victoria and Albert Museum
Conservation Department
Cromwell Road, South Kensington
GB- SW7 2RL London

Készült: 90 1296 Pannon Nyomda, Veszprém
Megjelent 12 (A/5) ív terjedelemben
Felelős vezető: Fekete István igazgató